Springer Proceedings in Earth and Environmental Sciences

Series Editor

Natalia S. Bezaeva, The Moscow Area, Russia

The series Springer Proceedings in Earth and Environmental Sciences publishes proceedings from scholarly meetings and workshops on all topics related to Environmental and Earth Sciences and related sciences. This series constitutes a comprehensive up-to-date source of reference on a field or subfield of relevance in Earth and Environmental Sciences. In addition to an overall evaluation of the interest, scientific quality, and timeliness of each proposal at the hands of the publisher, individual contributions are all refereed to the high quality standards of leading journals in the field. Thus, this series provides the research community with well-edited, authoritative reports on developments in the most exciting areas of environmental sciences, earth sciences and related fields.

More information about this series at http://www.springer.com/series/16067

Tatiana B. Yanovskaya · Andrei Kosterov ·
Nikita Yu. Bobrov · Andrey V. Divin ·
Alexander K. Saraev · Nadezhda V. Zolotova
Editors

Problems of Geocosmos—2018

Proceedings of the XII International Conference and School

Springer

Editors
Tatiana B. Yanovskaya
Saint Petersburg State University
St. Petersburg, Russia

Andrei Kosterov
Saint Petersburg State University
St. Petersburg, Russia

Nikita Yu. Bobrov
Saint Petersburg State University
St. Petersburg, Russia

Andrey V. Divin
Saint Petersburg State University
St. Petersburg, Russia

Alexander K. Saraev
Saint Petersburg State University
St. Petersburg, Russia

Nadezhda V. Zolotova
Saint Petersburg State University
St. Petersburg, Russia

ISSN 2524-342X ISSN 2524-3438 (electronic)
Springer Proceedings in Earth and Environmental Sciences
ISBN 978-3-030-21787-7 ISBN 978-3-030-21788-4 (eBook)
https://doi.org/10.1007/978-3-030-21788-4

This Springer imprint is published by the registered company Springer Nature Switzerland AG
The registered company address is: Gewerbestrasse 11, 6330 Cham, Switzerland

Preface

This volume is a collection of papers presented at the 12th conference and school "Problems of Geocosmos" held on October 8–12, 2018, at Peterhoff Campus, St. Petersburg State University. Organized primarily by faculty of the Department of the Earth Physics, this conference is one of the few in Russia to cover a broad range of geophysical research. In 2018, 140 scientists from Russia and abroad attended, including about 45 young scientists, graduate and undergraduate students.

The conference was organized into four parts: Exploration Geophysics (EG), Paleomagnetism, Rock Magnetism and Geomagnetism (P), Seismology (S), and Solar-Terrestrial Physics (STP).

Part EG has covered two broad areas of environmental and engineering geophysics: near-surface research and deep geoelectric studies. Five papers by EG part attendees are presented in this volume. Vorobev et al. discuss the problem of geomagnetically induced currents (GIC) and their influence on stable operation of electric power networks. The treatment is based on a statistical relationship between GIC and variations of the geomagnetic field. Ermolin and Ingerov introduce new methodological approaches to processing audio-magnetotelluric (AMT) and magnetovariational (MVP) data in a heterogeneous environment. Ermolin et al. consider possibilities for improving AMT data quality in the so-called dead band when measuring the AMT field at high latitudes. New theoretical developments in the field of magnetotelluric (MT) measurements and interpretation of MT data are presented in two papers by Plotkin and Mogilatov and Plotkin et al. The first paper considers a possibility to measure the vertical electrical component of electromagnetic field with the help of the circular electric dipole (CED). In the second paper, the possible influence of the Hall effect in MT sounding is discussed.

Part P provided a forum to discuss problems of paleomagnetism, rock magnetism, and geomagnetism. Nine papers in this volume cover the whole range of these topics. Grishchenko et al. and Guzhikov et al. discuss magnetostratigraphy of Valanginian parts in Crimea and Maastrichtian in the Lower Volga area, respectively. New paleomagnetic data are presented for the Lower Carboniferous of the South Urals (Golovanova et al.) and the Lower Cretaceous of the Amur terrane at the junction zone between Sikhote-Alin and Mongol-Okhotsk orogenic belts (Peskov et al.).

Kosterov et al. analyze magnetic hysteresis loops measured in strong, up to 7 Tesla, magnetic field on high-coercivity rock samples. Study of magnetic properties of fallow soils in Tatarstan is presented by Fattakhova et al. Two papers by Merkuriev et al. discuss applying marine magnetic anomaly data to geochronological and geohistorical analysis and to studies of non-dipole component of the Earth's magnetic field, respectively. Finally, Starchenko presents a method to obtain a simple estimate of the relationship between planetary convection and magnetism using scaling laws.

Presentations in Part S covered a wide range of problems, from global seismicity and new earthquake precursors to study of peculiarities of the Earth structure by seismic methods at different scales. Results of the development of new measuring devices were also presented. The papers in this volume reflect all the above-mentioned lines of investigation. Characteristics of the spatial–temporal distribution of the aftershocks are presented by Guglielmi et al. and Zotov et al.; distribution of seismicity in the Crimea is described by Gobarenko and Yegorova. A relationship between the ionospheric anomalies and strong earthquakes is studied by Korsunova et al. Three papers (Gobarenko et al., Yanovskaya et al., Krasnoshchekov and Ovtchinnikov) discuss new studies of the Earth structure covering, respectively, the Earth's crust, upper mantle, and core. New seismic instruments having an advantage of higher precision compared to the existing ones are described by Kislov and Gravirov, and Gravirov et al. The paper by Sharov presents a comprehensive overview of the results of superdeep drilling in Fennoscandia.

Part on Solar-Terrestrial Physics is traditional for "Problems of Geocosmos" conferences and is attended by participants from leading Russian research institutes and universities, and foreign research centers. Topics covered at STP Part include a wide range of magnetospheric, ionospheric, climatic, and solar studies. This volume contains contributions dealing with temporal relations in characteristics of the lower atmosphere and solar activity at the multi-decadal time (Veretenenko and Ogurtsov); pitch-angle diffusion coefficient of energetic protons on the basis of data of the THEMIS spacecraft (Popova et al.); temporal and spatial variability of geomagnetic and telluric electric fields in the Eastern Fennoscandia (Kozyreva et al.); the effect of auroral particle precipitation and polar cap patches on scintillations of the GPS signals in the polar ionosphere (Belakhovsky et al.). Based on the aa (and/or ap) indices, Sokolov et al. presented an improved catalogue of magnetic disturbances including storms with aa >30. Kopytenko et al. discussed possible origins of the displacement of the position of the auroral oval by more than 1500 km over the past 50 years. Remenets et al. reported results of the terrestrial electromagnetic sounding of the lowest part of ionosphere during the natural disturbances.

St. Petersburg, Russia
March 2019

Tatiana B. Yanovskaya
Andrei Kosterov
Nikita Yu. Bobrov
Andrey V. Divin
Alexander K. Saraev
Nadezhda V. Zolotova

Contents

Part I
Exploration Geophysics

Distortion of Local Magnetovariational Anomalies by Effect of Regional Structures

E. Ermolin and O. Ingerov

Abstract In this paper the authors consider the influence of the host medium on the morphology of the magnetovariational (MVP) anomaly from a local object with a relative conductivity of the section 10,000 Sm·m and 500 m depth. Conclusions are based on mathematical modeling. The results have shown that the covered rocks are the main cause of a local anomaly distortion. Resistivity changing of bottom layers have a weak effect on local anomaly disturbing. The anomaly of a local object can be isolated as a result of subtracting a regional component from the Wiese matrix of the original field. The express interpretation of MVP data and inversion codes can be applied for the described anomaly.

Keywords Magnetovariational method (MVP) · Induction vector · Tipper

1 Introduction

Two groups of methods are based on measuring the natural alternating electromagnetic (EM) field of the Earth: magnetotelluric (MT) [1] and magnetovariational (MVP) [2]. In the MT sounding method 4 horizontal components (Ex, Ey, Hx, Hy) are measured. In the MVP method 3 orthogonal magnetic components (Hx, Hy, Hz) are studied. Both methods can be operated independently, but it is more expedient to use them together. This 5-component EM method is an effective tool for solving a wide range of geological exploration tasks.

The response functions in the MVP method are the induction vector and the tipper [2]. Earlier it was shown that by means of using the singular points on the pseudo-sections of the tipper, it is possible to quickly estimate the parameters of 2D and 3D anomalous objects of a simple geometric shape [3, 4]. The algorithms

E. Ermolin (✉)
GM Service, Saint-Petersburg, Russia
e-mail: e.ermolin.gms@gmail.com

O. Ingerov
Phoenix-Geophysics, Toronto, Canada
e-mail: olexandr_ingerov@yahoo.ca

T. B. Yanovskaya et al. (eds.), *Problems of Geocosmos—2018*, Springer Proceedings in Earth and Environmental Sciences, https://doi.org/10.1007/978-3-030-21788-4_1

developed by the authors can be called "the method of express-interpretation of magnetovariational data". This technique has been developed for a homogeneous half-space. But in real conditions we deal with a complex environment and the responses of various geological objects are superimposed on each other. Large lateral inhomogeneities of the Earth's crust and the upper mantle create anomalies. The MVP method can find these anomalies at a distance of up to several hundred kilometers. As a result, the response from the local searched object is considerably distorted, and it becomes impossible to apply the express interpretation methods. At the same time, it is interesting to separate the local component from the regional influence. The current state of this problem is described by Berdichevsky and Dmitriev [1]. On the qualitative level, for the first time the division of the regional and local components of the real part of the induction vector was performed in the work [5]. The quantitative interpretation of the separated local component is an unresolved issue nowadays. The purpose of this work is to estimate the influence of different variants of the horizontally layered host medium and regional 2D conductors on the parameters of the anomaly created by a local body of a simple geometric shape. The most important goal is the answer to the question: is it possible to use the separated local part for further quantitative interpretation?

2 Methods and Modeling Results

We have performed mathematical modeling by means of the WinGLink software [6]. 2D anomalous body of a regular geometric shape is taken as a basic geoelectrical model. Within the framework of this paper the parameters of the anomalous local object have not been changed (the cross section of the body (200 × 200 m), the resistivity (4 Ω·m), the depth to the upper edge (500 m). The object was placed in a horizontally layered medium, shown in Fig. 1. The model consists of several key elements: a sedimentary cover with the thickness of 200 m and the resistivity of 20 Ω·m (total longitudinal conductivity of 10 Sm), a conductive layer of 1 km thickness (resistivity 20 Ω·m and 500 m depth to the upper edge), the boundary of the Earth's crust (M) and the boundary of the lithosphere sole (L).

The studies were divided into 3 stages. At the first stage, the response from the model presented in Fig. 1 was analyzed. At the second stage the influence of the host medium on the local anomaly was estimated. At the third stage, the influence of a 2D regional conductor located at a distance from 40 to 0.4 km from a local object was estimated. As a regional conductor, the suture zone model is taken. Similar objects are often found in the Earth's crust [7].

At the first stage the most significant pseudo-sections of the MT and MVP responses, calculated from the basic model (Fig. 1a) were analyzed. The object clearly appears in the MT and MVP response in spite of the presence of an overlapping cover. In the impedance phase of the TE-mode, the object is displayed on 2 period ranges (Fig. 1c). The first range is around 0.02 s (the amplitude of the anomaly is 6°). The second range is around 2 s (the amplitude of the anomaly is 3°). On the

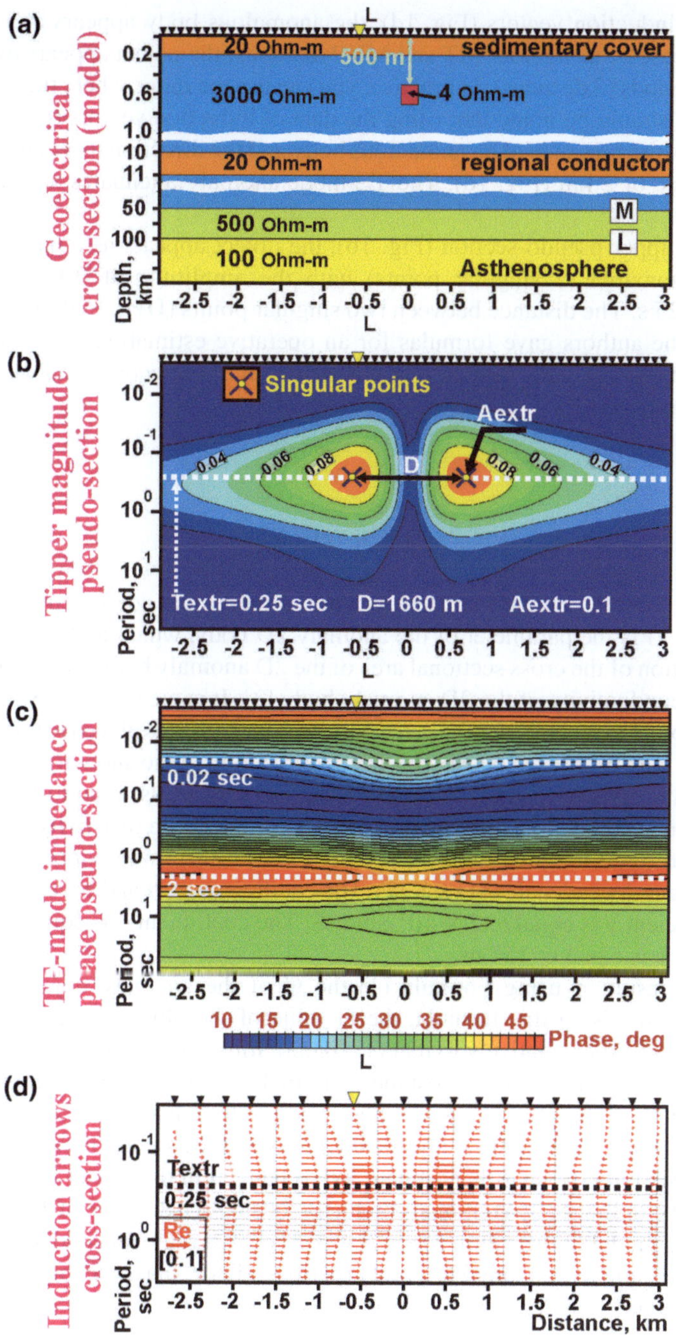

Fig. 1 Geoelectrical basic model (**a**), Pseudo-sections of the response functions (**b–d**)

section of induction vectors (Fig. 1d), the anomalous body appears at a period of 0.25 s. In case of using both MT and MVP methods, the object appears over a wide range of periods (3 decades). In practice, you can use the range where the data quality is better. It should be noted that using the data of induction vectors (pointing to the conductor), the presence of an anomalous body and the direction to it can be detected at a distance of 2 km (Fig. 1d). The TE-mode anomaly attenuates at a distance of 0.7 km.

On the tipper pseudo-section (Fig. 1b), the object appears as a paired anomaly having 2 maxima (2 singular points) with the amplitude of 0.1 at the period Textr = 0.25 s. The distance between two singular points (D) is 1660 m. In previous work [3] the authors gave formulas for an operative estimation of the anomalous conductivity of the cross section (G) and the depth of the geometrical center of the anomaly body (H):

$$G = \gamma \cdot T, \tag{1}$$

$$H = \alpha \cdot D + c, \tag{2}$$

where γ is a coefficient in Sm·m/s; α is a non-dimensional coefficient; c is a coefficient in m. G is the parameter of the anomaly 2D body, which is determined as the multiplication of the cross sectional area of the 2D anomaly body and the anomalous electrical conductivity of the 2D anomaly body [3]. It is measured in Sm·m.

The anomaly object presented in the previous work [3] was placed in homogeneous half-space with 3000 Ω·m resistivity. The cross section of the anomaly body was 200 × 200 m, the resistivity was 4 Ω·m. The G of that object was equal to 10,000 Sm·m. The thin layer (thickness 25 m) at the top of the model with resistivity 100 Ω·m has been added (total longitudinal conductivity of 0.25 Sm). In previous work authors proved linear dependencies (1) and (2). The coefficients γ, α and c were determined. The coefficient γ is equal to 2×10^5 Sm·m/s. The coefficient α is equal to 0.46. The coefficient c is equal to 154 m.

As the result of using Formula (1) the G of the anomaly object is equal to 50,000 Sm·m. Using this formula, the G value of the object presented in Fig. 1a increased five times relatively to real G (10,000 Sm·m). Obviously, the difference is caused by the difference in the host medium. In the second stage of the investigation the influences of the several types of host-medium have been studied.

At the second stage of the study, the thickness and resistivity of the covered rocks and the resistivity of the asthenosphere were changed in the model from Fig. 1a. In addition, the influence of a regional conductor located at a distance of 40 km was estimated. The model of the suture zone [7] was taken as a regional conductor. We compared the parameters of singular points (period—Textr, maxima amplitude—Aextr and the distance between the maxima—D) on the tipper cross-sections calculated from different geoelectrical models. The results of the comparison are shown in Table 1. It can be seen that if total conductivity of overlapping rocks changes, the parameters of the singular points (Textr, D, Aextr) on the tipper cross-section alter

Table 1 Changes of the singular points parameters under the influence of varying the model shown in Fig. 1a

a Difference in singular points for models with 0.25 Sm and 10 Sm

Period (**Textr**)	Distance (**D**)	Amplitude (**Aextr**)
66.94 %	358 m	0.287

b

		Local body	Local body, asthenosphere resistivity 1 Ohm-m	Local body and suture zone
25 m cover, 100 Ohm-m (0.25 Sm)	**Textr** (change)	0.248 s	0.250 s (0.81 %)	0.249 s (0.20 %)
	D (change)	1276 m	1283 m (7 m)	1280 m (4 m)
	Aextr (change)	0.098	0.098 (0)	0.096 (0.002)
200 m cover, 200Ohm-m (10 Sm)	**Textr** (change)	0.082 s	0.082 s (0.37%)	0.081 s (0.91 %)
	D (change)	1665 m	1656 m (9 m)	1634 m (31 m)
	Aextr (change)	0.385	0.385 (0)	0.393 (0.008)

considerably (Table 1a). There are no fundamental changes of singular points parameters in the case the asthenosphere resistivity changes. It is more important that there is no change in the case of adding the suture zone, situated 40 km away from the local object, in the cross-section (Table 1b).

At the third stage the suture zone was located at a different distances from the local anomaly. The influence of the regional conductor on the local anomaly was analyzed more detailed. The geoelectrical models for the distances of 40, 7.5 and 0.4 km are shown in Fig. 2a. The tipper magnitude pseudo-sections calculated from the models are shown in Fig. 2b. The parts of the tipper magnitude pseudo-sections where the local anomaly appears are shown in Fig. 2d. The white isolines illustrate the response of the local anomalous body without the suture zone presence. The results of the calculation in case of the local body absence are shown in Fig. 2c. If the regional structure is located at a distance of 40 km from the local anomalous object (Fig. 2a-1), the anomaly morphology from the local object is practically not distorted. This can be seen from a detailed examination of the pseudo-section of the tipper (Fig. 2d-1). White isolines reflect the shape of the anomaly. Therefore, we can use express-interpretation methods.

When an anomalous object is located at a distance of 7.5 km from the suture zone (Fig. 2a-2), the low-period part of the anomaly of the local object is distorted (Fig. 2d-2). We can not determine the parameters of singular points, but the presence of a local object can be assessed at a qualitative level by the presence of a less distorted high-frequency part of the anomaly. Figure 2 in section e-2 shows the projections of

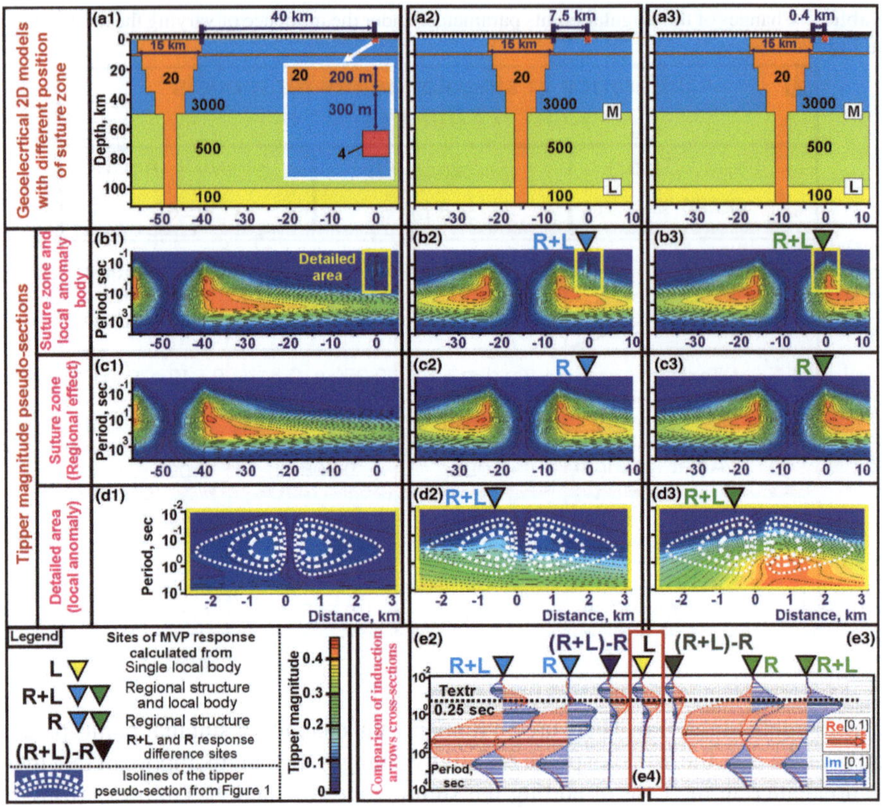

Fig. 2 2D geoelectrical models (**a**); pseudo-sections calculated from models containing: a local body and a regional structure (**b, d**), only a regional structure (**c**); projections of real (red vectors) and imaginary (blue vectors) parts of induction vectors in the Parkinson convention (pointed to conductor) (**e**)

the real and imaginary parts of induction vectors calculated for the model shown in Fig. 2a-2. Curves R + L correspond to the models where both a regional conductor and a local object present. The R curves are calculated only for the regional conductor (there is no local object). The curve L (Fig. 2e-4) is calculated only for a local object (there is no regional structure). It can be seen that the local object on the R + L curve is clearly visible at the period of 0.25 s. Moreover, the value of the maximum period and the amplitude of the vectors coincide with the curve L. We performed a subtraction of the values of the components of the matrix R from the components of the Wiese matrix R + L for all frequency range. The final curve is presented in the same section in Fig. 2e. This is the curve of the vectors (R + L) − R. If the curves of the induction vectors (R + L) − R and L have to be compared, it is difficult to visually distinguish them. Consequently, after dividing the regional and

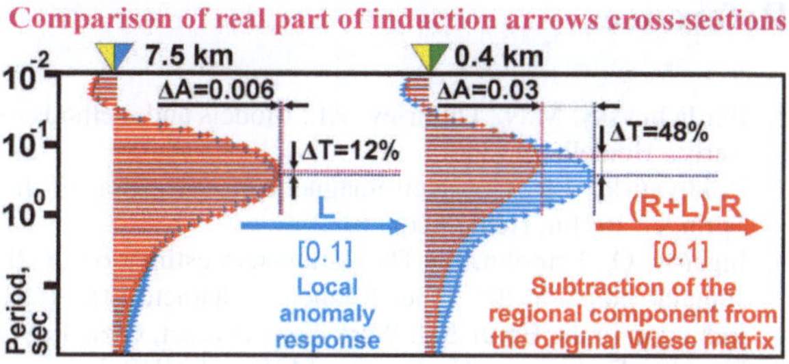

Fig. 3 Comparison of induction arrows

local parts by subtracting the components of the Wiese matrix, we can apply the express-interpretation or inversion technique to the magnetovariational data.

If the regional object is located at a distance from a local object less than its depth (Fig. 2a-3), it is practically impossible to determine the presence of a local object even at a qualitative level. Nevertheless, with the formal subtraction (R + L) − R, we obtain a curve of the vectors that does not in principle differ from the vectors of the local object (L). A more detailed comparison of the real part of induction arrows between L and (R + L) − R curve for 7.5 and 0.4 km separation of the local body and the suture zone is shown in Fig. 3.

If the distance between the regional structure and the local object is less than the depth to the upper edge, the difference of (R + L) and L curve of induction arrows is greater than the field measurement error.

3 Conclusions

The modeling results have shown that for the local anomalous object with a relative conductivity of the cross section of 10^4 Sm·m and a depth to the upper edge of 500 m, the following conclusions are valid:

1. The host medium affects the morphology of the magnetovariational anomaly. The morphology of the anomaly is primarily related to the parameters of the sedimentary cover.
2. If the local and regional anomaly can be visually separated, the presence of large remote regional conductors in the studied area has a very weak effect on the shape of the anomaly from the local 2D object.
3. The local magnetovariational anomaly, distorted by regional anomalies, can be highlighted by simple subtracting the regional constituent of the components of the Wiese matrix from the original field.

The possibility to separate local and regional parts of anomalous signal in magnetovariational method was demonstrated for synthetic data, calculated from current geoelectrical model. Application this approach for real field data is the target of future investigations.

References

1. Berdichevsky, M.N., Dmitriev, V.I.: Models and methods of magnetotellurics, 564 p. Springer, Berlin, Heidelberg (2008)
2. Rokityansky, I.I.: Geoelectromagnetic Investigation of the Earth's Crust and Mantle, 381 p. Springer, Berlin, Heidelberg (1982)
3. Ingerov, O., Ermolin, E.: The parameters estimation of 2D conductive isometrical bodies by singular points at the tipper frequency characteristic. In: IAGA WG 1.2 on Electromagnetic Induction in the Earth 20th Workshop Abstract, Giza, Egypt (2010)
4. Ermolin, E., Ingerov, O., Fox, L.: MTS and MVP data integration to estimate the 2D anomaly bodies parameters situated away from measuring profile. In: 22-th EM Induction Workshop Weimar, Germany (2014)
5. Lozovoy, A.L., Mendry, Y.V., Ingerov, I.: Localization of the position of conducting anomalous objects with the help of local and regional real induction vectors. In: 14th EAGE Conference & Exhibition, Almaty, Kazakhstan (2018)
6. WinGLink Homepage. https://www.slb.com/services/seismic/seismic-reservoir-characterization/electromagnetics/emsoftware/winglink.aspx. Accessed 2019/02/13
7. Feldman, I., Erinchek, Yu., Egorov, A.: Scientific and methodological problems of the geological interpretation of a digital database of materials of deep electrical exploration in the territory of the European part of Russia. Zapiski Gornogo Instituta **176**, 154–158 (2008). (in Russian with English abstract)

The Features of Dead Band AMT Signal in Chukotka Region

E. Ermolin, O. Ingerov, A. A. Yankilevich, N. N. Pokrovskaya, T. V. Davytkina and V. Melnikov

Abstract In this paper we analyze the amplitude spectra of the audiomagnetotelluric (AMT) natural-source signal. Special attention is given to the signal in the frequency range from 1 to 5 kHz known as "dead band". We study the measurements of the base sites that were used during summer fieldwork in 2013, 2014 and 2017 in Chukotka Region (Russian Far East). The area is located above the Arctic Circle. As it has been described by previous researchers, the stable signal in the dead band can be acquired only at nighttime. We would like to point out that in the daytime the local maximum at 2.4 kHz frequency is observed within the dead band in Chukotka. The 2.2 and 2.6 kHz frequencies can be reconstructed by using more than 3 h of daytime data acquisition. These frequencies can be used as the ranging marks. These ranging marks sometimes allow reconstructing the AMT curve by using the relationship between the amplitude and phase. We present a proposal for improving data quality in the dead band during the daytime.

Keywords Magnetotellurics · Magnetovariational profiling · EM monitoring · Amplitude spectra · Dead band

1 Introduction

Magnetotelluric sounding (MT) and magnetovariational profiling (MVP) methods [1] are based on measuring five components of the natural alternating electromagnetic field of the Earth. This technique is a reliable instrument for solving a large number

E. Ermolin (✉) · V. Melnikov
GM Service, Saint-Petersburg, Russia
e-mail: e.ermolin.gms@gmail.com

O. Ingerov
Independent Expert, Toronto, Canada
e-mail: olexandr_ingerov@yahoo.ca

A. A. Yankilevich · N. N. Pokrovskaya · T. V. Davytkina
Saint-Petersburg Mining University, Saint-Petersburg, Russia
e-mail: yankilevich@mail.ru

T. B. Yanovskaya et al. (eds.), *Problems of Geocosmos—2018*, Springer Proceedings in Earth and Environmental Sciences, https://doi.org/10.1007/978-3-030-21788-4_2

of exploration tasks. For investigation at small depths (less than 2 km), the high-frequency modification of magnetotelluric method—AMT (audiomagnetotellurics) is applied. In the AMT modification, the frequency range from 10 kHz to 1 Hz is used depending on the modification of magnetic sensors. It is considered that the source of the field in AMT is thunderstorms in equatorial regions. The peculiarity of this source is that the minimum of the signal amplitude is presented in the frequency range from 1 to 5 kHz. The specified range is called "dead band". An extensive study of this problem is found in Xavier Garcia and Alan G. Jones article [2]. For enhancing data quality in the dead band they propose the following: 1. Perform a sounding at night; 2. Increase the sensitivity of magnetic sensors; 3. Use more than one reference point for processing. At present, a new approach has appeared: using dispersion relationship in the magnetotelluric tensor in the first stage of data processing [3]. Unfortunately, it is hardly ever feasible to apply all the given recommendations. However, the information about the curve in dead band is very important for many exploration tasks.

We performed AMT survey for gold exploration in Chukotka [4]. The geological task demanded obtaining AMT data of a very high quality, because the search objects (gold-bearing quartz veins) are small (thickness from 1 to 5 m). The veins are covered with a layered volcanic rock cover with a thickness of 80–300 m.

The investigated area is located above the Arctic Circle where the signal in the dead band is much weaker. All regular measurements were done by using remote reference station technique [5]. Despite this, we could not get good quality AMT data in the dead band during the daytime. Nighttime measurements were the only way to acquire adequate data in the high-frequency band. That decreased the productivity of the fieldwork significantly. Nevertheless, some measurements were carried out in the daytime in the hope of obtaining AMT curves of an adequate quality. In July and August, it can be done with a 50–70% chance and in October with a 10% chance.

The purpose of this investigation was to search for the features of the amplitude spectra of 5 field components (Ex, Ey, Hx, Hy, Hz). Some features will probably help to find new ways of enhancing the AMT and MVP data quality in the dead band.

2 Methods

We used the records of the remote reference sites in the periods of fieldwork in Chukotka: August–October 2013; June–September 2014; July–August 2017. The natural field was registered daily for 15 h from 12:00 to 8:00 local time. The averaged coordinates of the area are: Longitude: 170° E, Latitude: 70° N. MTU-5A receivers, AMTC-30 magnetic coils and the non-polarized lead electrode by "Phoenix-Geophysics Ltd" were used. The magnetic sensors were installed on the tripods manufactured by "AGCOS Inc". All the cables and tripods were buried in the base site. At the 24 kHz sampling frequency, 4 records in each of 10 s times-lot were used. The results of data acquisition (time series) were divided into 30-min intervals. SSMT2000 software was used for data processing. Estimates of impedance (apparent resistivity, magnetic or electric field spectra) with auto electric reference (local E) [5] were used. In this article we consider only the year 2017 data, because the main conclusions of three seasons data analysis have not got principle differences.

3 The Data Analysis and Results

We visualized the results of data processing. The analysis has been done for: I—amplitude spectra as the function of time; II—the curves of amplitude spectra (for a more detailed analysis); III—amplitude at the frequencies 3 and 1.5 kHz; IV—analysis of AMT apparent resistivity curves. All considered data refer to one site.

I. The amplitude spectra for 5 components as the function of time are shown in Fig. 1 (August, 2017). In the dead band, the clear periodicity of the signal amplitude is observed which is connected with a certain time of the day. This blue-purple band of the low level of all components in the frequency band from 6 to 0.9 kHz contracts and expands over time. It gets widest (the lowest amplitude) from 12:00 to 15:00 (in the daytime) and the most narrow (the maximum amplitude) from 00:00 to 03:00 (at night). The day of August, 11, stands out in the average regularity. On this day an abnormally large amplitude spectral value is observed throughout the day. The dead band is shown more detailed in Fig. 2 from 8 to 13 of August in the frequency range from 10 to 0.1 kHz.
   In Fig. 2 the general minimum of the amplitude spectra in the dead band is observed in the daytime. But at the same time, the local maximum is observed at frequency 2.4 kHz within the general minimum. This local maximum is more visible in the vertical magnetic component. Over here the two local minimums of the dead band are clearly visible: 3 and 1.5 kHz.
   We also visualized the amplitude of signal curves as a function of frequency to analyze the features of amplitude spectra in the daytime more detailed.
II. Frequency characteristics of a 4-h daytime record on August, 8 (from 12:30 to 16:30), and a short 30-min night record on August, 9 (from 02:00 to 02:30), are shown in Fig. 3. These are typical spectra in the field seasons of 2013, 2014 and 2017. In Fig. 3 it is clearly seen that the amplitude spectra at night is higher than those in the daytime. In Fig. 3 the local maximum is clearly observed on daytime spectra in the dead band at the frequencies 2.2 and 2.6 kHz. This local maximum is marked by a transparent blue square in Fig. 3. The table in Fig. 3 demonstrates that the amplitude spectra of the local maximum in the dead band are higher by 85% (relative to the values of two nearby local minimums). Thus, there is a strong possibility that the AMT curve can be rebuilt at the local maximum using the daytime record. The presence of the local maximum in the dead band can be an indicator of sufficient data quality in the daytime. Consequently, for daytime measurements it is essential to carry out an operative analysis of the

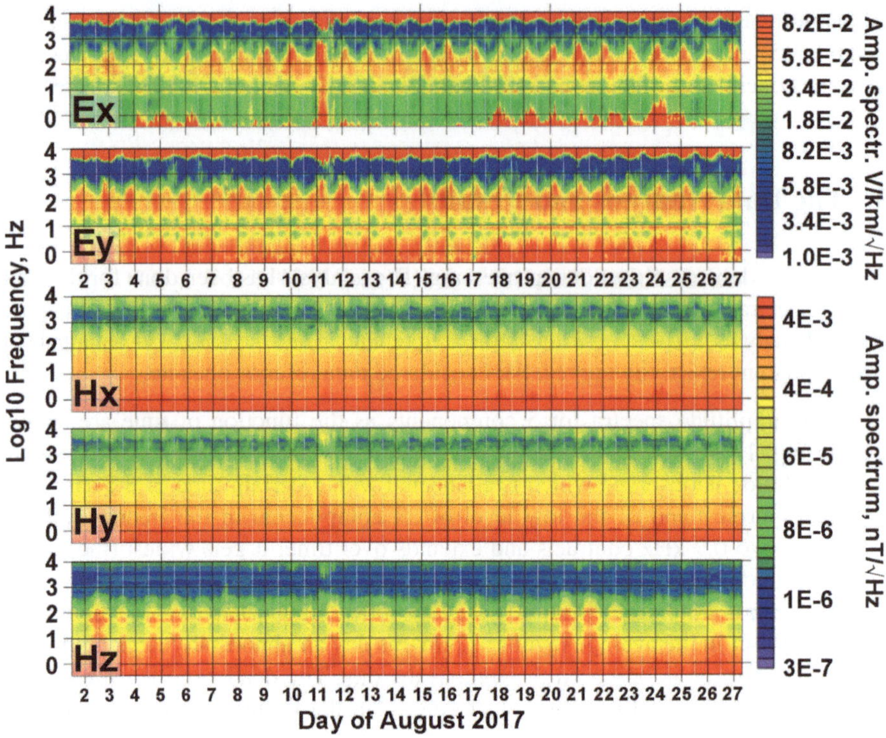

Fig. 1 AMT amplitude spectra in August, 2017

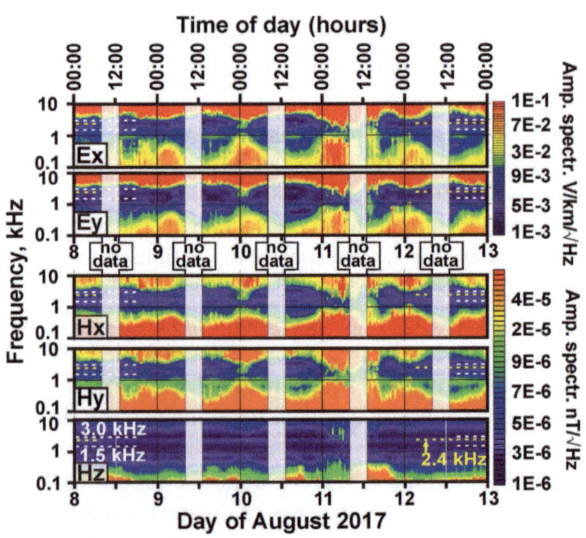

Fig. 2 Amplitude spectra from 8 to 13 of August

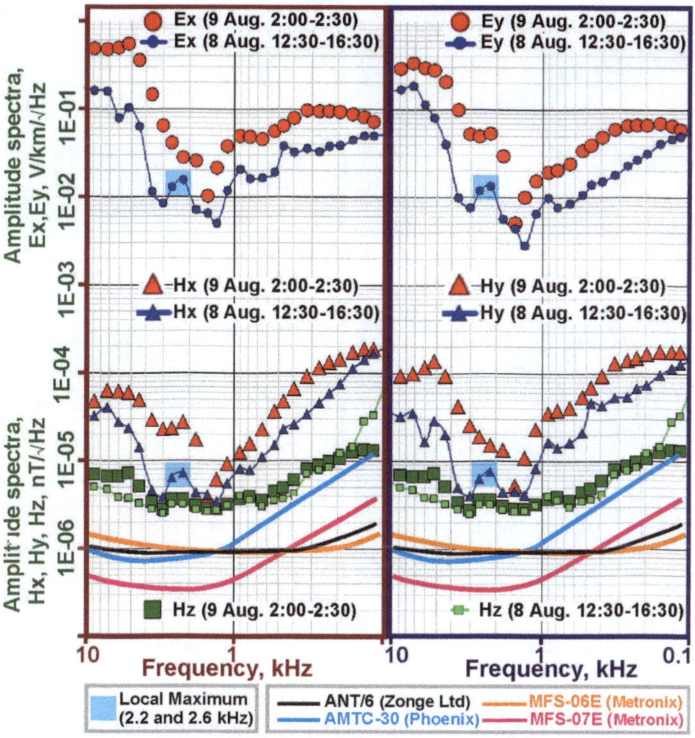

Hx, nT·√Hz			Ex, V/km/√Hz		
3 kHz (Local Min)	2.4 kHz (Local Max)	1.5 kHz (Local Min)	3 kHz (Local Min)	2.4 kHz (Local Max)	1.5 kHz (Local Min)
3.9E-6	6.8E-06	4.1E-06	8.4E-03	1.4E-02	6.5E-03

Fig. 3 The comparison of the amplitude spectra measured at night and in the daytime. Sensors noise characteristics are shown as color lines (the manufacturer's website official data). Hx and Ex local maximum and two local minimums amplitude values in the daytime are shown in the table

amplitude spectra and other impedance functions immediately after recording in field conditions.

It should be noted that in the dead band the amplitude of magnetic components is lower than that of electric components (relative to 100 Hz frequency). Therefore, the noise in magnetic channels is greater than noise in electric channels. For this reason, in the investigated area for the dead band we used the estimates of impedance with remote electric reference (Remote E) [5]. The comparison between the natural-signal amplitude levels and official characteristics of sensors noise of the commonly used sensors are shown in Fig. 3. This figure illustrates that the own noise of modern magnetic sensors is significantly lower than the natural-signal amplitude in the dead band. Consequently, the dead band problem is not connected with the sensor noise but in a greater degree with external noise. As a result, a crucial task is to improve the ways of sensor installation

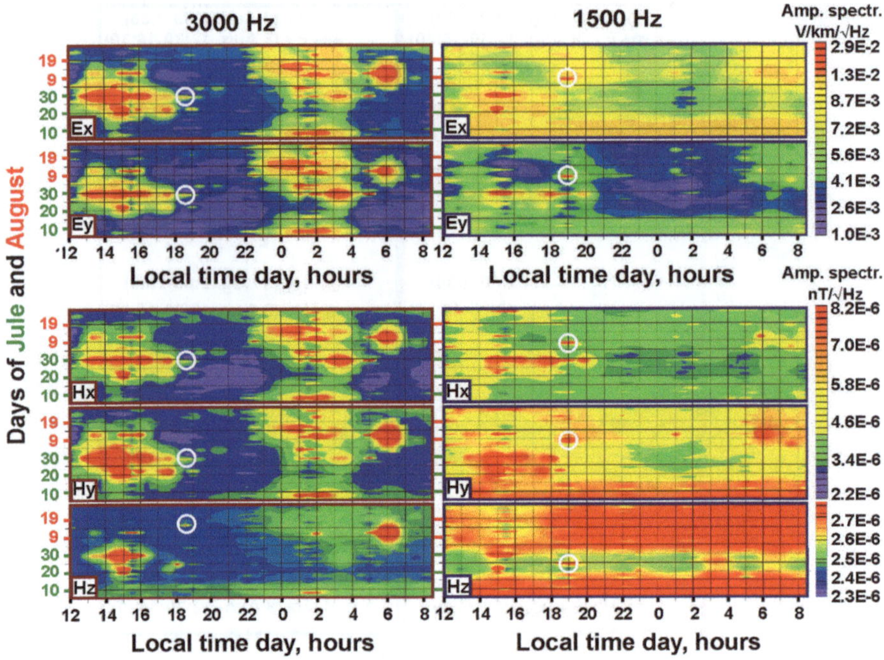

Fig. 4 Dependence of the signal amplitude in the local time of the days from July, 4, to August, 26, at frequencies 1.5 and 3 kHz. Local maximums are shown with white circles

to decrease external noises. Due to high noises in the magnetic channels, the values of apparent resistivity appear to be substantially smaller. This is what we observe in the dead band.

III. The signal amplitudes at 1.5 and 3 kHz have minimal values in the dead band. For this reason, we performed a detailed analysis of these frequencies specifically. The results are shown in Fig. 4. At 3 kHz frequency the signal is higher at night from 0:00 to 4:00 for all components. Also, from July, 15, to August, 15, the increased values are observed from 13:00 to 17:00 local time. At 1.5 kHz frequency the signal level is higher in the daytime. At night from 22:00 to 5:00 the values are low. That is why the frequency 1.5 kHz is the most challenging problem concerning the dead band.

Single local anomalies of the increased amplitude values deserve special attention (the white circles in Fig. 4). They chaotically appear throughout the day. We used the MTU-5A receivers, where at a sampling frequency of 24 kHz, in fact only 2–5% of the measurement time is recorded. The increasing of the sampling frequency and the recording of continuous time series will probably help to measure the rare "pulses" of the signal. It is likely that this technique will help to partially solve the problem with the data quality of AMT curves in the dead band.

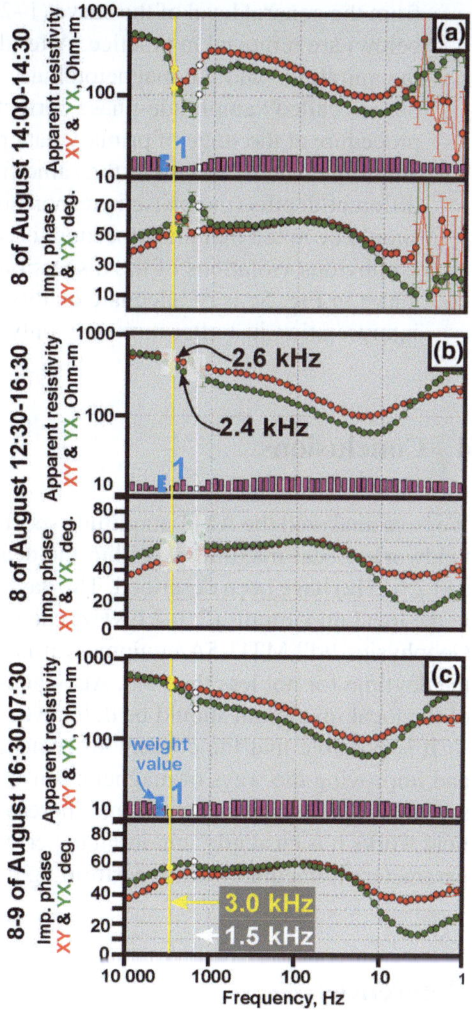

Fig. 5 Apparent resistivity curves and impedance phases for: a 30-min (**a**) and 4-h (**b**) daytime record on August, 8; 15-h night record on August, 8–9 (**c**). Chukotka region

IV. The apparent resistivity curves and impedance phase curves of AMT are shown in Fig. 5 for a short daytime record (5a), 4-h daytime record (5b), and 15-h record (5c). The remote-reference processing was applied.

The curve in Fig. 5a is a rejected take because the amplitude and phase values at 11 consecutive frequencies in the dead band deviate significantly from the real curve.

The curve in Fig. 5b is of much better quality. Its main feature is that the amplitude and phase values at frequencies of the local maximums (2.2 and 2.6 kHz) can be restored as a result of the interactive rejection of solutions with a low weight. Due to the presence of these benchmarks, we can restore the apparent resistivity curve in the dead band. As a rule, the values that significantly deviate from the general level of the curve (1–2 frequencies above and 2–3 frequencies below) are removed in practice. After that, the dispersion relationship between the amplitude and the magnetotelluric impedance phase are used. This procedure is called "amplitude-phase correction" [6]. Some authors implement this procedure at the stage of primary data processing [3]. The basic idea is that the amplitude and the phase in the same frequency range carry information about different depths (the phase is responsible for the deeper part of the section). It should be noted that this procedure is correct only for the areas where there are no obvious violations of the dispersion relations in the impedance tensor. The curve in Fig. 5c is of adequate quality. We can use this curve for analysis and interpretation just after removing only the value at 1.5 kHz.

4 Conclusions

We have analyzed the AMT amplitude spectra in the dead band in Chukotka summer fieldwork. Local maximums within the dead band in the daytime (frequencies 2.2 and 2.6 kHz) have been described. The daytime records can be used only if the curve in the local maximums (2.2–2.6 kHz) can be reconstructed. As a rule, for "Phoenix-Geophysics ltd" MTU-5A equipment it is necessary to carry out measurements in the daytime for not less than 3 h. Also, no more than 2–3 frequencies on both sides of the local maximum should be defective.

It is possible that the continuous registration of time series at high frequencies and improving the ways of magnetic sensor installation will allow decreasing measurement length in the daytime and increasing data quality in the dead band. In the field works it is currently important to perform operative visualization of impedance parameters immediately after performing measurements on the site.

References

1. Berdichevsky, M.N., Dmitriev, V.I.: Models and methods of magnetotellurics, 564 p. Springer, Berlin, Heidelberg (2008)
2. Garcia, X., Jones, A.: Atmospheric sources for audio-magnetotelluric (AMT) sounding. Geophysics **67**, 448–458 (2002)
3. Jones, A.: Magnetotellurics: status quo and quo vadimus. DMEC Exploration 2017 Abstr 11, 139–158 (2017)
4. Ermolin, E., Ingerov, O., Savichev, A.: Gold exploration in Chukotka region by using audio-magnetotellurics. In: 22-nd EM Induction Workshop, Weimar, Germany Abstr (2014)
5. Gamble, T.D., Goubau, W.M., Clark, J.: Magnetotellurics with remote reference. Geophysics **44**, 53–68 (1979)
6. Feldman, I., Ermolin, E.: Amplitude-phase correction of magnetotelluric impedance curves. Zapiski Gornogo Instituta **194**, 200–210 (2011). (in Russian with English abstract)

Role of Hall Effect in Magnetotelluric Sounding

Valery V. Plotkin, Vladimir S. Mogilatov and Vladimir V. Potapov

Abstract Many minerals belong to the semiconductor class from the point of view of the mechanism of electrical conductivity. One of the experimental methods for determining the electrical conductivity of semiconductor minerals is based on the Hall effect. Due to the presence of the Earth's magnetic field it is possible to expect the manifestation of this effect in natural conditions during electromagnetic soundings. To detect manifestations of the Hall effect during magnetotelluric (MT) sounding, a polarization analysis method is proposed, based on a data processing algorithm that divides the time spectrum of the MT field into normal mode spectra with right and left circular polarization. Numerical estimates and results of a special field experiment are presented.

Keywords Magnetotelluric sounding · Hall effect · Normal modes · Right and left circular polarizations

1 Introduction

Host rocks overlaying oil and gas deposits penetrated by a stream of hydrocarbon fluids, represent a semiconductor medium in which manifestation of the Hall effect is possible [1]. Interest in this phenomenon arose from the results of electromagnetic soundings with controlled sources in hydrocarbon-rich areas [2]. Similar phenomena may be observed during electromagnetic soundings of porous geological media due to the effect of the Earth's magnetic field on currents in fluids. Here the role of viscosity is large, and viscosity coefficient can be determined from the experiment.

V. V. Plotkin (✉) · V. S. Mogilatov · V. V. Potapov
Institute of Petroleum Geology and Geophysics SB RAS, Koptug Ave. 3, Novosibirsk 630090, Russia
e-mail: PlotkinVV@ipgg.sbras.ru

T. B. Yanovskaya et al. (eds.), *Problems of Geocosmos—2018*, Springer Proceedings in Earth and Environmental Sciences, https://doi.org/10.1007/978-3-030-21788-4_3

2 The Polarization Analysis Method

So far, for geological media the characteristics of microprocesses are insufficiently known, therefore it is better to address an experiment. For its planning, it is possible to consider the influence of the Earth's magnetic field by introducing the Hall conductivity σ_H into the electrical conductivity tensor. In an anisotropic medium, the field splits into two components, called normal modes, which differ in attenuation coefficients and phase velocities. The difference between modes is connected with its polarization and direction of rotation of the field vector; in the first mode the field rotates clockwise, in the second counterclockwise. For physical reasons, it is clear that due to Hall effect the response of the medium may be different in the case of excitation of the medium by only one of the normal waves.

Analysis of MTS data shows that generally the direction of rotation of MT-field vectors (right or left polarization) is defined by characteristics of frequency spectrum. It is well known that polarization is generally elliptic, and at different frequencies, the direction of rotation of field vector is determined by the phase difference of spectral elements of horizontal components at different frequencies. So, if we assume the field time dependence in the form $e^{i\omega t}$ and introduce the designations $H_x(\omega) = A_x(\omega)e^{i\varphi_x}$ and $H_y(\omega) = A_y(\omega)e^{i\varphi_y}$, then the horizontal component of magnetic field vector rotates clockwise when $\sin(\varphi_y - \varphi_x) > 0$, and counterclockwise when $\sin(\varphi_y - \varphi_x) < 0$. This can be explained by the ratio of amplitudes of normal modes with circular polarization representing the MT-field.

The mode amplitudes can be determined using a polarization analysis of time spectrum of MT-field. For this purpose, it is necessary to know the polarization coefficients for a given mode. The calculations carried out in [3] for the horizontally layered medium with the parameters $\sigma_n, h_n, n = 1, \ldots, N, h_N \to \infty$ showed that in middle latitudes mode polarizations coefficients in all layers are close to $\pm i$ when $\sigma_H \ll \sigma_n$. It means that under these conditions modes with right or left circular polarization are normal. Then, for any elliptic polarization, the decomposition of MT-field into modes with right and left circular polarizations can be performed using the following representation for the real part of horizontal components:

$$Re\{H_x(\omega)e^{i\omega t}\} = Re\{A_x(\omega)e^{i(\varphi_x+\omega t)}\} = A_1\cos(\varphi_1 + \omega t) + A_2\cos(\varphi_2 + \omega t),$$

$$Re\{H_y(\omega)e^{i\omega t}\} = Re\left\{A_y(\omega)e^{i(\varphi_y+\omega t)}\right\} = A_1\sin(\varphi_1 + \omega t) - A_2\sin(\varphi_2 + \omega t),$$

where the mode with index 1 is characterized by the counterclockwise rotation of magnetic field horizontal components, and the mode with index 2 by the clockwise rotation. Amplitudes and phases of modes are then calculated using the polarization analysis formulas:

$$H_{1,2}(\omega) = A_{1,2}e^{\varphi_{1,2}},$$

$$A_1^2 = \frac{\left[A_x^2 + A_y^2 + 2A_xA_y\sin(\varphi_x - \varphi_y)\right]}{4}, \quad A_2^2 = \frac{\left[A_x^2 + A_y^2 + 2A_xA_y\sin(\varphi_y - \varphi_x)\right]}{4},$$

$$\tan\varphi_1 = \frac{A_x\sin\varphi_x + A_y\cos\varphi_y}{A_x\cos\varphi_x - A_y\sin\varphi_y}, \quad \tan\varphi_2 = \frac{A_x\sin\varphi_x - A_y\cos\varphi_y}{A_x\cos\varphi_x + A_y\sin\varphi_y}$$

$$\tan\varphi_1 = \frac{A_x\sin\varphi_x + A_y\cos\varphi_y}{A_x\cos\varphi_x - A_y\sin\varphi_y}, \quad \tan\varphi_2 = \frac{A_x\sin\varphi_x - A_y\cos\varphi_y}{A_x\cos\varphi_x + A_y\sin\varphi_y}$$

According to the similar formulas, a transition to the modes is carried out for the horizontal components of the electric field. Further on, for any time harmonic, instead of standard impedance ratios for horizontal MT-field components, impedance ratios are introduced for the modes with right and left circular polarizations:

$$E_1(\omega) = Z_{11}(\omega)H_1(\omega) + Z_{12}(\omega)H_2(\omega), \quad E_2(\omega) = Z_{21}(\omega)H_1(\omega) + Z_{22}(\omega)H_2(\omega).$$

The difference between mode impedances $Z_{11}(\omega)$ and $Z_{22}(\omega)$ will indicate a possible occurrence of the Hall effect. This difference will also manifest itself in the values of apparent resistivities and impedance phases introduced by conventional formulas.

3 A Numerical Model of the Hall Effect in MTS Data

We will confirm the above by numerical calculations using the synthetic MTS data obtained as follows. First, using the normal distributions of random values, initial time spectra for the horizontal magnetic field components $H_{x,y}(\omega)$ were synthesized. The parameters of distributions were chosen to achieve similarity with examples of available experimental data. Such approach allows a multiple repetition of numerical experiments without carrying out real MTS. Then, for the given model of horizontally layered medium, the impedance tensor was calculated using the method described in [4]. The time spectra of horizontal components of electric field are determined using the standard impedance ratios for the magnetic field time spectra:

$$E_x(\omega) = Z_{xy}(\omega)H_y(\omega) + Z_{xx}(\omega)H_x(\omega), \quad E_y(\omega) = Z_{yy}(\omega)H_y(\omega) + Z_{yx}(\omega)H_x(\omega).$$

Using such a scheme for the preparation of synthetic MTS data, we next consider the results of their processing using the proposed version of the MT field polarization analysis.

Numerical calculations were carried out for the horizontally layered medium model consisting of four layers with thicknesses from top to down of 0.7, 5, 2 and 9 km and specific electric resistivities (SER) 100, 1000, 300 and 100 of Ω m respectively, resistivity of underlying homogeneous subsurface was 20 Ω m.

It is interesting to compare the calculation results for the cases when the Hall conductivity of the medium is zero, and when it is present in all layers of the medium.

We consider that $\sigma_H = 1/1000$ S/m, the Earth's magnetic field is deflected from the vertical by an angle of 25°, the *OX* axis is set to the plane of magnetic meridian and directed northward, the *OZ* axis is directed down to the medium. The MTS curves obtained by modal impedances $Z_{11}(\omega)$ and $Z_{22}(\omega)$ are presented in Fig. 1 (on the left when the Hall conductivity is absent, on the right when it is not equal to zero). Naturally, the observed difference in the modal curves increases with increasing Hall conductivity. It should be noted that this difference practically does not depend on the ratio of amplitudes $A_{x,y}(\omega)$ in the original spectra of horizontal components of magnetic field (Fig. 2, $\sigma_H = 1/1000$ S/m).

The question arises whether the observed effect is due to the influence of the Hall conductivity. Would not the manifestations be similar, for example, in the case of apparent anisotropy arising in a thin-layer medium?

In order to reveal differences in the results of processing data by the proposed method for a medium with the Hall conductivity and for a thin-layer medium, respectively, calculations were also carried out for a transversal anisotropic medium. We assume that in this case in the coordinate system where the tensor $\hat{\sigma}'$ is diagonal (in the main axes) alternating thin layers with different SERs, the normal to which is directed along the axis OZ', lead to an anisotropy, and it was considered that $\sigma'_{xx} = \sigma'_{yy}$, $\sigma'_{zz} = \sigma'_{xx}/k_a$, where k_a is the anisotropy coefficient.

Note that when taking into account the Hall conductivity, it was assumed that the geomagnetic field is directed along the axis OZ' in the same coordinate system, and the tensor $\hat{\sigma}'$ has identical diagonal components $\sigma'_{xx} = \sigma'_{yy} = \sigma'_{zz}$, but additional antisymmetric components $\sigma'_{xy} = -\sigma'_{yx} = \sigma_H$ appear. Differences of tensors $\hat{\sigma}'$ are

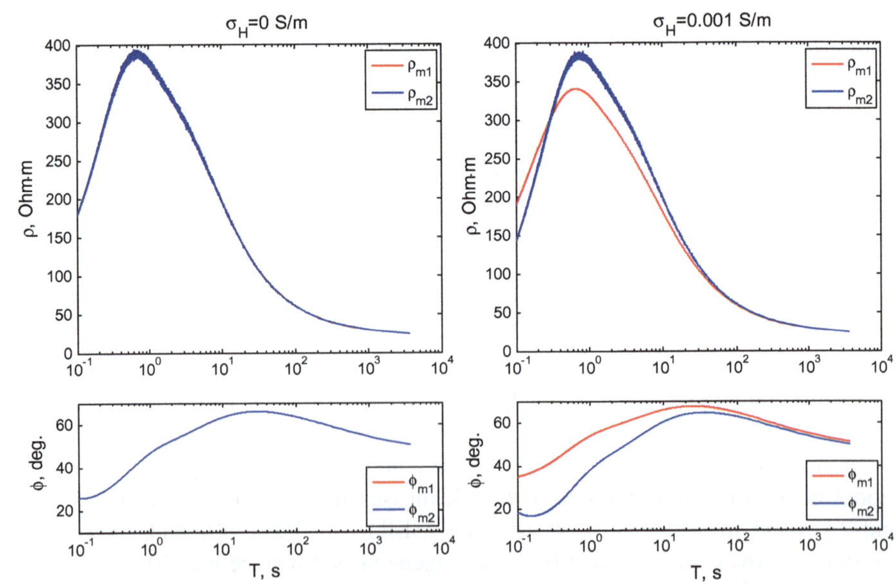

Fig. 1 The MTS modal curves in the case when the Hall conductivity is presented in the medium

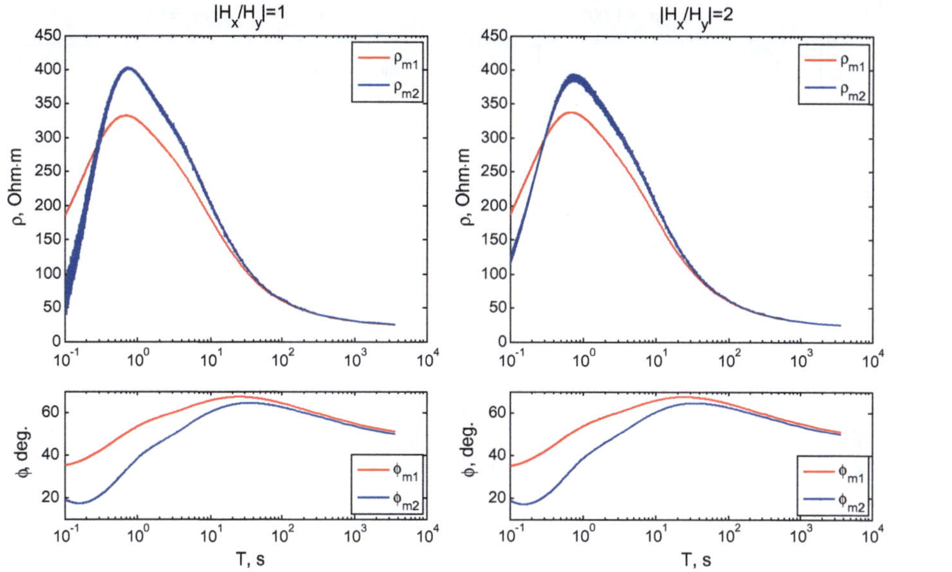

Fig. 2 The MTS modal curves for different ratios of $H_{x,y}(\omega)$ amplitudes

associated with these circumstances when studying the Hall effect and the apparent anisotropy of a thin-layer medium. The same transition matrix was used in the calculations of tensors σ′ in the laboratory coordinate system in both cases.

As an example of differences in conductivity tensors, we present the values of its components in the laboratory coordinate system at the same depth (on the left—in case of the Hall's effect, on the right—for the thin-layer medium):

$$\hat{\sigma} = \begin{pmatrix} 0.05 & 0.0009 & 0.0004 \\ -0.0009 & 0.05 & 0 \\ -0.0004 & 0 & 0.05 \end{pmatrix}, \quad \hat{\sigma} = \begin{pmatrix} 0.05 & 0 & 0 \\ 0 & 0.0485 & 0.0032 \\ 0 & 0.0032 & 0.0432 \end{pmatrix}$$

Certainly, the calculations were performed for the same selected model of horizontally layered medium consisting of four layers described above. The results are presented in Fig. 3. It can be seen that for the thin-layer medium (even with the bigger coefficient of anisotropy $k_a = 1.2$) the difference of modal curves is significantly less than in the case of presence of the Hall effect.

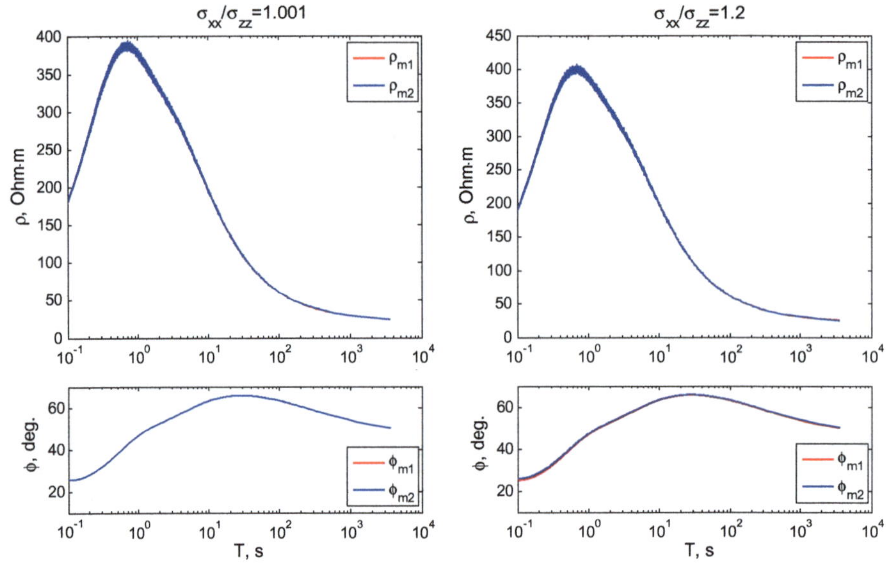

Fig. 3 The MTS modal curves in the case of a thin layer anisotropic medium

4 The Results of Experiment to Detect the Hall Effect in Real MTS Data

The experiment to detect the Hall effect is easiest to carry out in the case of a horizontally layered medium. The suitable site for such an experiment is available in the Tatar area of West-Siberian lowland, near Orlovka village (54°51′ N, 76°02′ E). MTS was performed in July 2018 using the MTU-5 station (Phoenix Geophysics). Signal registration lasted 40 h. The results obtained by a standard processing of experimental data are presented in Fig. 4.

On the short periods, the selected area quite corresponds to a horizontally layered medium. Therefore, the polarization analysis was performed for these periods (Fig. 5) with various selections of time windows lasting 10^4 s and sample rate of 15 Hz. For each window the time spectra of horizontal component variations of the field were transformed to the spectra of amplitudes of normal modes with right and left circular polarization. The similar procedure was carried out in addition to the processing of spectra. In Fig. 5, dots show the estimates of mode apparent resistivities obtained from different data samples; the solid lines are their approximations by polynomials. The dotted lines present the modal curves calculated for two medium models (shown with dotted lines in the left part of the figure) and the specified values σ_H. The differences of modal curves are also shown for clarity. The search of medium models and σ_H allows approximately bring together all curves. The estimates show that the σ_H values do not exceed ~10^{-3} S/m.

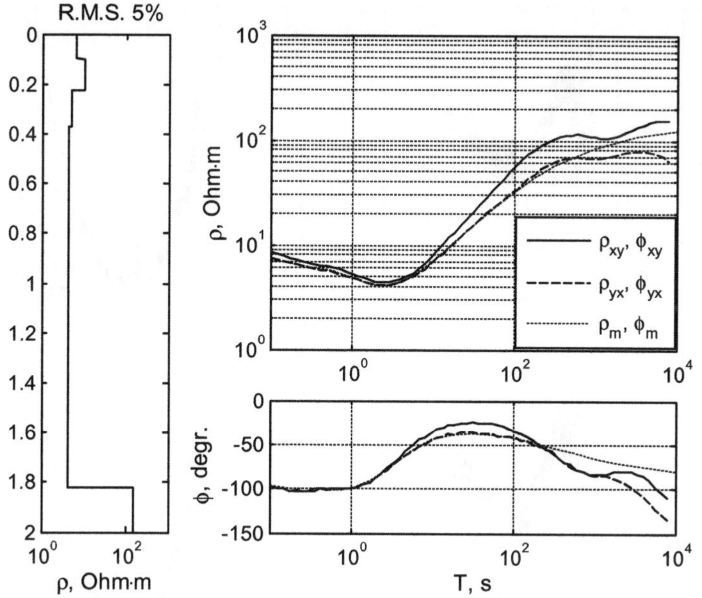

Fig. 4 MTS curves and the model of medium obtained by a 1D inversion

It should be noted that a possibility exists that differences of modal curves are connected with the lateral inhomogeneity of the medium. To estimate its role, it is convenient to construct polar diagrams of impedances. In case of a 2D heterogeneity the polar diagrams of additional impedances taking into account the Hall effect and without it considerably differ. An example of polar diagram for the discussed experimental data is shown in Fig. 6. Apparently, the role of lateral inhomogeneity in this case is small. It is yet another argument favoring a possibility of manifestation of the Hall effect.

5 Conclusion

To find manifestations of the Hall effect in the MTS data, it is possible to apply the method of polarization analysis based on the data processing algorithm with the division of time spectra of MT-field into spectra of normal modes with right and left circular polarization. It is established that with the use of the proposed method it is possible to distinguish effects for a medium with the Hall conductivity and for a thin-layer medium, respectively. First experimental estimates showed that σ_H values do not exceed ~10^{-3} S/m.

Acknowledgements This work was executed with a financial support of Russian Foundation for Basic Research, grant 17-05-00083.

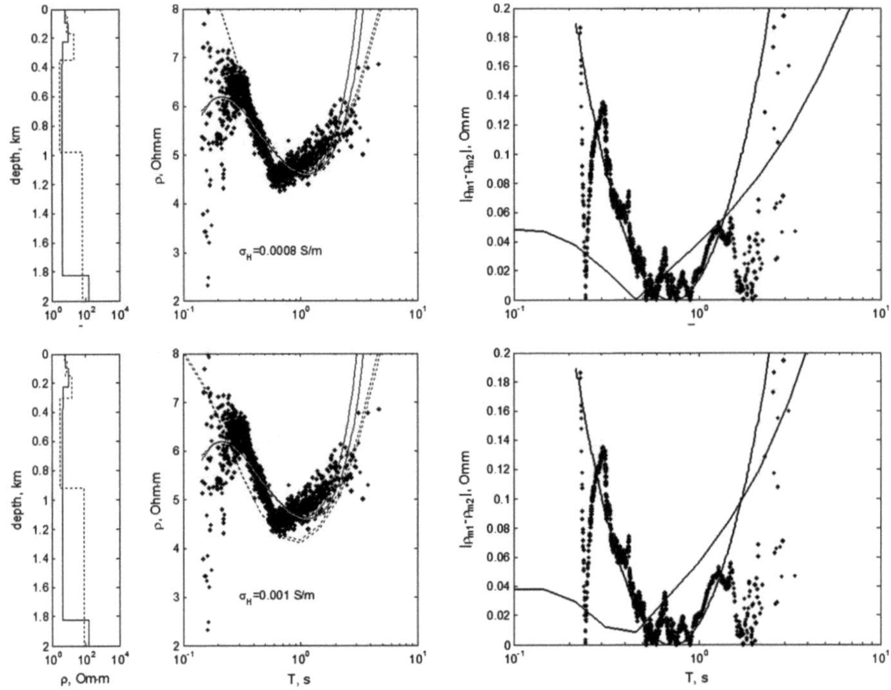

Fig. 5 The models of medium (left), the modal curves (center) and their distinction (right). Dots are the experimental estimates, solid lines are the approximations by polynomials, dotted lines are the results of calculation of modal curves for the model shown by dotted line and the given Hall conductivity

Fig. 6 The polar diagram of impedances for the time period T = 1 s

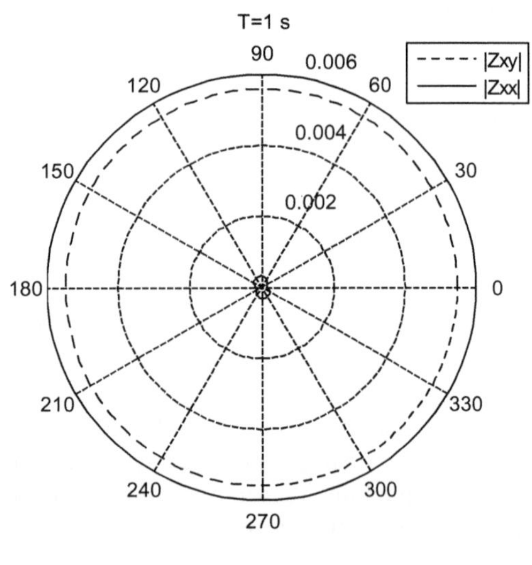

References

1. Gololobov, D.V., Malevich, I.Yu.: Physical and electrochemical processes in the environment above the carbon-hydrogen deposit. Doklady BGUIR (1), 22–27 (2005) (in Russian with English abstract)
2. Mogilatov, V.S.: Influence of geomagnetic field on the transient process of the secondary currents in the Earth. Geofizika (4), 70–75 (2013) (in Russian with English abstract)
3. Plotkin, V.V.: Technique of contribution definition of the Hall's effect by magnetotelluric sounding. In: XIII International Scientific Congress and GEO-SIBIR-2017 INTEREKSPO Exhibition. International Scientific Subsurface Use Conference. Mining. New Directions and Technologies of Search, Investigation and Development of Mineral Deposits. Geoecology. Collection of Materials, vol. 2, no. 3, pp. 187–192. Novosibirsk (2017) (in Russian)
4. Aleksandrov, P.N.: The forward problem of geoelectrics in a one-dimensional bianisotropic medium. Izv. Phys. Solid Earth **37**(4), 324–333 (2001)

About Measurement of Vertical Component of Electric Field During Magnetotelluric Sounding

Valery V. Plotkin and Vladimir S. Mogilatov

Abstract In magnetotelluric sounding (MTS), variations of five components of electromagnetic field are usually registered. With the standard approach in experiment transfer functions are defined as ratios between horizontal components of electric and magnetic fields. Deviations from the Tikhonov-Cagniard basic model are defined by the fifth registered component—the vertical component of magnetic field. In order to increase the reliability of results obtained by the inversion and to gain more information from MTS, it is proposed to carry out the additional registration of vertical component of the electric field in the medium. At present, for registration of variations of vertical electric component the method of measurement of potential difference arising on the vertical line immersed in the sea is applied. On the land, for this purpose it is necessary to drill a hole or use already available wells, which may be difficult in practice. It is proposed to apply a circular electric dipole (CED) for registration of vertical electric component. Presently, CED is successfully used as an emitter in electromagnetic sounding methods. Its feature to excite in the layered medium a single TM-mode whose electric field has a vertical component, is used. In practice, CED is usually implemented with eight long radial lines. Taking into account this feature, CED can also be used as the MTS receiving system for registration of variations of a TM-mode field. Possibilities of using this schema in MTS are analyzed.

Keywords Magnetotelluric sounding · Vertical component of electric field · Circular electric dipole

1 Introduction

Variations of five components of electromagnetic field are usually registered during magnetotelluric soundings (MTS). It is connected with the fact that the MTS method

V. V. Plotkin (✉) · V. S. Mogilatov
Institute of Petroleum Geology and Geophysics SB RAS, Koptug Ave. 3,
Novosibirsk 630090, Russia
e-mail: PlotkinVV@ipgg.sbras.ru

T. B. Yanovskaya et al. (eds.), *Problems of Geocosmos—2018*, Springer Proceedings
in Earth and Environmental Sciences, https://doi.org/10.1007/978-3-030-21788-4_4

is based on the Tikhonov-Cagniard model in a form of a horizontally-layered medium excited by the vertically falling plane wave. With the standard approach, transfer functions are determined in experiment as ratios between horizontal components of electric $E_{x,y}$ and magnetic fields $H_{x,y}$. Deviations from the Tikhonov-Cagniard basic model are defined by the fifth registered component—the vertical component of magnetic field H_z.

Since the electrical conductivity of the atmosphere $\sigma_a \sim 10^{-14}$ S/m is very small, and the condition of continuity of vertical current is satisfied, it turns out that on the Earth's surface the vertical component of electric field E_z is close to zero, and there is no need for its registration.

However, in the real case the component E_z becomes zero only at the boundary with the atmosphere, while gradually decreasing in the adjacent skin layer. Moreover, if a medium within this layer is anisotropic, the E_z component may be non-zero even at the boundary, since the transverse current in this case does not necessarily coincide in direction with the E_z component.

On the other hand, in order to determine the component E_z or components $E_{x,y}$, in the field experiment the registration of potential differences for lines of finite length is carried out. It is possible to claim that anyway in practice the specified potential differences will always be non-zero and depend on the line length.

Considering, that in the process of solution of inversion problem for 3D medium the component E_z for each actual model will also be determined, a comparison of the calculated values of E_z with the experimental data is very desirable for higher reliability of the results. With a complex geoelectric structure of the medium and the presence of distortions in MTS curves, difficulties arise in interpreting the MTS data in practice. Unless possible distortions are taken into account, unreliable results of such an interpretation are quite likely: lateral near-surface heterogeneities can be perceived as deep "conductive" layers.

To eliminate these difficulties, various data processing algorithms and non-standard MTS options are being developed. Registration of vertical component E_z could become one of those options. The other one proposed in [1] consists of the spatial averaging of electric field horizontal components by registration of these components with long lines along the studied profile (electromagnetic array profiling).

In this regard, it is interesting to analyze a possibility of registering the vertical component E_z using the circular electric dipole (CED) located on the surface. The CED configuration was proposed in [2] as a power supply for the controlled-source electromagnetics (CSEM). At present, CED is successfully applied as an emitter in electromagnetic sounding methods [3], using its feature to excite only one TM-mode with the electric field having a vertical component in the horizontally layered medium. Since the magnetic field of TM-mode is equal to zero in the nonconductive atmosphere, registration of data on the Earth's surface eliminates the contribution of the direct field, and all observed variations of the magnetic field are associated only with deviations from the horizontally layered medium.

The ideal external current circuit and the real source are shown in Fig. 1. In practice, a CED is usually implemented with eight long radial lines. Considering the above-mentioned feature, CED can also be used as an MTS receiver for registration

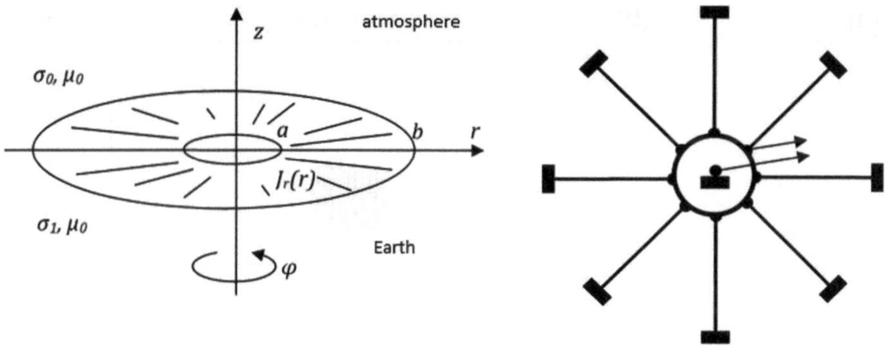

Fig. 1 Ideal (on the left) and real (on the right) circular electric dipole on the Earth's surface

of variations of TM-mode field, namely, of the vertical component E_z of electric field in the medium under the point of sounding. The method of measurement of potential difference on the vertical line immersed in a water reservoir is used for registration of E_z variations [see, for example, 4, 5]. On the land, it is necessary to drill a hole for this purpose, or use already available wells, which may be impractical. Therefore, search for new ways to register E_z is of great interest. The importance of obtaining new experimental results on the E_z variations for many areas of geophysics should also be noted. Possibilities of using CED in MTS are analyzed in this article.

2 Results of Numerical Calculations of Potentials of U_z and U_{CED}

It has been already established in [3] that when using CED as a source eight lines evenly distributed in azimuth are sufficient. To use CED as a receiver, it is necessary to define the ratio of potential differences registered on the vertical line in the medium U_z and on the Earth's surface by means of CED U_{CED}. The CED radius is an important characteristic that in practice determines the ratio of these potential differences. Influence of various factors when using CED as the MTS receiver can be investigated using numerical modeling.

For numerical simulation, the field model based on the Trefftz method [6] considering distortions of MTS curves caused by the three-dimensional heterogeneity of the medium was used. We consider calculation results for the medium model presented by seven layers with resistivities of 25, 50, 100, 1000, 10, 1000 and 1 Ω m (from top to bottom) and thicknesses of 0.1, 0.1, 0.1, 15, 10, 45 and 15 km, respectively, and the underlying homogeneous subsurface of 100 Ω m (Fig. 2a). Such normal geoelectric section includes two conductive layers: the Earth's crust at depths of 15–25 km and the asthenosphere at depths of 70–85 km. Three near-surface layers with thickness of 0.1 km can be used to simulate MTS curve distortions. In particular, influence

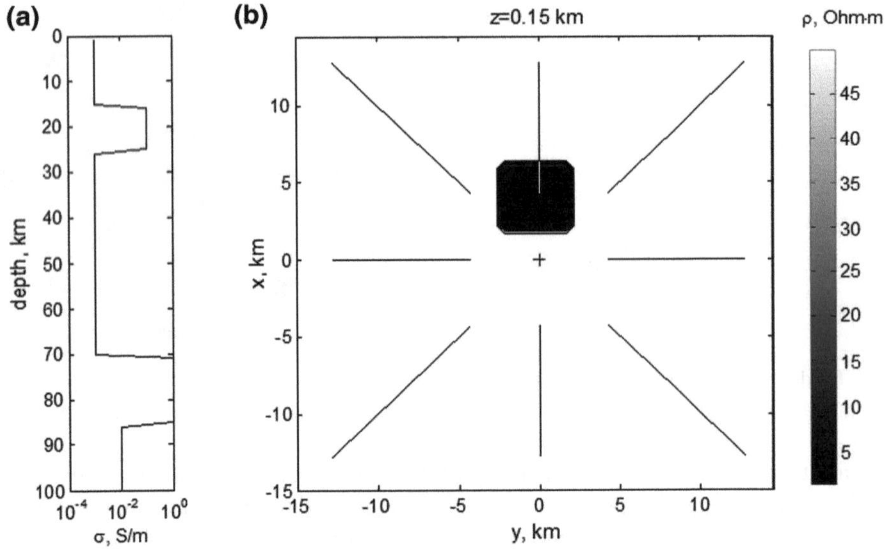

Fig. 2 3D model of the medium: the normal geoelectrical section (**a**), the horizontal section of second near-surface layer (**b**). CED radial lines on the Earth's surface are shown, the cross denotes its center

of a separate heterogeneity placed in the second layer from the surface (Fig. 2b) is studied here in detail.

It was assumed that MTS is carried out at the polygon with sizes of 30 × 30 km, with CED placed in its center. Their radial rays are shown in Fig. 2b by straight lines. When calculating using the numerical model based on the Trefftz method, each of seven medium layers is represented by 49 identical parallelepipeds with heights equal to the layer thickness and horizontal sections with sizes of 4.29 × 4.29 km (7 parallelepipeds along each of horizontal coordinate axes), in which the medium is uniform. In each parallelepiped, solutions of the Maxwell equations are used in the form of transverse counter propagating waves along each of coordinate axes. Considering the polarization of all waves, there are 12 such waves in each parallelepiped. All unknown amplitudes are solutions of the general system of equations obtained from conditions of matching tangential components of fields on all sides of the mentioned parallelepipeds and from boundary conditions. Using obtained wave amplitudes, we determine all components of the electromagnetic field in the centers of parallelepipeds and on the Earth's surface. Potential differences U_z and U_{CED} are determined further by integrals of the longitudinal component of electric field along vertical and radial registering lines respectively. To calculate all impedance tensor components, the calculations are repeated for two independent polarizations of vertical plane wave falling on the medium. Potential differences U_z and U_{CED} are also determined for each of two independent polarizations of primary wave. For this medium model we consider potential differences that are non-zero for chosen CED configuration, and correspond to the polarization of magnetic field of the primary wave along the *OY* axis with the amplitude of 1 nT.

In Fig. 3 the curves of apparent resistivity are shown obtained for the center of the polygon and the chosen medium model for two different resistivity values ρ_n in a heterogeneity. As can be seen, in the presence of heterogeneity in the second layer of the medium from the surface (Fig. 2b) there are noticeable galvanic distortions. On the long time periods, the curves of the apparent resistivities ρ_{xy} and ρ_{yx} are also shifted along the ordinate axis relative to normal curve ρ_l. The more conductive the heterogeneity is, the greater this shift becomes.

Figure 4 shows the calculated dependences of potential differences U_z and U_{CED} on the time period in the center of the polygon (marked with the cross in Fig. 2b). Calculations were performed for different sets of values of lateral heterogeneity parameters: resistivity ρ_n, total longitudinal conductance Σ_0, characteristic scale r_0, volume conductivity $r_0^2\Sigma_0$ and distance r_n from the center of heterogeneity to the CED center. Figure 4a displays changes in curves depending on resistivity values in a heterogeneity with identical other parameters. The greater are these deviations of resistivity in heterogeneity from background, the larger are values of potentials U_z and U_{CED}.

Due to controlled differences in the real CED from the ideal (in quantity and lengths of radial lines), the shift between curves U_z and U_{CED} along ordinate axis appears. Curves obtained by removal of such offset over long time periods are displayed in Fig. 4a. For this purpose, all values of U_{CED} curves are increased by a factor of 3.5. Note that this coefficient is the same for different resistivity values in the heterogeneity and under identical other conditions.

As can be seen, small differences between the U_z and U_{CED} curves remain for short time periods. This is confirmed by analyzing the data of ideal CED. With decreasing the distance to heterogeneity while preserving its volume conductivity $r_0^2\Sigma_0$, the coincidence of U_z and U_{CED} curves is also observed over an increased range of long

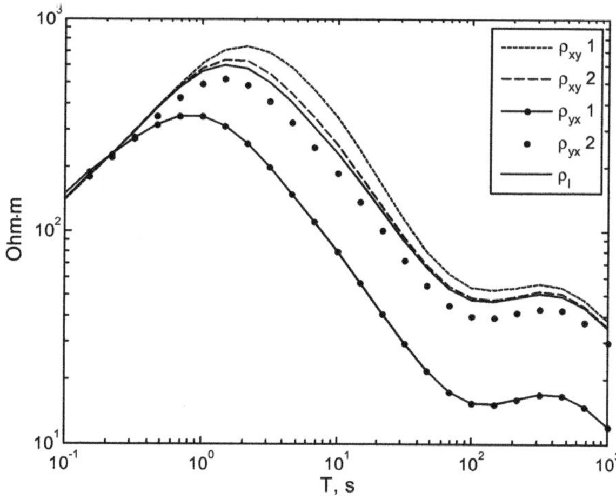

Fig. 3 Curves of apparent resistivities ρ_{xy} and ρ_{yx} in the polygon center for two resistivity values ρ_n in heterogeneity (1—$\rho_n = 1\ \Omega$ m, 2—$\rho_n = 10\ \Omega$ m, ρ_l—the normal curve for the background medium)

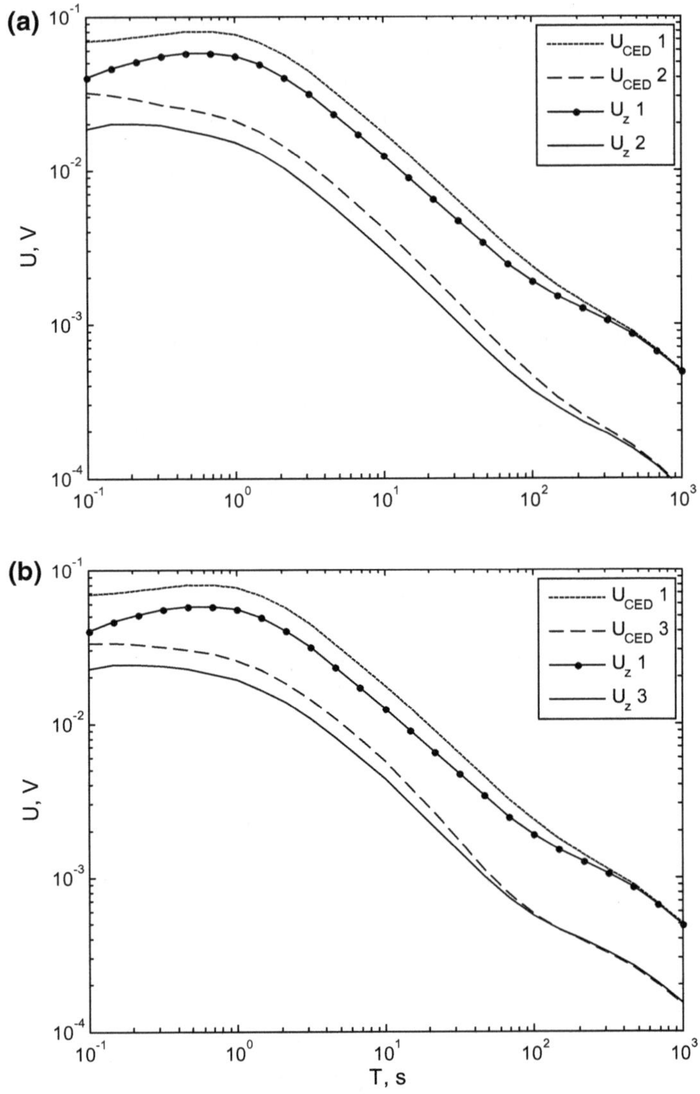

Fig. 4 Curves of the potential differences arising on the vertical line in the center U_z and totally on all CED radial lines U_{CED} for different parameters of lateral heterogeneity: volume conductivity $r_0^2\Sigma_0$ and removal of center of heterogeneity from CED center r_n: **a** 1—$\rho_n = 1\ \Omega$ m, $r_n = 4.29$ km, $r_0^2\Sigma_0 = 1.84 \times 10^9$ Sm^2, 2—$\rho_n = 10\ \Omega$ m, $r_n = 4.29$ km, $r_0^2\Sigma_0 = 1.84 \times 10^8$ Sm^2; **b** 1—$\rho_n = 1\ \Omega$ m, $r_n = 4.29$ km, $r_0^2\Sigma_0 = 1.84 \times 10^9$ Sm^2, 3—$\rho_n = 1.1\ \Omega$ m, $r_n = 1.43$ km, $r_0^2\Sigma_0 = 1.84 \times 10^9$ Sm^2

periods (Fig. 4b). Curves for the case 3 were obtained in calculations with changes in parameters: for a 10 × 10 km polygon, the resistivity value in a heterogeneity $\rho_n = 1.1\ \Omega$ m and other equal conditions. The normalization coefficient for U_{CED} curves for the case 3 turned out to be 1.12.

3 The Application of CED in the Arctic Region

Application of CED at its placement on the ice floe drifting in the Arctic is of interest. Because of nonconductive atmosphere, the medium excitation by TM-mode is excluded. The field recorded by CED is bounded only with 3D-inhomogeneities of the medium. To estimate CED application opportunities for MTS in the Arctic, we again carried out numerical calculations of the electromagnetic field by the Trefftz method [6]. Let us consider results of these calculations for the model of the medium presented in Fig. 5.

For the position of object and CED shown on Fig. 5, examples of results of calculation of the apparent resistivity and the vertical component of magnetic field are given in Fig. 6. Well conducting ocean shields manifestations of a target object in the behavior of magnetotelluric curves. Apparently, the object is shown in changes of apparent resistivity on the 100-th shares of percent. Variations of vertical component of magnetic field on day surface are also very small ~0.001 nT.

In Fig. 7 dependences of the potential of the TM-mode on different time periods, which are registered by the CED in its profile movement across the target, are given.

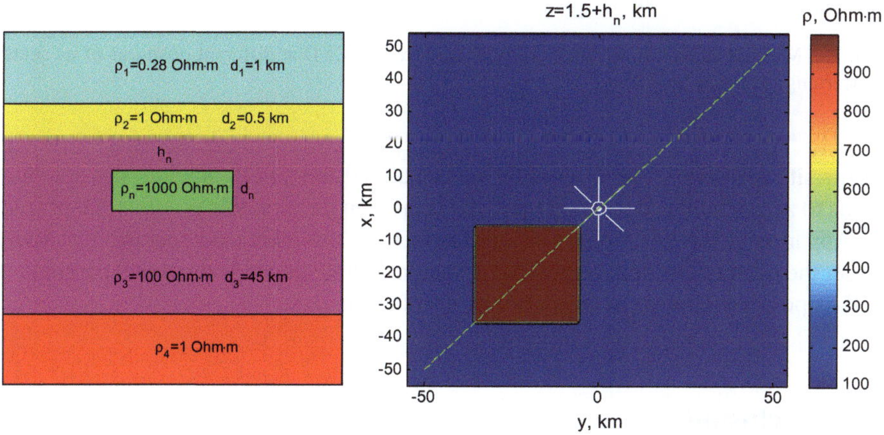

Fig. 5 Geoelectrical section (on the left) and the plan of heterogeneity that is a target object under the ocean floor (on the right). In the center, there is CED shown white. Dotted line shows the profile along which it moves in drift. Values ρ_1, d_1 correspond to the sea water, ρ_2, d_2—to the sediments, ρ_3, d_3—to the crust, ρ_4—to the mantle and ρ_n, d_n, h_n—to the target object

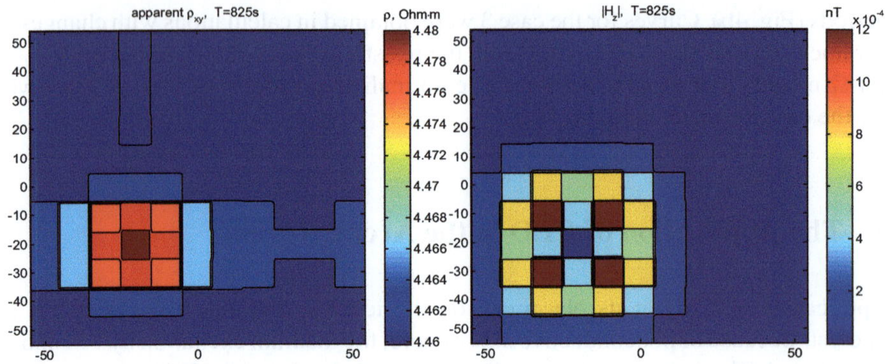

Fig. 6 Maps of apparent resistivity (on the left) and vertical component of magnetic field (on the right) for the time period of T = 825 s. Object parameters are $d_n = 2$ km, $h_n = 2$ km

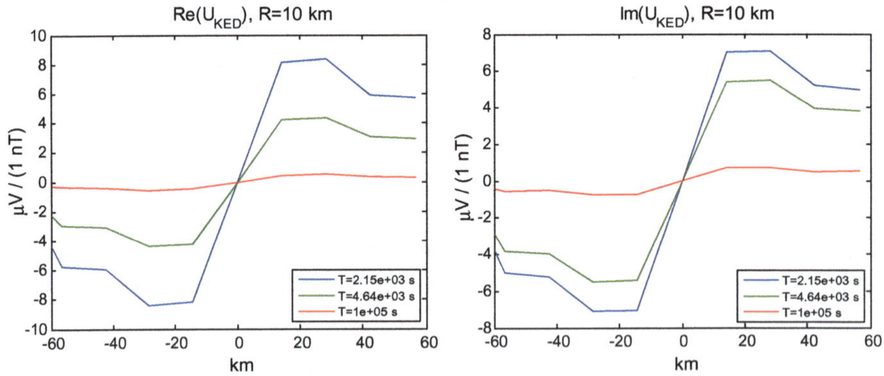

Fig. 7 TM-mode potential for three periods registered by CED at different positions of its center concerning the target object

In the situation when the centers of the target and the CED coincide, the potential tends to zero, changing the sign upon the transition through this point on the profile.

Apparently, the CED can find the target with the increased specific resistivity under the ocean floor. It is clear that CED signal value depends on object parameters and its position relative to the center of installation.

4 Conclusion

In magnetotelluric sounding (MTS), variations of five components of electromagnetic field are usually registered. With the complex geoelectric structure of the medium and the presence of distortions in MTS curves, difficulties arise in practical interpretation of the MTS data. In order to increase the reliability of results obtained by inversion and to extract more information from MTS, it is proposed to carry out an additional registration of vertical component of the electric field in the medium applying a circular electric dipole (CED) for this purpose.

The potential difference registered by ideal CED completely coincides with the potential difference registered by the vertical line in the medium, if thickness of skin layer is much more than a characteristic size of lateral inhomogeneities (at low frequencies or long time periods). Similar situation takes place for galvanic distortions of MTS curve in the electrostatic field, which arise in a layer with lateral inhomogeneities. An application of CED allows to obtain the dependence of potential on the vertical line in the virtual well in the medium from the time period.

The numerical modeling of CED application in the case of near-surface inhomogeneities in the horizontally layered medium was carried out. Owing to controlled differences of real CED from ideal one (by quantity and lengths of radial lines), we observe a shift between curves of U_z and U_{CED} along the coordinate axes. The coefficient of the shift does not depend on the resistivity values in a heterogeneity under identical other conditions. After accounting for such shifts on the long time periods, small differences between U_z and U_{CED} curves still remain on the short time periods.

Thus, the potential difference measured by means of CED in the electric field excited in the 3D medium by a vertically falling incident plane wave is equivalent to the potential difference arising on the long vertical line in the CED center. In practice, it means an opportunity to register the potential difference along the vertical line without drilling the well necessary for it.

Targets with the increased specific resistivity under the ocean floor can be found by the CED installation.

Acknowledgements This work was supported by the Russian Science Foundation under grant 18-17-00095.

References

1. Torres-Verdin, C., Bostick Jr., F.: Principles of spatial surface electric field filtering in magnetotellurics: electromagnetic array profiling (EMAP). Geophysics **57**(4), 603–622 (1992)
2. Mogilatov, V.S.: Method of geoelectric exploration. Patent RF N 1062631. Bulletin N 47, publish 23.12.83 (1982) (in Russian)
3. Mogilatov, V.S.: Pulse Geoelectrics, 181 p. Novosibirsk State University, Novosibirsk (2014) (in Russian)
4. Moroz, Yu.F., Moroz, T.A., Buglova, S.G.: Vertical and horizontal components of the lake Baikal electrotelluric field and their relation to the electric conductivity. Izv. Phys. Solid Earth **44**(3), 239–248 (2008)
5. Shneer, V.S., Gaidash, S.P., Trofimov, I.L., Korotaev, S.M., Kuznetsova, T.V., Tsirulnic, L.B., Panfilov, A.I., Budnev, N.M., Mirgazov, R.R.: Long-term observations of the electric field vertical component in lake Baikal (preliminary results). Izv. Phys. Solid Earth **43**(4), 331–335 (2007)
6. Plotkin, V.V., Gubin, D.I.: Accounting for near-surface inhomogeneities over a horizontally layered section in magnetotelluric sounding. Russ. Geol. Geophys. **56**(7), 1083–1090 (2015)

Statistical Properties of the Geomagnetic Field Variations and Geomagnetically Induced Currents

A. V. Vorobev, V. A. Pilipenko, Ya. A. Sakharov and V. N. Selivanov

Abstract The relationships between variations of geomagnetic field and geomagnetically induced current (GIC) are studied using data from the station for registration of GIC in the electric transmission line and the IMAGE magnetic observatory. A highest correlation of GIC intensity ($R > 0.7$) is found with the field variability $|d\mathbf{B}/dt|$, whereas the correlations of GIC with the time derivatives of X and Y components are about the same. The diurnal variations of hourly values of geomagnetic field variability and GIC intensity have a wide night maximum associated with the substorm activity, and a wide morning maximum presumably caused by geomagnetic pulsations of the Pc5-Pi3 type. A regression linear model is constructed to estimate GIC magnitude from the local time derivative of geomagnetic field. The statistical distributions of the probability density of $|d\mathbf{B}/dt|$ and GIC correspond to the log-normal law. On the basis of the constructed distributions the probabilities of extreme values of GIC and $|d\mathbf{B}/dt|$ are estimated.

Keywords Geomagnetically induced currents · Geomagnetic field variations · Space weather

---

A. V. Vorobev (✉)
Ufa State Aviation Technical University, Ufa, Russia
e-mail: geomagnet@list.ru

V. A. Pilipenko
Institute of Physics of the Earth, Moscow, Russia

Geophysical Center of the Russian Academy of Sciences, Moscow, Russia

Ya. A. Sakharov
Polar Geophysical Institute, Apatity, Russia

V. N. Selivanov
Center for Physical and Technical Problems of Energy of the North, Apatity, Russia

T. B. Yanovskaya et al. (eds.), *Problems of Geocosmos—2018*, Springer Proceedings in Earth and Environmental Sciences, https://doi.org/10.1007/978-3-030-21788-4_5

1 Introduction

Research on space weather is stimulated, on one hand, by the fundamental scientific interest in the Earth's environment comprising magnetosphere, ionosphere, atmosphere, and lithosphere. On the other hand, understanding and monitoring of space weather is vitally important to ensure the stable operation of technological systems. One of the most significant manifestations of space weather is the excitation of geomagnetically induced current (GIC) in conducting technological structures (power transmission lines, pipelines, transformers, cable networks) during magnetic storms and substorms [4]. With the development of technology, energy systems are becoming increasingly susceptible to space weather perturbations. Modern energy electric power networks with complex geometry in fact operate as a giant antenna which is electromagnetically coupled to the currents of the Earth's ionosphere. In grounded networks, GICs were recorded during magnetic storms up to 200–300 A [8], while currents with an intensity of only a few Amperes are capable to shift some types of transformers from the linear regime [15]. Although the most powerful disturbances of geomagnetic field, which lead to the excitation of intense GICs, occur in auroral latitudes, dangerous GICs can also be observed at middle and low latitudes [16].

Diagnostics and prediction of GIC levels during various types of geomagnetic disturbances, which can be used by network operators to take the necessary measures to mitigate the risk of catastrophic failures, is an extremely important task. At the same time, the solution of such a problem is not reduced simply to the "engineering" application of the space physics results for calculating GIC in a specific technological system, but also requires clarification of the physical nature of some magnetosphere and ionosphere phenomena. Greatest perturbations of magnetic field on the Earth's surface are caused by an extended auroral electrojet which creates magnetic perturbations oriented predominantly in the N–S direction [2]. However, small-scale ionospheric current structures can make a significant contribution to the rapid changes in the magnetic field, which are essential for the excitation of GIC [1, 12]. The nature of such structures has not yet been fully clarified.

Geophysical literature describes many individual events, when a close connection between geomagnetic field variations and GICs has been revealed during such space weather events as interplanetary shocks [7] and magnetic storms [3, 14]. At the same time, statistical studies of the relationship between geomagnetic field variations and GIC are lacking. Research of space weather effects on technological systems is limited by the lack of databases of GIC recording available for scientific analysis.

This paper is largely based on the data of unique Russian system of GIC registration in power transmission lines, deployed at the Kola Peninsula [10, 11]. The paper examines the statistical characteristics of geomagnetic disturbances, geomagnetic field variability $d\mathbf{B}/dt$, and GIC during 2015. In the case of a closed circuit with Ohmic resistance, the GIC magnitude would be completely determined by the electromagnetic induction, i.e. time derivative of geomagnetic field $d\mathbf{B}/dt$. In reality, even in the simplest case, GIC occurs in a spatially distributed system formed by the power lines, substations with poorly known characteristics, and underlying Earth surface with frequency-dependent anisotropic geoelectric properties. Thus, it is not a priori evident whether magnetic field variability $d\mathbf{B}/dt$ totally controls GIC intensity. For practical applications, it is important to estimate what GIC magnitude can be expected for various disturbances. Knowledge of such empirical relationship is necessary to build a diagnostic model for GIC based on parameters of geomagnetic field. A statistical distribution of the probability density of the geomagnetic field derivative and GIC may help to outline responsible mechanisms and to estimate probabilities of extreme values of GIC and $|d\mathbf{B}/dt|$ [9].

2 Data and Pre-processing

The GIC registration system measures with a time resolution of 1 min the quasi-DC current flowing in grounded neutral of transformer (http://eurisgic.org). For this study the "Vyhodnoi" (VKH) station with geographic coordinates 68.83° N, 33.08° E has been selected, in which the registration of the GIC is carried out on the 330 kV line. This station terminates the power line, so a measured current matches much better GIC in the line.

To characterize the local geomagnetic field activity, data of nearby magnetic stations from the IMAGE network (www.geo.fmi.fi/image) are considered. The 1-min data from closest to the GIC station observatories have been used: IVA (geographical coordinates 68.56° N, 27.29° E, separation 236 km), KEV (69.76° N, 27.01° E, separation 260 km) and SOD (67.37° N, 26.63° E, separation 313 km). Station IVA is at the same geomagnetic latitude with VKH, so global geomagnetic disturbances at those sites are expected to be similar. About 93.3% of the total data set is available for cross-station analysis (490,385 samples) [17].

The statistical relationships between GIC and planetary characteristics of space weather, such as AE (http://wdc.kugi.kyoto-u.ac.jp/aedir) and PC (http://www.geophys.aari.ru) indices have been examined in [16]. Apparently, planetary indices cannot reveal the conditions under which extreme values of currents occur at a selected station.

The perturbations of geomagnetic field horizontal component $\Delta\boldsymbol{B} = \{\Delta X, \Delta Y\}$ (X and Y are the N–S and E–W field components, correspondingly) and the time derivative $d\mathbf{B}/dt = \{dX/dt, dY/dt\}$ are calculated. Under horizontal homogeneity of geoelectric properties of underlying medium, the orientation of vector $d\mathbf{B}/dt$ is orthogonal to orientation of excited telluric field $\boldsymbol{E}$. The values of ΔB are calculated relative to the background level of B_0, which has been taken as the average value of $B(t)$ per day. To avoid difficulties with the changing sign of magnetic disturbance, the absolute values of disturbance components $|\Delta X|$, $|\Delta Y|$ are used. As a characteristic of field variability, absolute values of derivatives of horizontal components $|dX/dt|$ and $|dY/dt|$, and the full derivative $|d\boldsymbol{B}/dt| = \sqrt{(dX/dt)^2 + (dY/dt)^2}$ are used.

3 Diurnal Variation of Geomagnetic Disturbances and GIC

To examine diurnal variation of geomagnetic disturbances and GIC (dependence on local time LT) we construct histograms of yearly-averaged values of various characteristics of geomagnetic disturbance and GIC for 2015. Daily variation of disturbance $|\Delta X|$ at station IVA (Fig. 1a) show the presence of midnight (LT ~ 24) and late afternoon (LT ~ 15) maxima. These maxima are caused by the intensification of the western and eastern auroral electrojets above the station during the substorm activations. Though the difference between minimum and maximum values is not large, the dispersion of yearly-averaged estimates is significant.

Daily variation of hourly-averaged geomagnetic field variability $|d\mathbf{B}/dt|$ (Fig. 1b) has a different character: the wide night (LT ~ 24) and morning (LT ~ 5–6) maxima are observed. The night maximum is obviously associated with substorm activity. The increased field variability in the morning hours is presumably caused by intense geomagnetic pulsations of Pc5-Pi3 range, which are observed most often in the early morning hours [5]. The appearance of large values of geomagnetic field time derivative associated with Pi3 pulsations was noted in [18].

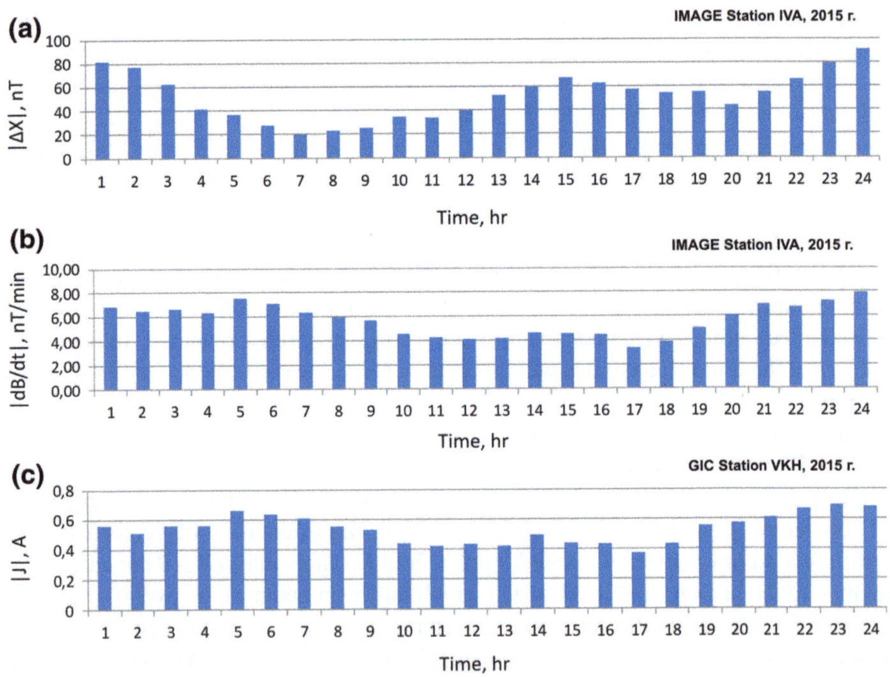

Fig. 1 Diurnal variations during 2015 of: **a** average magnetic perturbation $|\Delta X|$ at the IVA station; **b** average magnitude of geomagnetic field variability $|d\mathbf{B}/dt|$ at IVA, and **c** average GIC intensity $|J|$ at VKH station

The daily variation of the average intensity of the GIC at VKH station follows the variation of geomagnetic field variability |$d\mathbf{B}/dt$| with the morning and midnight maxima (Fig. 1c). Our results confirm the conclusion made in [13] that the morning maximum in |$d\mathbf{B}/dt$| does not have a counterpart in the distribution of geomagnetic disturbances intensity |ΔX|, and in the region of maximum eastern electrojet there is no increase in the level of variability |$d\mathbf{B}/dt$|.

4 Correlations Between GIC and Magnetic Field Variability

Knowledge of statistical relationships is necessary as a first step to build diagnostic models of GIC based on the general characteristics of space weather. The problem how well the geomagnetic indices characterizing the substorm activity (AE, PCN) can predict the GIC magnitude was addressed in [16]. The maximum correlation between the absolute value of GIC |J| recorded by VKH station was found for the indices AE ($R = 0.56$) and AL ($R = 0.55$), while the correlation with the PCN index was lower ($R = 0.44$).

Here we examine whether the magnitude of local geomagnetic disturbance and field variability are sufficient to predict the GIC magnitude. Table 1 shows the Pearson correlation coefficient R between the GIC absolute value |J| and geomagnetic field perturbations |ΔX|, |ΔY|, and the rate of change of field components |dX/dt|, |dY/dt| at IVA, KEV and SOD stations for 2015.

Correlations of |J| with the variability of horizontal components |dX/dt| and |dY/dt| are higher than with the field perturbation magnitude |ΔX|, |ΔY| by ~30%. Therefore, |$d\mathbf{B}/dt$| is a more promising parameter to characterize a GIC. Nonetheless, though the correlation coefficient is rather high, $R \sim 0.7$, the determination coefficient $D = R^2 \sim 0.5$ indicates that $d\mathbf{B}/dt$ is responsible for ~50% of GIC variations only.

Correlations |J| with variations of X- and Y-component derivatives are close (the difference in second digits). This result confirms that field derivative $d\mathbf{B}/dt$ fluctuates not only in magnitude but also in direction, which can indeed be caused by the presence of rapidly varying local vortex-like structures superimposed on the electrojet magnetic field [12].

Table 1 Correlation coefficients R between |J| at VKH and geomagnetic variations at near-by magnetic stations

IVA				SOD				KEV			
|ΔX|	|dX/dt|	|ΔY|	|dY/dt|	|ΔX|	|dX/dt|	|ΔY|	|dY/dt|	|ΔX|	|dX/dt|	|ΔY|	|dY/dt|
0.49	0.70	0.44	0.67	0.49	0.68	0.43	0.63	0.48	0.68	0.43	0.68

5 Regression Model of GIC

For applied assessments, it is important to know what magnitude of GIC can be expected under current level of geomagnetic field variability. To answer the question, a linear regression model has been synthesized, which in general has the form:

$$|J| = w_0 + w_1 \cdot |d\boldsymbol{B}/dt| \pm \Delta \quad (1)$$

where w_1, w_0 are weight coefficients, and Δ is the average simulation error. A model of the form (1) makes it possible to estimate statistically GIC value $|J|$ by the values of local magnetic field variability.

Linear regression model has been constructed for magnetic field variability $|d\mathbf{B}/dt|$ at IVA station for 2 months (from March 1 to April 30, 2015), as the largest interval for which there are no gaps in all analyzed parameters. The weights w_1, w_0 in (1) are calculated by the gradient descent method. The calculation with the model with restrictions (neglecting $|d\mathbf{B}/dt| \leq 1$ nT/min) gives the following coefficients: $w_0 = 0$, $w_1 = 0.074$ A min/nT. Figure 2 shows the result of comparing the simulation of GIC values with actual observations for the period of magnetic storm on March 17, 2015 with a series of substorm activations. A comparison of the predictions of models (2) with the measured values shows that the model based on $|d\mathbf{B}/dt|$ well predicts the moments of GIC intensifications and their magnitude in general, but underestimates extreme values. Model error $\Delta = \pm 0.9$ A proves that the regression model using the parameter $|d\mathbf{B}/dt|$ is adequate.

In general, the statistical model works well (small Δ) for intermediate values of $|d\mathbf{B}/dt|$ with the occurrence ~1% (which statistically corresponds to

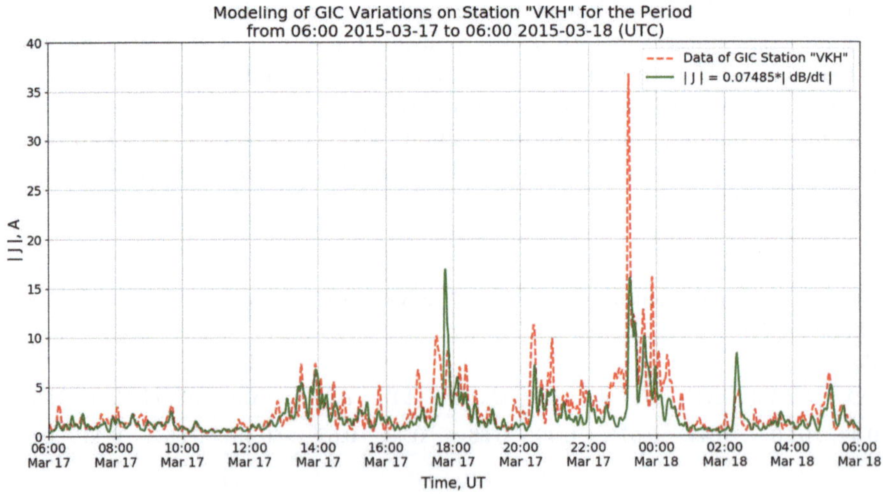

Fig. 2 Results of GIC modeling for the storm period from 00 UT to 24 UT on March 17, 2015

|*d***B**/*dt*| < 40 nT/min and the level |*J*| < 3 A). For large GIC values (|*J*| > 20 A), the regression model has $\Delta = \pm 2.3$ A, $w_0 = 11.677$ A, $w_1 = 0.11$ A min/nT.

6 Statistical Distributions of GIC and Geomagnetic Variations

The probability density function *F*(*x*) of perturbation amplitude *x* is determined by a physical mechanism of the process. For example, under the action of random independent effects, a normal (Gaussian) distribution is formed; in a confined system, energy of its components is distributed according to the exponential Boltzmann/Laplace law; the power-law distribution (Pareto-type) is often attributed to self-organized criticality; a random multiplicative action of several factors results in a log-normal distribution, etc. The presence of heavy tails of distribution is important. With such power distributions, variance of a studied quantity is determined mainly by rarely intense deviations, rather than by frequent small deviations. In geophysical studies the following distributions are commonly encountered:

- the log-normal distribution

$$F(x,\sigma) = \frac{1}{\sigma x\sqrt{2\pi}} \exp\left(-\frac{1}{2}\left(\frac{\ln(x)}{\sigma}\right)^2\right);$$

- power-law distribution

$$F(x,\alpha) = F_0(x/x_0)^{-\alpha};$$

- the generalized Pareto power distribution

$$F(x,c) = (1+cx)^{-1-\frac{1}{c}}$$

We have calculated the normalized histogram *n*(*A*) as a proxy of the probability density distribution, i.e. the probability of amplitude detection in the interval *A*, *A* + *dA* as follows $n(A) = N(A)/Nt$, where *N*t is the sample size. Also, we have calculated the amplitude distribution (survival function) *P*(*A*) as follows

$$P(>A) = \int_A^\infty n(A)dA$$

This function is a probability to observe a magnitude >*A*. Often many natural processes are described by the power-law distribution $P(>A) \propto A^{-\alpha}$. This dependence in the log-log scale looks as a straight line.

Figure 3 shows normalized histograms of X-component disturbances at IVA station for 2015. According to Table 2, the probability density distribution of $|\Delta X|$ values most consistent with the generalized Pareto distribution with $c = 0.47$. The tail ($|\Delta X|$ > 400 nT) of the amplitude distribution is well approximated by the power-law function.

Histograms of distribution of values $|d\mathbf{B}/dt|$ and $|J|$ are given in Figs. 4 and 5. The probability density distributions of $|d\mathbf{B}/dt|$ and $|J|$ are best approximated by the log-normal distribution with $\sigma = 1.15$ and $\sigma = 1.19$, correspondingly. The tails of

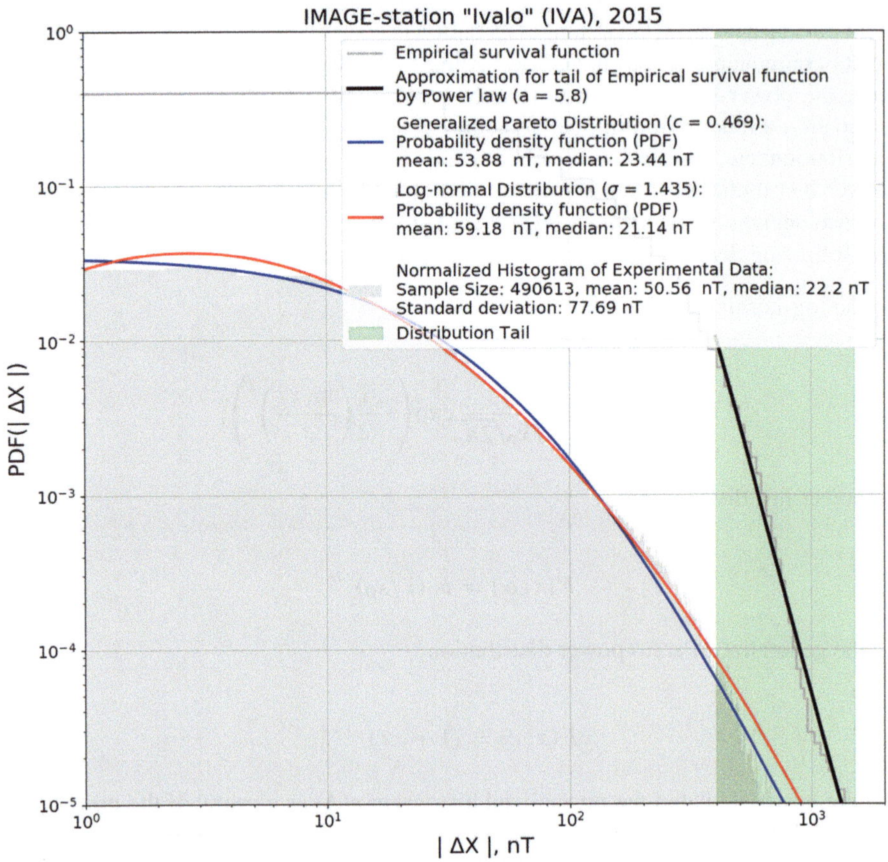

Fig. 3 Statistical distributions of magnetic field perturbation $|\Delta X|$ at IVA

Table 2 Kolmogorov criterion for the distribution of magnetic variations and GIC

Distribution	Time series		
	$\|X\|$	$\|d\mathbf{B}/dt\|$	$\|J\|$
Log-normal	0.0286	0.0331	0.0185
Generalized Pareto	0.0196	0.0763	0.0649

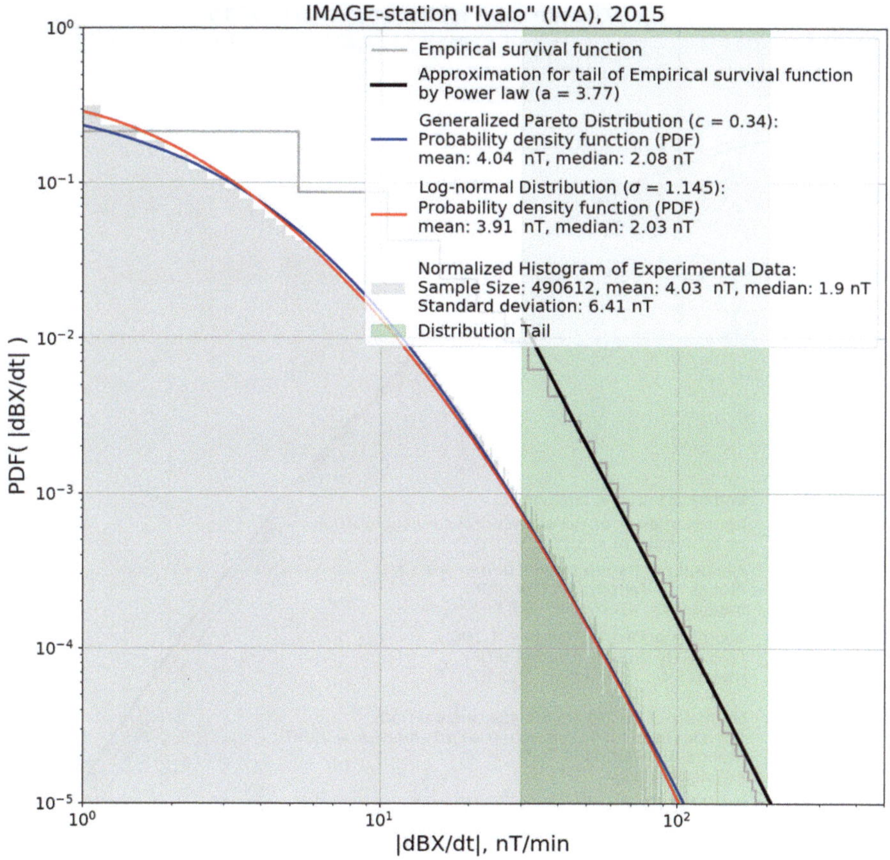

Fig. 4 Statistical distributions of the field variability $|d\mathbf{B}/dt|$ at IVA

the amplitude distributions of the field variability ($|dX/dt| > 30$ nT/min) and of the GIC ($|J| > 2$ A) are also well approximated by the power-law functions.

The obtained non-Gaussian distributions allow us to correctly determine the median, expected value and the probability of observing the analyzed parameters in a given range, as well as to evaluate the association of recorded values with anomalous events. Knowledge of such statistical distribution makes it possible to estimate probability of an extreme event, which during the observation period may not even be observed (assuming that it obeys the same law) [9]. From the probability curve (Figs. 3, 4 and 5), it is possible to estimate statistically which maximum perturbation of $d\mathbf{B}/dt$ and J is possible for a given observation period. In 2015 $|J| > 10$ A is observed ~0.03% of the time, and $|d\mathbf{B}/dt| > 60$ nT/min is observed ~0.2% of the time. With a probability of ~0.01% (approximately 50 times a year), regional perturbations of GIC and magnetic field may exceed $|J| > 13$ A, $|d\mathbf{B}/dt| > 113$ nT/min, and $|\Delta X|$

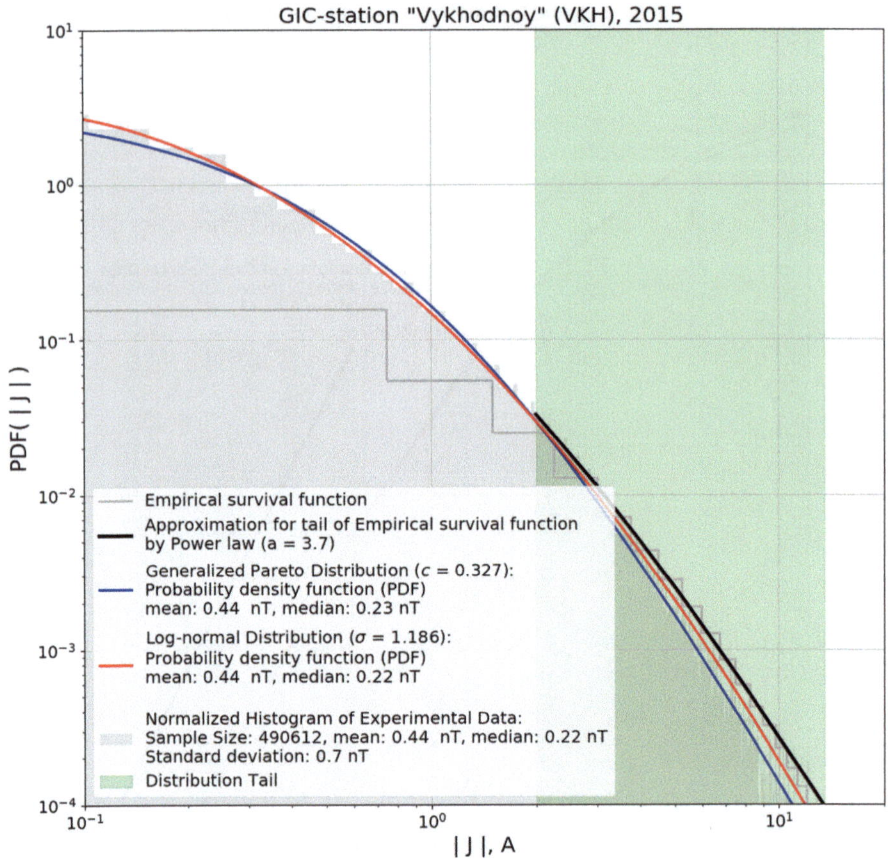

Fig. 5 Statistical distributions of GIC at VKH station

> 880 nT. Significant GIC variations ($|J| > 1$ A) are observed with a probability of ~9.7%.

After evaluating and analyzing the statistical characteristics of the time series, one can speculate about the similarity of their physical mechanisms. To test the hypothesis that the analyzed sample belongs to a particular known distribution, the Kolmogorov criterion is used, which characterizes the absolute maximum discrepancy between the experimental curves and the expected known distribution. A distribution with the minimum value of this criterion describes best the statistics of the experimental sample (Table 2).

According to Table 2, one may conclude that the statistics of $|\Delta X|$ values distribution is somewhat better described by the generalized Pareto distribution, whereas field variability and GIC correspond match better the log-normal distribution. The proposed hypotheses can be rejected with a significance level not exceeding 0.01%.

The fact that the probability distribution of both F ($|J|$) and F ($|d\mathbf{B}/dt|$) are described by heavy-tailed log-normal distribution indicate that this distribution is formed as a result of a multiplicative stochastic effect. It is interesting, that according to many observations, turbulence of near-Earth plasma often has a log-normal form [6]. Thus, such a coincidence may indicate that the turbulence of the near-Earth plasma is largely responsible for the variability of geomagnetic field and, therefore, for the appearance of GIC.

7 Conclusions

The yearly-averaged correlation between GIC and variability of geomagnetic field components $|dX/dt|$ and $|dY/dt|$ was found to be rather high, $R \sim 0.7$, higher than that between GIC and magnetic perturbations $|\Delta X|$, $|\Delta Y|$, $0.5 < R < 0.7$. The correlations $|J|$ with variation's of the derivatives of X- and Y-components are close, which confirms the quasi-isotropy of rapid variations of geomagnetic field derivative $d\mathbf{B}/dt$ [1, 12].

Daily variations of average values of geomagnetic field variability $|d\mathbf{B}/dt|$ and GIC intensities have a wide night maximum associated with the electrojet, and a wide morning maximum, presumably caused by intense geomagnetic pulsations of the Pc5-Pi3 type.

Regression linear diagnostic model with input parameter $|d\mathbf{B}/dt|$ predicts a GIC of moderate magnitude with an error ± 0.9 A. Large GICs ($20\ \text{A} < |J| < 45\ \text{A}$) can be predicted based on the parameter $|d\mathbf{B}/dt|$ with an accuracy of ± 2.3 A.

The statistical probability density distributions of values $|d\mathbf{B}/dt|$ and $|J|$ are most consistent with the log-normal distribution, whereas the probability density of the values $|\Delta X|$ somewhat better corresponds to the generalized Pareto distribution. On the basis of the constructed distributions the probabilities of extreme values of GIC and $|d\mathbf{B}/dt|$ can be estimated.

Acknowledgements This study was supported by the grant 16-17-00121 from the Russian Science Foundation. The authors are grateful to the research teams operating the IMAGE magnetometer network. The maintenance of the GIC recording system by the Polar Geophysical Institute and Center of Physical and Technical Problems of the Northern Energetics is supported by the grant 17-48-510199 from the Russian Foundation for Basic Research and the Murmansk region government.

References

1. Belahovsky, V.B., Pilipenko, V.A., Sakharov, Ya.A., Selivanov, V.N.: Characteristics of the variability of a geomagnetic field for studying the impact of the magnetic storms and substorms on electrical energy systems. Izv. Phys. Solid Earth **54**, 52–65 (2018)
2. Boteler, D.H., Pirjola, R.J., Nevanlinna, H.: The effects of geomagnetic disturbances on electrical systems at the Earth's surface. Adv. Space Res. **22**, 17–27 (1998)

3. Kappenman, J.G.: An overview of the impulsive geomagnetic field disturbances and power grid impacts associated with the violent Sun-Earth connection events of 29–31 October 2003 and a comparative evaluation with other contemporary storms. Space Weather **3**, S08C01 (2005). https://doi.org/10.1029/2004SW000128
4. Kelly, G.S., Viljanen, A., Beggan, C.D., Thomson, A.W.P.: Understanding GIC in the UK and French high-voltage transmission systems during severe magnetic storms. Space Weather **15**, 99–114 (2017)
5. Kleimenova, N.G., Kozyreva, O.V.: Spatio-temporal dynamics of geomagnetic pulsations Pi3 and Pc5 during extreme magnetic storms in October 2003. Geomag. Aeron. **45**, 71–79 (2005)
6. Kozak, L.V., Savin, S.P., Budaev, V.P., Pilipenko, V.A., Lezhen, L.A.: Character of turbulence in the boundary regions of the Earth's magnetosphere, Geomag. Aeron. **52**, 445–455 (2012)
7. Pilipenko, V.A., Bravo, M., Romanova, N.V., Kozyreva, O.V., Samsonov, S.N., Sakharov, Ya.A.: Geomagnetic and ionospheric responses to the interplanetary shock wave of March 17, 2015. Izv. Phys. Solid Earth **54**, 721–740 (2018)
8. Pirjola, R., Pulkkinen, A., Viljanen, A.: Studies of space weather effects on the Finnish natural gas pipeline and on the Finnish high-voltage power system. Adv. Space Res. **31**, 795–805 (2003)
9. Pulkkinen, A., Bernabeu, E., Eichner, J., Beggan, C., Thomson, A.W.P.: Generation of 100-year geomagnetically induced current scenarios. Space Weather **10**, S04003 (2012). https://doi.org/10.1029/2011SW000750
10. Sakharov, Ya.A., Danilin, A.N., Ostafiychuk, R.M.: Registration of GIC in power systems of the Kola Peninsula. In: Proceedings of 7th International Symposium on Electromagnetic Compatibility and Electromagnetic Ecology, pp. 291–293 (2007)
11. Sakharov, Ya.A., Danilin, A.N., et al.: Geomagnetically induced currents in the power systems of the Kola peninsula at solar minimum. In: Proceedings of 8th International Symposium on Electromagnetic Compatibility and Electromagnetic Ecology, pp. 237–238 (2009)
12. Viljanen, A.: The relation between geomagnetic variations and their time derivatives and implications for estimation of induction risks. Geophys. Res. Lett. **24**, 631–634 (1997)
13. Viljanen, A., Tanskanen, E.: Climatology of rapid geomagnetic variations at high latitudes over two solar cycles. Ann. Geophys. **29**, 1783–1792 (2011)
14. Viljanen, A., Tanskanen, E.I., Pulkkinen, A.: Relation between substorm characteristics and rapid temporal variations of the ground magnetic field. Ann. Geophys. **24**, 725–733 (2006)
15. Vakhnina, V.V., Shapovalov, V.A., Kuznetsov, V.N., Kretov, D.A.: The influence of geomagnetic storms on thermal processes in the tank of a power transformer. IEEE Trans. Power Delivery **30**, 1702–1707 (2015)
16. Vorobev, A.V., Pilipenko, V.A., Sakharov, Ya.A., Selivanov, V.N.: Statistical relationships between variations of the geomagnetic field, auroral electrojet and geomagnetically induced currents. Sol. Terr. Phys. **5**, 35–42 (2019)
17. Vorobev, A.V., Vorobeva, G.R.: Approach to assessment of the relative informational efficiency of Intermagnet magnetic observatories. Geomag. Aeron. **58**, 625–628 (2018)
18. Yagova, N.V., Pilipenko, V.A., Fedorov, E.N., Lhamdondog, A.D., Gusev, Yu.P.: Geomagnetically induced currents and space weather: Pi3 pulsations and extreme values of time derivatives of the geomagnetic field's horizontal components. Izv. Phys. Solid Earth **54**, 749–763 (2018)

Part II
Paleomagnetism, Rock Magnetism and Geomagnetism

Magnetostratigraphy and Biostratigraphy of the Valanginian in the Crimean Mountains

V. A. Grishchenko, A. G. Manikin, Yu. N. Savelieva and A. A. Feodorova

Abstract Preliminary results of the study aiming to construct a magnetostratigraphic scheme of the Valanginian of the Crimean Mountains are presented. Paleomagnetic and petromagnetic data were obtained on oriented samples taken from 363 stratigraphic levels of 10 sections in East and Southwest Crimea. Studied Valanginian sediments are generally a rather unfavorable object for paleomagnetic research because of several factors such as landslide deformations, hypergenous rock alteration, presence of condensed series, etc. Despite that, careful analysis of obtained data allows us to evaluate their magnetic polarity and to propose a first draft composite magnetostratigraphic section of the Valanginian for the Crimean Mountains. For the Valanginian sediments of the Southwest Crimea, new data were obtained on the foraminifers and ostracods which control the magnetostratigraphic correlation paleontologically.

Keywords Berriasian · Valanginian · Crimean Mountains · Magnetostratigraphy · Biostratigraphy · Geomagnetic polarity · Magnetic chrons · Foraminifers · Ostracods

1 Introduction

Sediments of Berriasian and Valanginian age are widespread on the territory of the Crimean peninsula. The most complete sections of the Valanginian stage, in terms of sedimentation continuity, are found in East Crimea. However, the complete absence of index fauna (ammonites), insufficient resolution ability of micropaleontologic methods and intensive deformations of an unclear nature (syn- or post-sedimentation) prevent their detailed subdivision, substantiation of their exact stratigraphic age and the determining the Berriasian-Valanginian boundary position. Despite all above difficulties, in the last few years a reliable magnetostratigraphy has been established

V. A. Grishchenko (✉) · A. G. Manikin
Saratov State University, Astrakhanskaya Str., 83, Saratov 410012, Russia
e-mail: grishenko-vladimir@bk.ru

Yu. N. Savelieva · A. A. Feodorova
AO "Geologorazvedka", Fayansovaya Str. 20/10, Saint-Petersburg 192019, Russia

T. B. Yanovskaya et al. (eds.), *Problems of Geocosmos—2018*, Springer Proceedings in Earth and Environmental Sciences, https://doi.org/10.1007/978-3-030-21788-4_6

for the Berriasian-Valanginian boundary interval in the section Zavodskaya balka (Feodosia) [8] and the synsedimantary age of deformations in sediments has been substantiated [17]. Studies of two remaining Valanginian sections (Sultanovka Village and Koklyuk Mountain) in Feodosia region are now in progress.

Southwest Crimea is the only region of the peninsula, where the Valanginian age of sediments is substantiated by ammonite fauna. However, this does not remove the difficulties related to the definition of the Berriasian-Valanginian boundary in Crimea. This problem has exacerbated after the decision to transfer the T. otopeta zone/subzone, previously traditionally identified with the base of the Valanginian (e.g., [11]), to the Berriasian as an upper subzone of the S. boissieri zone (e.g., [1, 21]). The lower Valanginian substage in Southwest Crimea is severely condensed, so that the maximum thickness of the correspondent coarse-grained sandstones does not exceed 9 m, but the upper substage is represented by a thick (~45 m) series of alternating clay and sandstone. The detailed microfaunistic characteristic of the Valanginian in Southwest Crimea has been missing until present.

The above difficulties hamper the correlation of Valanginian sections across the Crimean Mountains, and their comparison with the key sections in Mediterranean. To overcome them, it is necessary to employ multiple methods, preferably those based on features having a global distribution all over the world. One of such methods is magnetostratigraphy, based on the registration of reversals of the Earth's magnetic field that are global by their nature. The magnetostratigraphic data are invaluable for distant correlations since the magnetic chrons boundaries serve as an isochronous marker, allowing, in combination with paleontological data, to precisely compare detailed scales of different regions. Obtaining the magnetic polarity characteristics of the Berriasian-Valanginian boundary interval in the Crimean Mountains would thus give a possibility for an isochronous correlation of Crimean sections with the coeval sections in other regions that are already characterized paleomagnetically.

In the framework of the project aimed on compiling a composite magnetostratigraphic section of the Valanginian in the Crimean Mountains we studied 9 sections expected to contain the Berriasian-Valanginian boundary interval, 3 sections in East Crimea (Zavodskaya balka, Koklyuk, Sultanovka) and 6 sections in Southwest Crimea (Dlinnaya, Sheludivaya, Patil-1, Patil-2, Selbukhra, Bolshoy Kermen) (Fig. 1). Paleomagnetism of Valanginian rocks of condensed sections in Southwestern Crimea was studied for the first time.

A key section of the upper Valanginian, the Rezanaya Mountain near the Verkhorechye Village of Bakhchisaray district has been paleomagnetically studied earlier [45]. However, bearing in mind a possibility of using a 2G Enterprises cryogenic magnetometer for magnetization measurements, in 2016–2018 we have conducted a detailed field and laboratory study of this section.

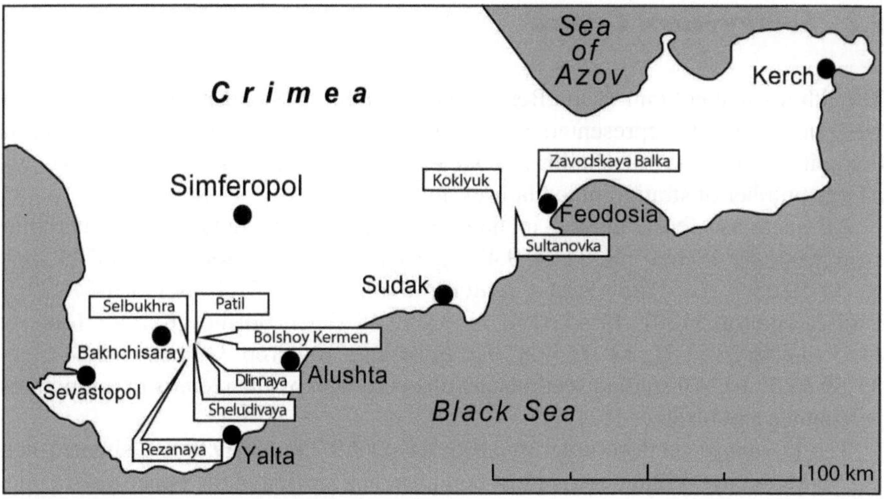

Fig. 1 Location of the studied sections

2 Characteristics of Sections

2.1 East Crimea

In East Crimea, within the Berriasian-Valanginian boundary sediments the Sultanovskaya formation and Nanikovskaya series are recognized (e.g., [9]). Sultanovskaya formation is composed of gray carbonate clay with rare thin layers of marl and limestone. Nanikovskaya series, according to our data, overlies concordantly the Sultanovskaya formation and is composed of similar clays with thin layers of siderite; however it is impossible to trace exactly the boundary between the two.

The Zavodskaya balka section is exposed by a clay extraction quarry in the northern outskirts of Feodosia. The results of complex paleontological and paleomagnetic studies of the Berriasian at Zavodskaya balka were published previously [4, 7, 20]. In 2015 sediments composing the Berriasian-Valanginian boundary interval (outcrop 3058, coordinates: 45°01′49.1″ N, 35°20′59.5″ E) were sampled. The Koklyuk section (outcrop 3030: 45°00′08.5″ N, 35°12′27.5″ E, and outcrop 3060: 45°00′08.6″ N, 35°12′31.3″ E) is situated near the Nanikovo Village in a ravine on the slope of Koklyuk Mountain. The Sultanovka section (outcrop 2926: 45°00′09.9″ N, 35°17′38.2″ E) is situated near the Sultanovka Village in the apex of Sultanovskaya syncline. All the studied sections are characterized by micropaleontological and ammonite data [8].

2.2 Southwestern Crimea

The lithological column of the Berriasian-Valanginian boundary interval in Southwestern Crimea is represented by an alternation of solid and loose calcareous dolomitic sandstones of shallow, shoal genesis. The sections are condensed and a large number of stratigraphic hiatuses are found [12].

All sections were sampled in natural outcrops at mountains Dlinnaya (outcrop 3065: 44°45′32.9″ N, 34°00′9.4″ E), Sheludivaya (outcrop 3107: 44°45′28.2″ N, 33°59′55.7″ E), Patil (Patil-1 (outcrop 3133): 44°45′51.4″ N, 33°59′46.7″ E; Patil-2 (outcrop 3139): 44°45′51.4″ N, 33°59′48.7″ E), Bolshoy Kermen (outcrop 3135: 44°46′22.3″ N, 34°00′54.8″ E), Selbukhra (outcrop 3134: 44°43′54.6″ N, 33°59′33.5″ E). All studied sections are characterized by the records of ammonites, belemnites and bivalves [12].

The Rezanaya section (outcrop 3106: 44°42′7.9″ N, 33°59′9″ E), situated near the Verkhorechye Village is the most complete key section of the upper Valanginian in the Crimean Mountains [12].

In the description of the section, made by E. Yu. Baraboshkin [12], five lithologic slices were detected. The first one, situated at the base of the section consists of "pudding" conglomerates and gravelstone. In the lower 1.5 m of the slice, the *T. otopeta* zone was recognized. Higher in the section an interval corresponding to the *Thurmanniceras pertransiens* zone follows. The sediments of the first slice are related to the Karatlykhskaya formation [46]. Overlying four slices in general are represented by alternation of clay with solid and loose sandstones of Rezanskaya formation [46]. The records of the species *Neohoploceras submartini* (Mallada) correspond to the second and third slices, which were assigned to the upper Valanginian. In the lower part of the fourth slice, the *Himantoceras trinodosum* zone is recognized, but most of the slice in terms of ammonite zonation is related to the *Teschenites callidiscus* zone of the upper Valanginian. The complex of fauna found in the fifth slice is interpreted as late Valanginian [12].

3 Methods

Oriented samples were collected from a total 363 stratigraphic levels in 10 sections described above. Sampling was conducted by two different methods: (1) using a mining pick (sections Zavodskaya balka, Koklyuk, Sultanovka, Dlinnaya and the Rezanaya Mountain); (2) using a specialized portable drill (model Pomeroy EZ core D261-C) with a diamond crown, made of non-magnetic metal (model Pomeroy BSS-1E Drill Bit), with the diameter of 1 in. (sections Sheludivaya, Patil-1, Patil-2, Selbukhra, Bolshoy Kermen and the Rezanaya Mountain). For the orientation of a drilled cylinder an orientation device (model OR-2) with an attached magnetic compass (model "Brunton Pocket Transit"). Sampling intervals were chosen in accordance with the thickness of studied section and varied between a minimum of

14 ccm for the condensed sections and 50–90 ccm for 'normal' sections. To exclude possible misinterpretations and errors in assimilating data obtained by different methods, sampling was carried out in bed-by-bed manner simultaneously with geological description and paleontological analysis. For further laboratory studies each sample, depending on the way of sampling, was sawn to the specimens of cubic (8 cm) or cylinder (10 cm) shape. The total quantity of oriented specimens was 1089.

In order to estimate the suitability of rocks for the paleomagnetic analysis and to obtain a supplementary geological information, each sample was explored for a number of petromagnetic characteristics. These included of magnetic susceptibility (K) and anisotropy of magnetic susceptibility (AMS), magnetic susceptibility after heating a specimen to 500 °C (Kt), natural remanent magnetization ($\mathbf{J_n}$), saturation remanent magnetization ($\mathbf{J_{rs}}$), coercivity of remanence ($\mathbf{H_{cr}}$). Out of these, parameters such as ratio $K/\mathbf{J_{rs}}$, increment of magnetic susceptibility after the heating of specimens up to 500 °C $dK = Kt - K$ and the Koenigsberger ratio Q were calculated. Rock magnetic experiments also included thermomagnetic analysis (TMA) carried out using a TAF-2 magnetic balance (Orion Ltd., Russia), and measurements of temperature dependences of magnetic susceptibility using an MFK1-FA kappabridge with an attached CS3 furnace (AGICO, Czech Republic).

Paleomagnetic studies were conducted using standard methods [29, 34] and consisted of consequential temperature demagnetization up to 350–500 °C with 50 °C increment or alternating field demagnetization up to 50–100 mT with 2–5 mT increment. Measurements of natural remanent magnetization were conducted using a spin-magnetometer JR6 (AGICO, Czech Republic) in the Petrophysics laboratory of SSU (Saratov) and a cryogenic magnetometer (2G Enterprises, US) in paleomagnetic laboratory (Institute of Physics of the Earth of the Russian Academy of Sciences, Moscow). To detect possible phase transformations of magnetic minerals during temperature demagnetization, after each heating step a magnetic susceptibility (K) of the sample has been measured using an MFK-1FB kappabridge.

The standard field tests [27, 28] and also a set of indirect characteristics (directions of different polarity are regularly grouped throughout the section, making large N- or R-magnetozones; polarity sign is indifferent to lithological composition; paleomagnetic structures of the examined sections conform to each other) [19] were used to verify the ancient nature of magnetization.

The samples for micropalaeontological studies were processed with the standard extraction technique for foraminifera and ostracods. The palynological samples were processed with the use of standard HF/HCl acid preparation method. The foraminifers were identified by A. Feodorova, ostracods by Y. Savelieva. Foraminifers and partially ostracods were photographed under a binocular microscope LOMO MCP-1 (Geologorazvedka); the rest of the ostracods were photographed on a scanning microscope JEOL-JSM-6390 LA (Botanical Institute and Palaeontological Institute, Russian Academy of Science). Foraminifer and ostracod collections are kept at the Petroleum Geology Department of the AO "Geologorazvedka", Saint-Petersburg, Russia.

4 Results

4.1 Micropaleontologic Data

Micropaleontologic data on the Berriasian-Valanginian boundary sections of East Crimea were published previously by V. V. Arkadiev and co-workers (e.g., [6–8, 35, 36, 37]). In the sections of Southwest Crimea the foraminifers and ostracods were studied in four sections, namely Patil-1, Patil-2, Dlinnaya and Rezanaya. In the powders that were investigated, we found: teeth, megaspores, fragments of echinoderms, belemnites, staghorns, shelly detritus, ichtyo detritus, blackened phytodetritus. The major part of samples contain medium- and coarse grained quartz, foraminifers and the fragments of echinoderms.

Foramnifers were found in all samples. In general they are represented by primitive forms, and also simple Haplophragmiidae, Trochammina and Lenticulina of different integrity degree. In two samples collected from the Rezanaya section, numerous Rotaliida, Ataxophragmiidae and singular planktonic forms were found.

Four assemblages of foraminifers were established.

The complex of "primitive" foraminifers, a typical "beach—shoal" complex, is represented by primitive forms (Hippocrepininae, Saccamminidae and Hormosinidae), simple Haplophragmiidae and singular Trochammina—agglutinated benthos foraminifers with coarse- and medium grained agglutinate. The presence of the species *Ammobaculites subasper* Bulynnikova, *Bulbobaculites inconstans* Bartenstein et Brand and *Trochammina* ex gr. *neocomiana* Mjatliuk (Patil-1 section; sample 3133/1) points at the early Cretaceous (Berriasian-Hauterivian) age of host rocks (Fig. 2). On the Dlinnaya Mountain (sample 3065/1), the age of stratigraphic level with *Reophax adaptatus* Dain, *Kutsevella* cf. *praegoodlandensis* (Bulynnikova), and *Haplophragmium* ex gr. *aequale* Roemer is assumed as Tithonian—lower Valanginian.

In the sections Patil-2, Dlinnaya and Rezanaya, the fossil complex with *Lenticulina espitaliei*, *Haplophragmium* ex gr. *Aequale* is established (Figs. 2, 3 and 4; Plates 1, 2). The particularity of this complex is the presence of rolled and split exemplars of secreting benthos characteristic for upper sublittoral inhabit, on the back of typical "shoreface" agglutinate forms with coarse- medium-grained agglutinate. The following forms are also identified here till the species rank: *Reophax* ex gr. *metensis* Franke, *Kutsevella* cf. *praegoodlandensis* (Bulynnikova), *Ammobaculites subasper* Bulynnikova, *Bulbobaculites inconstans* Bartenstein et Brand, *Haplophragmium monstratus* (Dain), *H.* ex gr. *aequale* Roemer, *Flabellammina* ex gr. *lidiae* Mjatliuk, *F. jurassica* Mjatliuk, *Triplasia emslandensis* Brat. et Brand, *Trochammina* ex gr. *neocomiana* Mjatliuk, *Verneuilinoides neocomiensis* (Mjatliuk), *Lenticulina macra* Gorbatchik, *L. espitaliei* Dieni et Massari, *L. insignita* Mjatliuk, *L.* aff. *akmetchetica* Mjatliuk, *L. andromede* Espitalie et Sigal, *L. nimbifera* Espitalie et Sigal, *L. nodosa* (Reuss), *Trocholina* cf. *alpine* (Leupold). These species are characteristic for the stratigraphic interval from the Tithonian to the Berriasian or from the Berriasian to the Hauterivian.

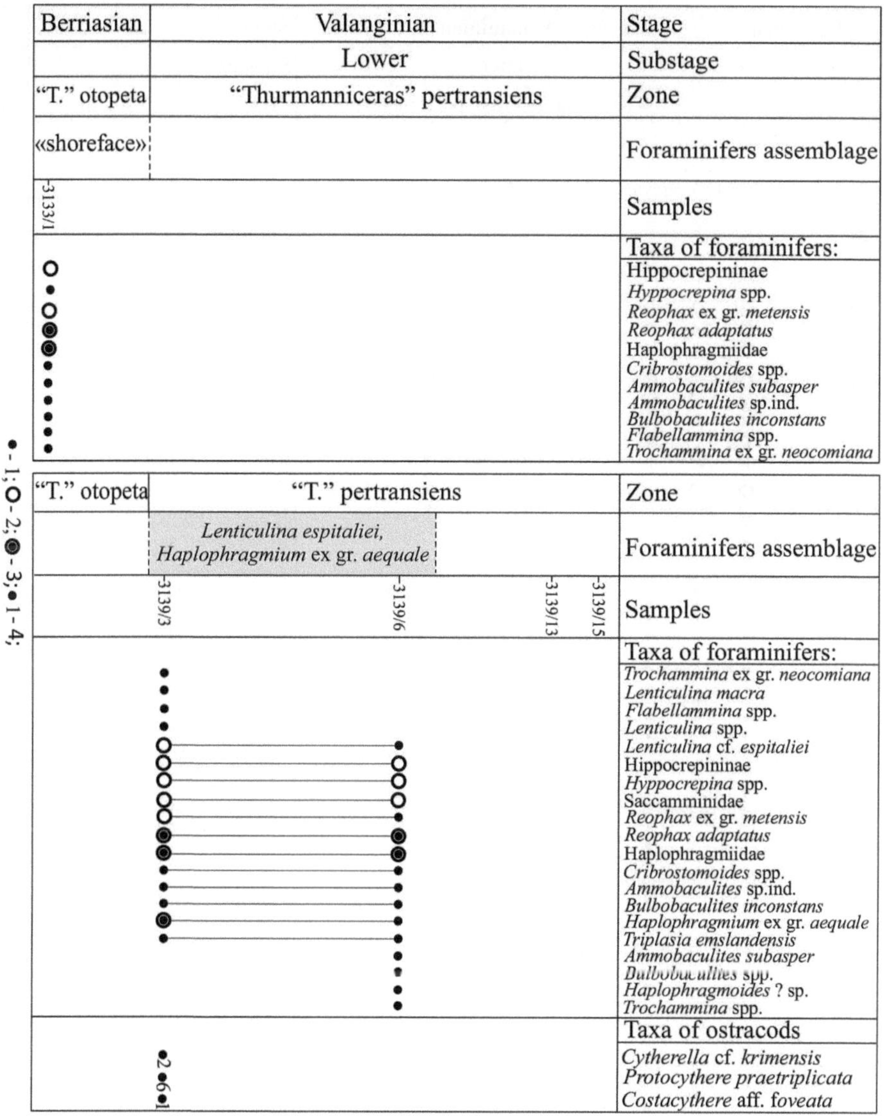

Fig. 2 Distribution of foraminifers and ostracods in outcrops Patil-3133 and Patil-3139 (Southwest Crimea). *Legend 1* 1–10; *2* 10–20; *3* 20–50—quantity of specimens in a sample; *4* occurrence and quantity of ostracods

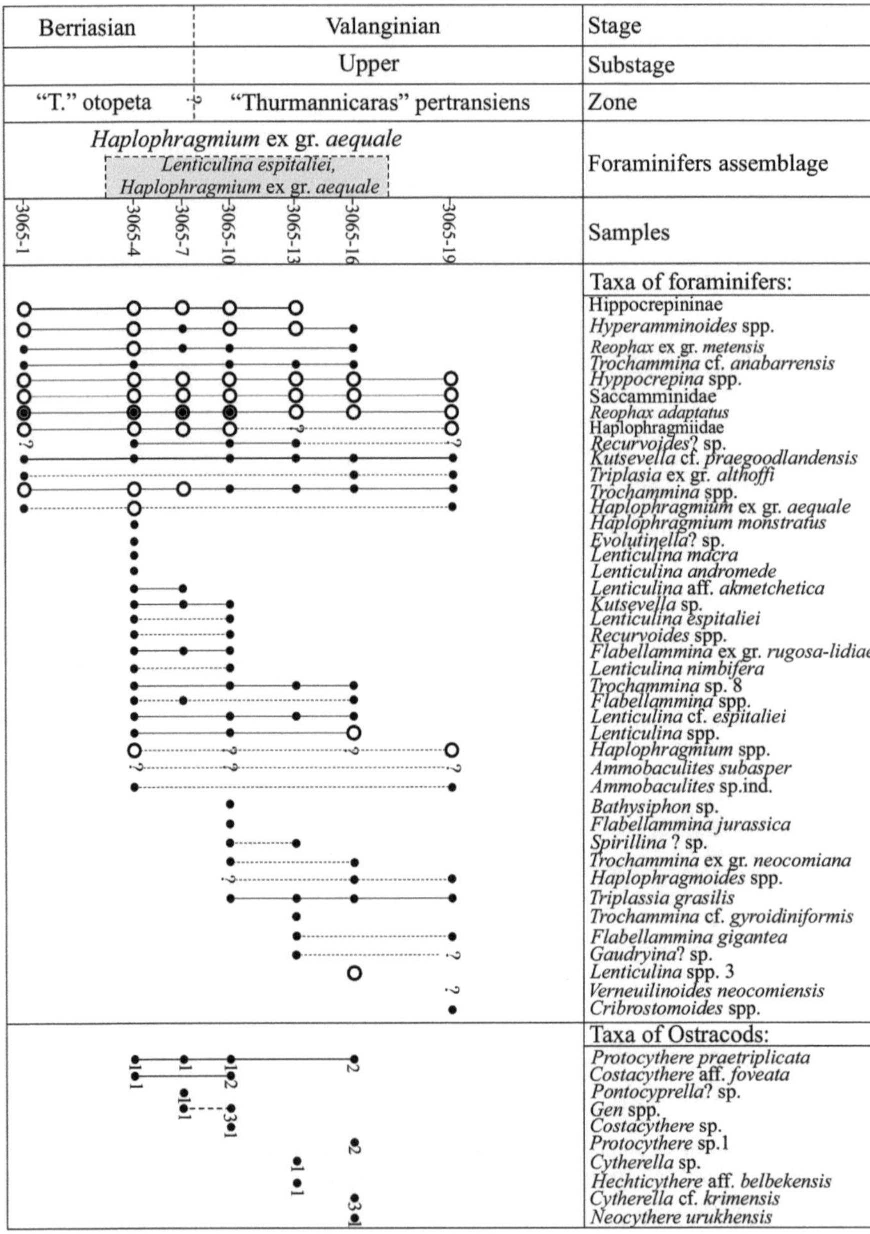

Fig. 3 Distribution of foraminifers and ostracods in the outcrop Dlinnaya-3065 (Southwest Crimea). Notation as in Fig. 2

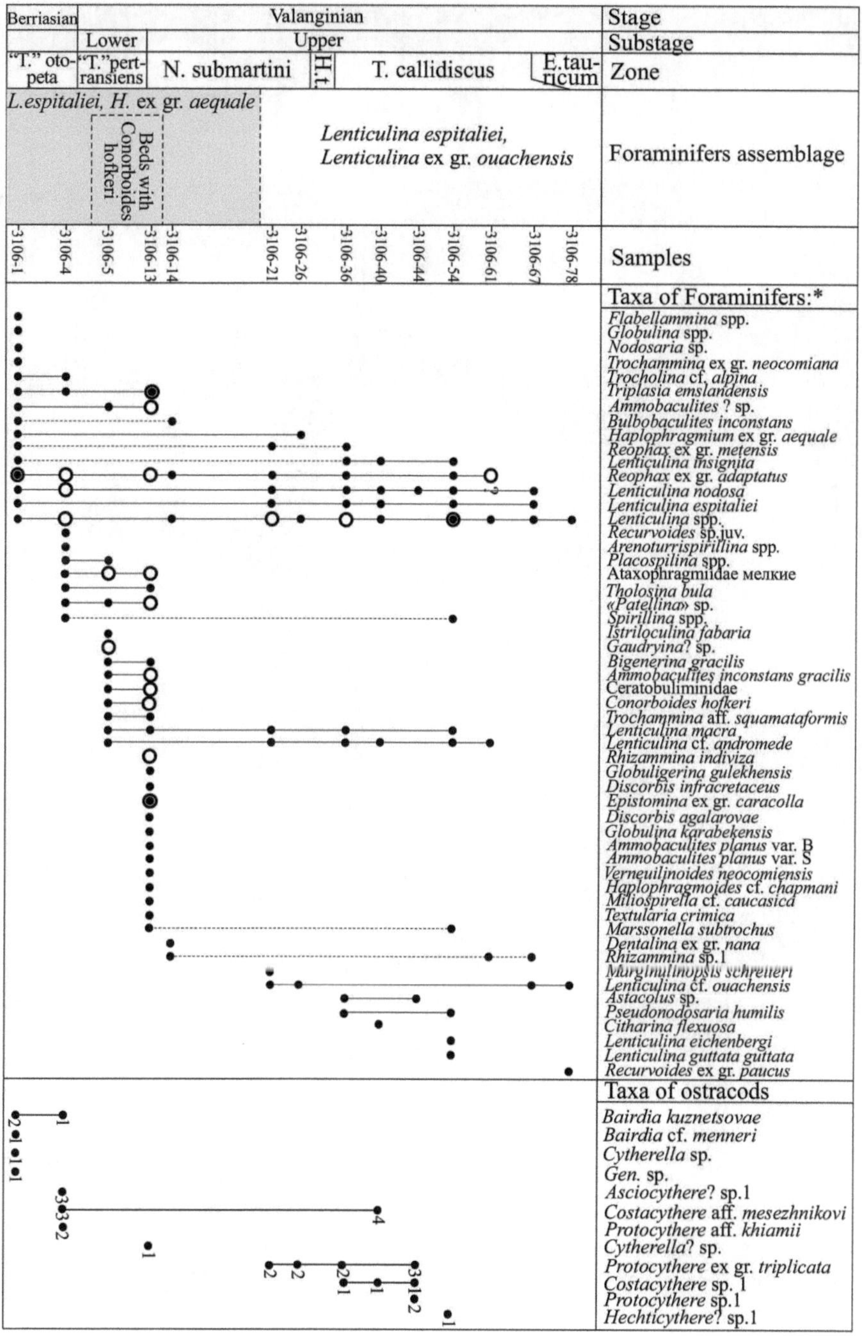

Fig. 4 Distribution of foraminifers and ostracods in the outcrop Rezanaya-3106 (Southwest Crimea). Numerous primitive forms Hippocrepininae, Saccamminidae and Hormosinidae found throughout the section are not shown in the figure. Notation as in Fig. 2

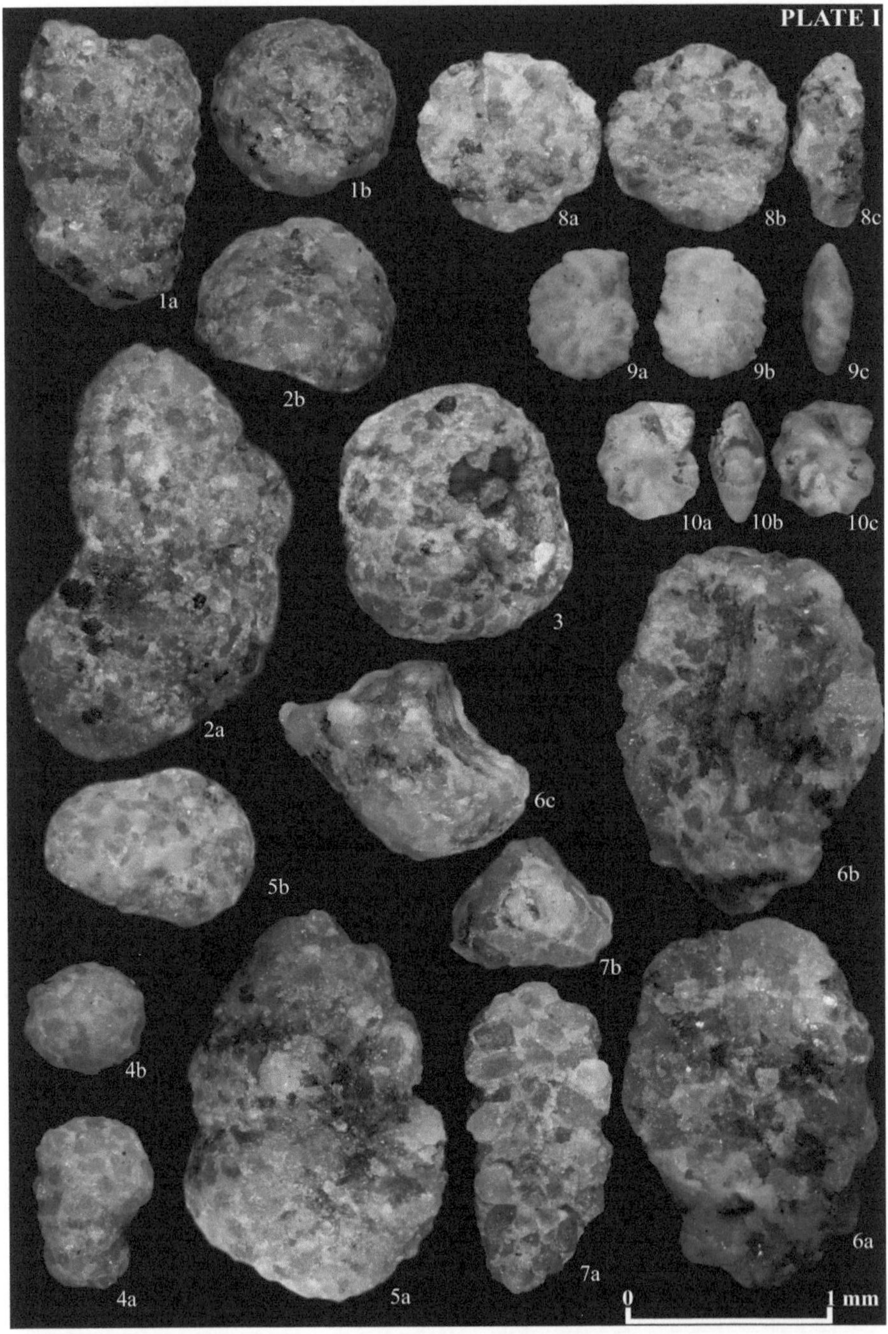
PLATE I
1a
1b
8a
8b
8c
9a
9b
9c
2b
10a
10b
10c
2a
3
6c
5b
6b
7b
4b
6a
7a
4a
5a
0
1 mm

◄**Plate 1** Foraminifers. Assemblage with *Lenticulina espitaliei, Haplophragmium* ex gr. *Aequale* Magnification X40. **1–4** *Haplophragmium* ex gr. *aequale* (Römer), **a** lateral view, **b** apertural view. Patil-2, sample № 3, Subzone Otopeta, Berriasian, **1a** lateral view, **1b** apertural view. **5** *Haplophragmium monstratus* (Dain), **5a** lateral view, **5b** apertural view. Patil section № 3139, sample № 3, Subzone Otopeta, Berriasian. **6** *Triplasia emslandensis* Brat. et Brand, **6a, b** lateral view, **6c** apertural view, Rezanaya section № 3106, sample № 4, Subzone Otopeta, Berriasian. **7** *Triplasia elegans* (Mjatluk), **7a** lateral view, **7b** apertural view, Dlinnaya section № 3065, sample № 4, Subzone Otopeta, Berriasian. **8** *Trochammina* ex gr. *neocomiana* Mjatliuk, **8a** dorsal view, **8b** ventral view, **8c** frontal view, Patil section № 3139, sample № 3, Subzone Otopeta, Berriasian. **9** *Lenticulina* cf. *espitaliei* Dieni et Massari, **9a, b** lateral view, **9c** frontal view. Patil section № 3139, sample № 3, Subzone Otopeta, Berriasian. **10** *Lenticulina nodosa* (Reuss), **10a** lateral view, **10b** frontal view, **10c** lateral view in water, Patil section № 3139, sample № 3, Subzone Otopeta, Berriasian

In the Rezanaya section, a complex with *Conorboides* cf. *hofkeri* (samples 3106/5 and 3106/13) was found (Fig. 4). In sample 3106/5, the complex is poor and represented by a small group of species: *Placopsilina neocomina* Bart et Brand., *Tholosina bula* (Brady), *Ammobaculites gracilis* Bartenstein et Brand, *Bulbabaculites inconstans* Bartenstein et Brand; downtrodden *Bigenerina gracilis* Antonova, *Istriloculina fabaria* Mats. et Temirb., *Trochammina* aff. *squamataformis* Kaptarenko, *Lenticulina macra* Gorbatchik, *L.* cf. *andromede* Espitalie et Sigal, *Conorboides* cf. *hofkeri* (Bartenstein and Brand); it also contains numerous dwarf forms of several Ataxophragmiidae species.

In sample 3106/13, a complex with *Conorboides* cf. *hofkeri* is more complete and variegated: besides the species listed above here there are *Rhizammina indiviza* Brady, *Haplophragmoides chapmani* Crespin, *Ammobaculites planus* Belousova, *Verneuilinoides neocomiensis* (Mjatliuk), *Marssonella* (*Dorothia*) *subtrochus* (Bartenstein), *Textularia crimica* (Gorbatchik), *Miliospirella* cf. *caucasica* Antonova, *Globulina karabekensis* Mjatluk, *Hoeglundina* ex gr. *caracolla* (Roemer), *Discorbis agalarovae* Antonova, *D. infracretaceus* Schokhina, and also the planktonic *Globuligerina gulekhensis* Gorbatchik et Poroshina. The relatively deep-water forms of Rotaliida dominate here.

The most numerous species of lower Cretaceous is *Hoeglundina* (*Epistomina*) *caracolla* (Roemer). The species *Conorboides hofkeri* (Bartenstein and Brand) is known from the lower part of the Berriasian to the Valanginian, indicating the Conorboides hofkeri—Conorbina heteromorpha zone comparable with the interval of ammonite zones from the top of the T. occitanica (D. dalmasi subzone) to the S. boissieri [24]. The species *Discorbis agalarovae* Antonova is known from the upper Tithonian—lower Valanginian, and *D. infracretaceus* Schokhina—from the topmost Tithonian-Berriasian.

The planktonic species *Globuligerina gulekhensis* Gorbatchik et Poroshinais is known from the Berriasian and Valanginian, however, in Crimea this species is characteristic for the upper Valanginian [24].

It is important to note that all the secretory forms are “sideriated”, and agglutinate forms have different degree of integrity with different “types” of wall: in such manner *Ammobaculites planus* Belousova is represented by exemplars with an agglutinate

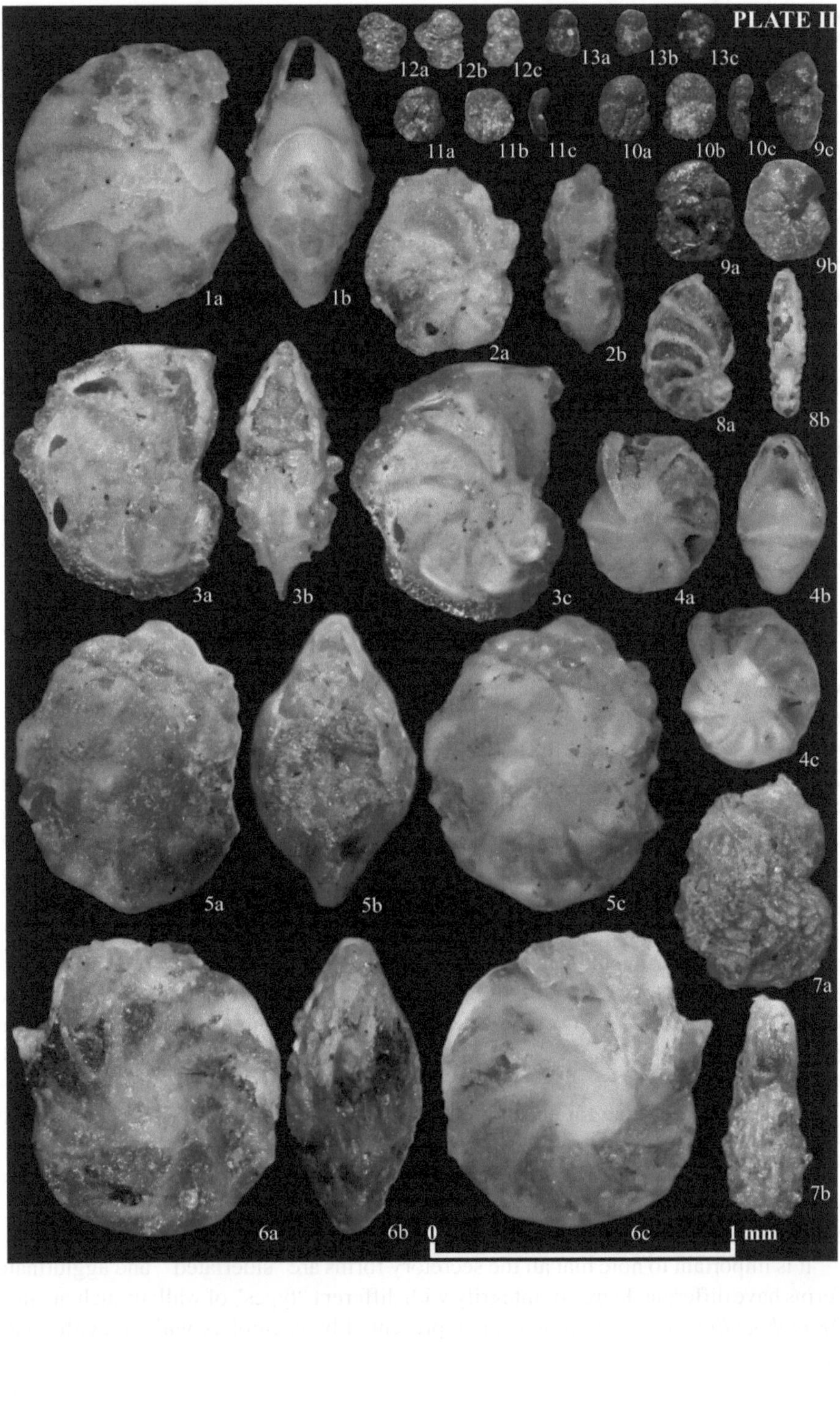
PLATE II
12a
12b
12c
13a
13b
13c
11a
11b
11c
10a
10b
10c
9c
9a
9b
1a
1b
2a
2b
8a
8b
3a
3b
3c
4a
4b
4c
5a
5b
5c
7a
6a
6b
6c
7b
0
1 mm

◂**Plate 2** Foraminifers. Assemblage with *Lenticulina espitaliei*, *Lenticulina* cf. *ouachensis*, Magnification X60. **1** *Lenticulina* cf. *ouachensis* Sigal, **1a** lateral view, **1b** frontal view, Rezanaya section № 3106, sample № 54, Zone T. callidiscus, Valanginian. **2** 3106_54, 2 *Lenticulina guttata* (Ten Dam), **2a** lateral view, **2b** frontal view, Rezanaya section № 3106, sample № 54, Sone T. callidiscus, Valanginian. **3** *Lenticulina eichenbergi* Bartenstein et Brand, **3a** lateral view, **3b** frontal view, **3c** lateral view in water, Rezanaya section № 3106, sample № 54, Zone T. callidiscus, Valanginian. **4** *Lenticulina nodosa* (Reuss), **4a** lateral view, **4b** frontal view, **4c** lateral view in water, Rezanaya section № 3106, sample № 54, Zone T. callidicus, Valanginian. **5** *Lenticulina espitaliei* Dieni et Massari, **5a** lateral view, **5b** frontal view, **5c** lateral view in water, Rezanaya section № 3106, sample № 21, Zone N. submartini, Valanginian. **6** *Lenticulina* cf. *andromede* Espitalie et Sigal, **6a** lateral view, **6b** frontal view, **6c** lateral view in water, Rezanaya section № 3106, sample № 54, Zone T. callidiscus, Valanginian. **7** *Marginulinopsis schreiteri* (Echrenberg), **7a** lateral view, **7b** frontal view, Rezanaya section № 3106, sample № 21, Zone N. submartini, Valanginian. **8** *Citharina flexuosa* (sp. D Espitalie et Sigal, 1963), **8a** lateral view, **8b** frontal view, Rezanaya section № 3106, sample № 54, Zone T. callidiscus, Valanginian. Assemblage with *Conorboides* cf. *hofkeri* Magnification X60. **9** *Hoeglundina* ex gr. *caracolla* (Roemer), **a** dorsal view, **b** ventral view, **c** peripheral view, Rezanaya section № 3106, sample № 13, Zone N. submartini, Valanginian. **10, 11** *Conorboides* cf. *hofkeri* (Bartenstein and Brand), **a** dorsal view, **b** ventral view, **c** peripheral view, Rezanaya section № 3106, sample № 13, Zone N. submartini, Valanginian. **12, 13** *Conoglobuligerina gulekhensis* (Gorbatchik et Poroshina), **a** spire view, **b** umbonal view, **c** peripheral view, Rezanaya section № 3106, sample № 13, Zone N. submartini, Valanginian

of quartz grains of small and medium size (*A. planus* var. B, characteristic for the shallow marine facies) and exemplars with a cryptocrystalline wall (*A. planus* var. S, characteristic for the moderate-marine conditions).

Bearing in mind dominating of characteristic species of the complex belonging to the Conorboides hofkeri—Conorbina heteromorpha zone [24] and the layers with Conorboides hofkeri [8, 35] comparable with the upper part of the Subthurmannia boissieri ammonite zone (Berriasella picteti (upper part) to Thurmanniceras alpillensis subzones), the host sediments may be dated as late Berriasian. However, considering the difference in common habit and the degree of integrity of associations of littoral and upper/middle sublittoral zone, this complex probably should be viewed as re-sedimented into the above described complex with *Lenticulina espital iei*, *Haplophragmium* ex gr. *aequale*.

In the Rezanaya section, a complex with *Lenticulina espitaliei*, *Lenticulina* cf. *ouachensis* was established (Fig. 4). In this complex, despite the sedimentation in littoral zone (shoreface and ebb-tide conditions), there is a large quantity of secretory foraminifers, characteristic for the upper sublittoral inhabit. The representatives of *Lenticulina* become more various, but *Haplophragmium* ex gr. *aequale*, represented by a large quantity of exemplars in the underlying sediments, was not found in this complex.

The following taxa are also recorded till the species rank: *Reophax adaptatus* Dain, *R.* ex gr. *metensis* Franke, *Recurvoides* ex gr. *paucus* Dubrovskaja, *Flabellammina* ex gr. *rugosa-lidiae* Mjatliuk, *Marssonella* cf. *subtrochus* (Bartenstein), *Dentalina* ex gr. *nana* Reuss, *Marginulinopsis schreiteri* (Echrenberg), *Citharina flexuosa* (var. D, Espitalie et Sigal, 1963), *Pseudonodosaria humilis* (Echrenberg), *Lenticulina eichenbergi* Bartenstein et Brand, *L.* cf. *andromede* Espitalie et Sigal,

L. cf. *ouachensis* Espitalie et Sigal, *L. macra* Gorbatchik, *L. insignita* Mjatliuk, *L. nodosa* (Reuss), *L. espitaliei* Dieni et Massari, *L. g. guttata* (Dam). The most part of species appear in the upper part of the Berriasian and are characteristic for the Valanginian or the Valanginian-Hauterivian. Similar increasing of diversity of nodozariides is characteristic for the terminal Berriasian—lower Valanginian of East Crimea [6–8, 24, 36].

Thus, despite the absence of short-lived characteristically Valanginian foraminifera species, this taxonomic composition indicates, with a high degree of confidence, the Valanginian age of the enclosing sediments.

Ostracods were found in three sections, namely Patil-2, Dlinnaya and Rezanaya, but they are not present in every sample. In total, 17 species belonging to 8 genera were found, the shells are mostly thick-walled, and the condition is satisfactory (Figs. 2, 3 and 4; Plate 3). Limited deep-water representatives of the genus *Bairdia* and singular *Pontocyprella* were found. Such assemblage is characteristic for the most shallow areas of sublittoral with an active hydrodynamic regime.

The most important stratigraphic meaning is borne by the representatives of the family *Protocytheridae* (*Protocythere*, *Costacythere*, *Hechticythere*) [32]. The species *Protocythere praetriplicata* Bartenstein et Brand is known from the upper part of the Berriasian on the Northern Caucasus (Urukh section) and from the Berriasian-Valanginian of Mangyshlak, Germany, France and Algeria [3, 10, 13, 22]. This species is an index of cognominal zone, determined in the middle part of the Berriasian of Central Asia [2]. It also composes the complex with *Palaeocytheridea fistulosa*, *Pal. bassovi*, determined in the upper part of the lower Valanginian of the Boyarka river in northern Siberia [25]. It composes the complex with *P. hannoverana*, *P. praetriplicata* of the lower Valanginian on Kolguev Island in the Barents Sea as well [23]. The close form, *Protocythere* ex gr. *triplicata* (Roemer), was found in the upper part of the Rezanaya section (complex of foraminifers with *Lenticulina espitaliei*—*L.* cf. *ouachensis*). The species *P. Triplicata* is one of the most widespread species, which is known from the Valanginian to the lower Barremian in central Asia [2], Precaspian [26], England [30, 39], Germany [14, 43] and France [10]. In Southwest Crimea, this species was recorded from the Hauterivian–Barremian at Verkhorechye [38]. It also has a status of an index species of cognominal zone, which was established in Central Asia in the range from the upper part of the upper Valanginian to the lower Hauterivian [2]. In the Hauterivian of Kolguev Island, a complex with *Protocythere triplicata*, *Eucythere neocomiana* has been found [23].

In the studied samples, *Costacythere* aff. *foveata* Tesakova et Rachenskaya, *C.* aff. *khiamii* Tes. et Rach., *Hechticythere* aff. *belbekensis* Tes. et Rach. were also found. Enumerated taxa are close forms of the species earlier known from the Berriasian sediments of the Crimean Mountains [5, 6, 8, 41]. We found *Costacythere* aff. *mesezhnikovi* Neale et Kolpenskaya, earlier reported from the middle and upper parts of the Berriasian of the Northern Caucasus (Urukh section), and *Hechticythere* aff. *belbekensis* Tes. et Rach., which is known from the upper part of the Berriasian of the same region and section [22]. The species *Cytherella krimensis* Neale is widespread in the Tithonian-Berriasian of the Crimean Mountains and in the middle and upper parts of the Berriasian in the Northern Caucasus (Urukh section, S. occitanica zone

PLATE III

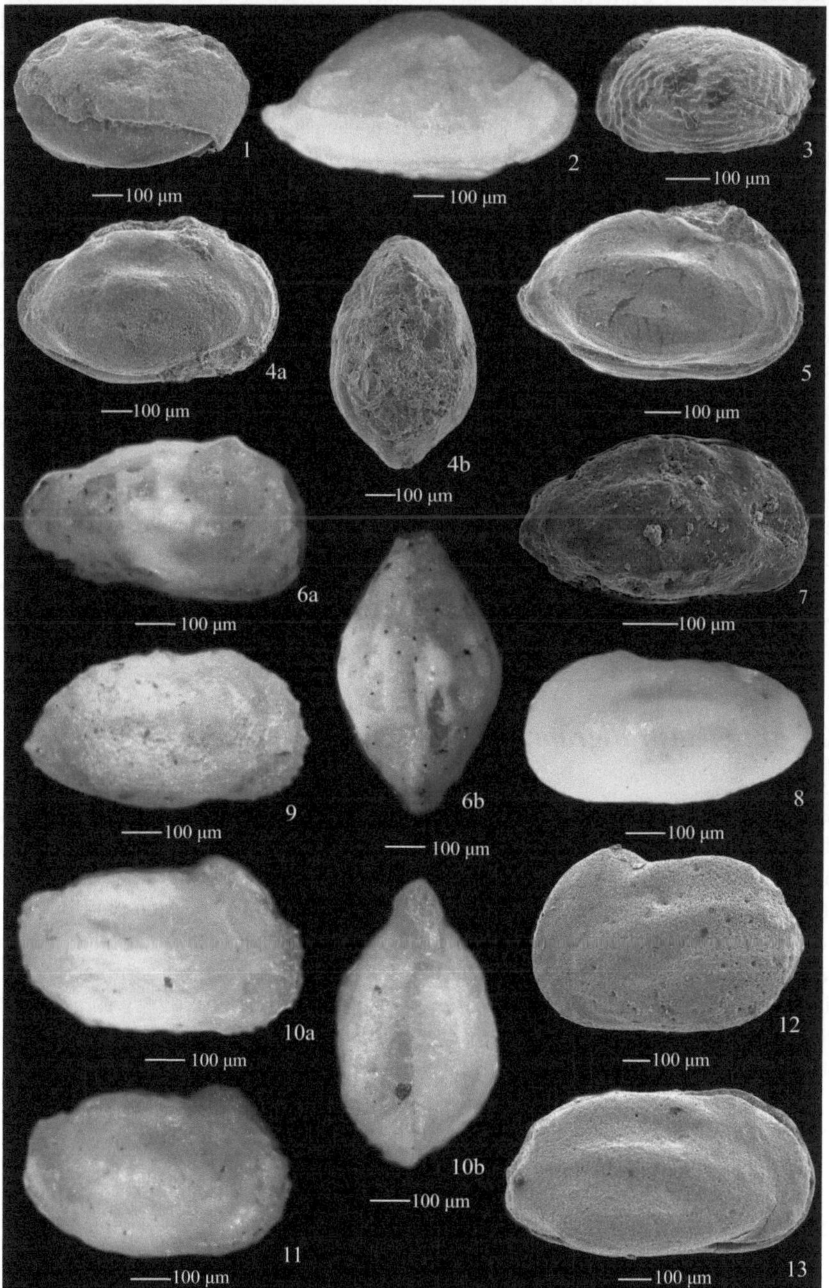
1
100 μm
2
100 μm
3
100 μm
4a
100 μm
4b
100 μm
5
100 μm
6a
100 μm
6b
100 μm
7
100 μm
8
100 μm
9
100 μm
10a
100 μm
10b
100 μm
11
100 μm
12
100 μm
13
100 μm

◂**Plate 3** Ostracods. **1** *Cytherella* cf *krimensis* (Neale, 1966), Dlinaya, sample 3065-16, carapace, left lateral view. **2** *Bairdia kuznetsovae* Tesakova et Rachenskaya, 1996, Rezanaya, sample 3106-4, carapace, right lateral view. **3** *Neocythere urukhensis* Neale et Koppenskaya, 2000, Dlinaya, sample 3065/16; left valve, lateral view. **4** *Protocythere praetriplicata* Bartenstein et Brand, 1959, Patily, sample 3139/3, carapace: **4a** right lateral view; **4b** dorsal view. **5** *Protocythere praetriplicata* Bartenstein et Brand, 1959, Dlinaya, sample 3065/16, carapace, right lateral view. **6** *Protocythere* ex gr. *triplicata* (Roemer, 1840), Rezanaya-3106, sample 3106-21, carapace: **6a** right lateral view; **6b** dorsal view. **7** *Protocythere* ex gr. *triplicata* (Roemer, 1840), Kermen, sample 3, carapace, right lateral view. **8** *Protocythere* sp. 1, "Dlinaya", sample 3065/16, left valve, lateral view. **9** *Costacythere khiamii* Tesakova et Rachenskaia, 1996, Rezanaya, sample 3106-4, right valve, lateral view. **10, 11** *Costacythere* aff. *mesezhnikovi* Neale et Kolpenskaya, 2000, Rezanaya, sample 3106-40, **10a** carapace, right lateral view; **10b** carapace, ventral view; **11** carapace, right lateral view. **12** *Costacythere* aff. *foveata* Tesakova et Rachenskaia, 1996, Patily, sample 3139/3, left valve, lateral view. **13** *Hechticythere* aff. *belbekensis* Tesakova et Rachenskaya, 1996, Dlinaya, sample 3065/13; carapace, right lateral view

(D. tauricum subzone) and S. boissieri zone) [5, 6, 22, 31]. The species *Neocythere urukhensis* Neale et Kolp. (= *Macrodentina melnikovae* Tes.) was earlier found in the upper part of the Berriasian in Central Crimea and the Northern Caucasus (Urukh section) [22, 42]. The species *Bairdia kuznetsovae* Tes. et Rach. is known from the Berriasian of Central and East Crimea; and species *B. menneri* Tes. et Rach. from the middle and upper parts of the Berriasian in the Crimean Mountains [5, 35, 42].

4.2 Magnetostratigraphy

Magnetite as a main carrier of $\mathbf{J_n}$ in rocks in East and Southwest Crimea was established earlier in the sections Zavodskaya balka, Dlinnaya and Sheludivaya [4, 7, 16]. Comparable results of magneto-mineralogical analysis were obtained for sections Patil-1, 2, Bolshoy Kermen, Selbukhra and Rezanaya. Fe_3O_4 is detected by the fall of magnetization on the TMA curves around 570–580 °C (Fig. 5a, c, d). The presence of a magnetically soft phase is confirmed by isothermal remanence acquisition curves (Fig. 5b). In several samples the presence of hydrous ferric oxides was established inferred from a characteristic fold on the graphs of TMA second derivative between 100 and 200 °C and a steady growth of isothermal remanence up to 700 mT (Fig. 5e, f). In a petromagnetic regard, all studied sections are well differentiated allowing to further subdivide the series into petromagnetic intervals [8, 15].

4.2.1 East Crimea

The total of 392 oriented samples, collected from 196 stratigraphic levels were studied.

The major amount of data has come from the magnetostratigraphic study of the Zavodskaya balka section. In the lower part of the section, two isolated groups of

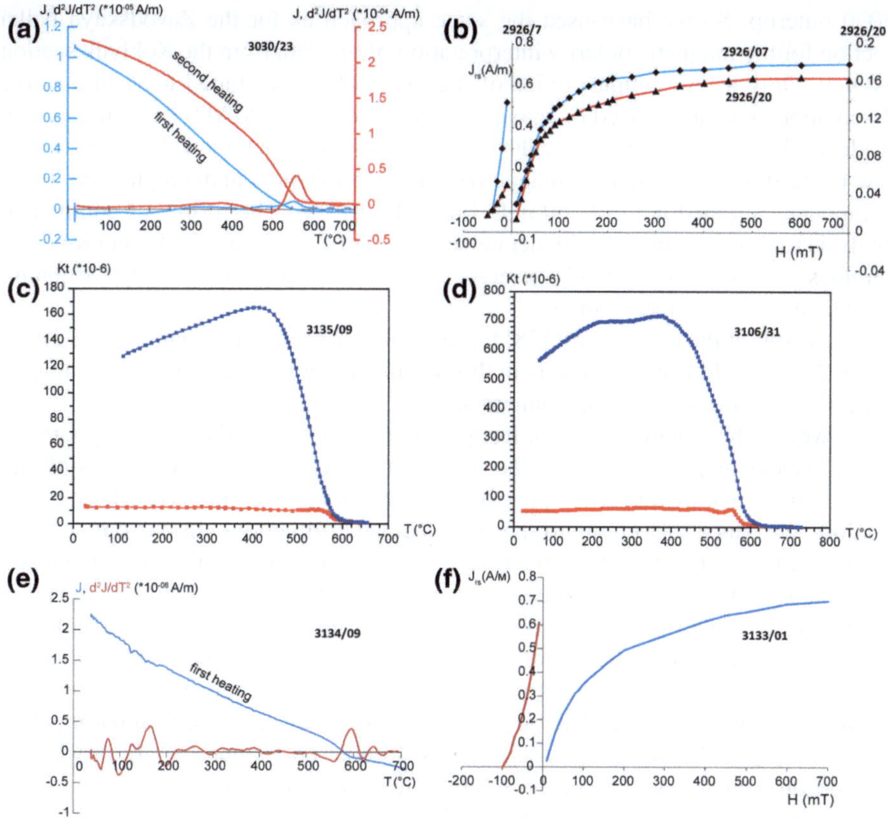

Fig. 5 Results of magnetomineralogical examinations: **a**, **e**—DTMA results; **b**, **f**—isothermal remanence acquisition curves; **c**, **d**—dependence of magnetic susceptibility on temperature

$\mathbf{J_n}$ directions appear to be present: in the N–NW quarter of lower hemisphere and in the SE sector of upper hemisphere, corresponding to the epochs of normal (N) and reverse (R) polarity of geomagnetic field, respectively. In the upper part of the section many paleomagnetic directions have an anomalous character (for example, negative inclination with northern declination) not allowing to determine the sign of polarity even hypothetically [8]. Further studies have detected the significant correlation (on the significance level $p = 0.001$) between the distribution of short axes of the ellipsoid of anisotropy of magnetic susceptibility (K3) and paleomagnetic directions of ChRM component. We attempted to use the anisotropy of magnetic susceptibility for correcting paleomagnetic data, and achieved a considerable improvement of paleomagnetic data in the upper part of the Zavodskaya balka section [17].

The Koklyuk section is characterized by a similar to that in Zavodskaya balka character of K3 (AMS) and $\mathbf{J_n}$ vectors distribution which are significantly correlated appears at the 95% level, and is likely caused by the same factors. In this section intensive landslide deformations are visible in the marl layers at the base of the

3030 outcrop. So we have used the same approach as for the Zavodskaya balka section for the magnetic polarity interpretation of the data from the Koklyuk section as well. Only a few samples (<7% of the total collection) failed to yield a stable $\mathbf{J_n}$ component characterized by maximum angle deviation (MAD) <15°. In general, Zijderveld diagram analysis reveals a two-component structure of $\mathbf{J_n}$ comprising a low-coercitivity component demagnetized in 5–25 mT field and a high-coercitivity component retained until 30–60 mT (Fig. 6 Ia, b, IIa). The results of component analysis for both thermal and AF demagnetization of Zavodskaya balka and Koklyuk samples generally show a good convergence thus raising the confidence to the quality of paleomagnetic determinations.

The result of the reversal test [28] for the Koklyuk section is negative ($A = 20.5°$; $A_c = 20.1°$). Fold test [27] results in this section are either inconclusive or indicate the presence of a post-folding component.

However, the negative reversal test per se does not contradict to the hypothesis of an ancient magnetization age, because it may be explained by viscous-plastic deformations in clay that are impossible to take into account properly. These deformations introduce distortions to the paleomagnetic directions. In our opinion, the important factor that designates the results of fold test is the ability of ferromagnetic grains to follow the direction of geomagnetic field during a synfolding deformation. Small particles may continue to be oriented along the field direction in a semifluid sediment while large particles are not. The degree of the sediment lithification, in turn, depends on a number of factors, e.g. carbonate content, organic matter content, degree of bioturbation etc. The described scenario may result in a situation when every sample will carry pre-, syn-, and post-folding components of $\mathbf{J_n}$ in variable proportion, which would be nearly impossible to resolve by demagnetization since their stability is expected to be quite similar.

In the Sultanovka section, it is impossible to determine a polarity sign in the ancient coordinate system. Directions of stable magnetization components show a more regular distribution in the present-day coordinate system than in the ancient one. The same is true for the projections of AMS ellipsoid axes: in the present-day coordinates they group into two clusters with average directions close to magnetization directions expected for normal and reverse polarity, respectively (Fig. 6 Ic, d, IIb, c). Similar behavior has been observed previously in other Crimean sections [17]. The fold test [27] clearly argues for the postfolding age of the magnetization at Sultanovka, but the presence of magnetozones with different polarity does not conform to a hypothesis of rocks remagnetization after their lithification. However, this contradiction may be erased considering a synsedimentational age of the syncline and magnetization acquisition in sediments dewatered after the fold development, simultaneously with the magnetic fabric. Therefore, we believe possible to use the directions of magnetization in geographic coordinate system to determine the polarity sign in this section.

The data thus acquired satisfy a number of indicators of primary magnetization [19, 40, 44]: (1) intervals of different polarity sign consistently group along the section, forming large N- and R-magnetozones; (2) the polarity sign does not depend on the lithological composition, because magnetozones of different polarity

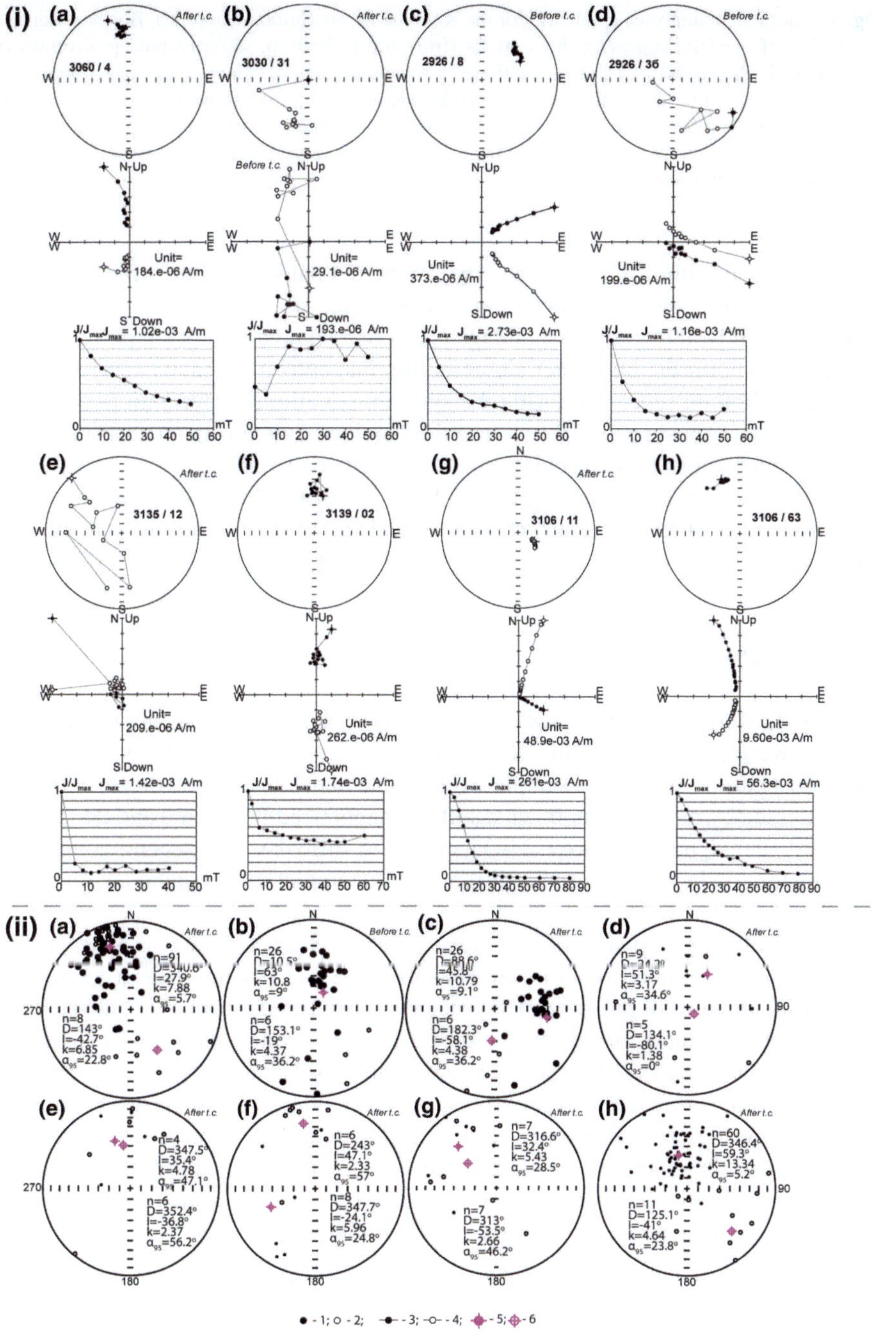
(i)
(a) After t.c. 3060 / 4
Unit= 184.e-06 A/m
J/J_{max} J_{max}= 1.02e-03 A/m
(b) After t.c. 3030 / 31
Before t.c.
Unit= 29.1e-06 A/m
J/J_{max} J_{max}= 193.e-06 A/m
(c) Before t.c. 2926 / 8
Unit= 373.e-06 A/m
J/J_{max} J_{max}= 2.73e-03 A/m
(d) Before t.c. 2926 / 36
Unit= 199.e-06 A/m
J/J_{max} J_{max}= 1.16e-03 A/m
(e) After t.c. 3135 / 12
Unit= 209.e-06 A/m
J/J_{max} J_{max}= 1.42e-03 A/m
(f) 3139 / 02
Unit= 262.e-06 A/m
J/J_{max} J_{max}= 1.74e-03 A/m
(g) After t.c. 3106 / 11
Unit= 48.9e-03 A/m
J/J_{max} J_{max}= 261e-03 A/m
(h) 3106 / 63
Unit= 9.60e-03 A/m
J/J_{max} J_{max}= 56.3e-03 A/m
N Up
S Down
W
E
mT
(ii)
(a) After t.c.
n=91 k=7.88 α_{95}=5.7°
n=8 D=143° I=-42.7° k=6.85 α_{95}=22.8°
(b) Before t.c.
n=26 I=63° k=10.8 α_{95}=9°
n=6 D=153.1° I=-19° k=4.37 α_{95}=36.2°
(c) After t.c.
n=26 I=45.8° k=10.79 α_{95}=9.1°
n=6 D=182.3° I=-58.1° k=4.38 α_{95}=36.2°
(d) After t.c.
n=9 I=51.3° k=3.17 α_{95}=34.6°
n=5 D=134.1° I=-80.1° k=1.38 α_{95}=0°
(e) After t.c.
n=4 D=347.5° I=35.4° k=4.78 α_{95}=47.1°
n=6 D=352.4° I=-36.8° k=2.37 α_{95}=56.2°
(f) After t.c.
n=6 D=243° I=47.1° k=2.33 α_{95}=57°
n=8 D=347.7° I=-24.1° k=5.96 α_{95}=24.8°
(g) After t.c.
n=7 D=316.6° I=32.4° k=5.43 α_{95}=28.5°
n=7 D=313° I=-53.5° k=2.66 α_{95}=46.2°
(h) After t.c.
n=60 D=346.4° I=59.3° k=13.34 α_{95}=5.2°
n=11 D=125.1° I=-41° k=4.64 α_{95}=23.8°
N
90
180
270
• - 1; ○ - 2; -•- - 3; -○- - 4; - 5; - 6

◂**Fig. 6** Component analyses results (**I**) for the Koklyuk (**a**, **b**), Sultanovka (**c**, **d**), Bolshoy Kermen (**e**), Patil-2 (**f**) and Rezanaya (**g**, **h**) sections (from top to bottom: stereographic projections of $\mathbf{J_n}$ changes in the course of demagnetization, Zijderveld diagrams, demagnetizations graphs) and Stereoprojections of the $\mathbf{J_n}$ stable components (**II**): **a** Koklyuk; **b, c** Sultanovka; **d** Bolshoy Kermen; **e, f** Patil-1 and 2; **g** Selbuhra; **h** Rezanaya. *Legend* 1, 2—on the lower and the upper semispheres respectively; 3, 4—on the horizontal and the vertical planes respectively; 5, 6—average direction of the $\mathbf{J_n}$ stable components

are established within monotonous clay series; (3) paleomagnetic structure of the studied sections shows many common features (Fig. 7).

Consequently, the whole assemblage of obtained data does not support a remagnetization hypothesis, but rather suggests that magnetization have been formed in a partially lithified sediment during the synsedimentational deformations [17]. For this reason, we consider possible to use the obtained data for magnetostratigraphic reconstructions despite the problematic paleomagnetic results.

The final composite paleomagnetic column constructed for East Crimea of contains 3 subzones of reverse, 2 subzones of normal and 1 subzone of mostly anomalous polarity (Fig. 7).

4.2.2 Southwest Crimea

In the paleomagnetic regard, the rocks of condensed sections at mountains Patil, Bolshoy Kermen and Selbukhra are analogues to Dlinnaya and Sheludivaya sections [16]. Most part of the studied samples were stable in paleomagnetic regard. They are characterized by two $\mathbf{J_n}$ components: soft (demagnetized between 0 and 24–32 mT) and hard (demagnetized above 24–32 mT) (Fig. 6 Ie–h). The second component is characteristic in some cases but it remains uncompletely demagnetized in many samples.

The results of group statistics on the established characteristic components have revealed that all studied sections are characterized by poor interbed grouping and shallow inclinations of paleomagnetic vectors (Fig. 6 IId–g) probably related to their coarse-grained composition originating from a shallow-water genesis and to presence of stratigraphic breaks. The standard tests used to substantiate the ancient age of magnetization, reversal test and fold test, either yielded a negative result (reversal test), or were impossible to apply because of too small variation of layers bedding elements. Despite all these factors, using several circumstantial attributes, employed while substantiating the ancient age of $\mathbf{J_n}$ in the magnetostratigraphic studies [19] it is possible to substantiate the primary nature of magnetization.

1. The established directions, corresponding to normal or reverse polarity, group in the sections consistently (Fig. 8).
2. The polarity sign does not depend on the lithological composition of studied rocks (Fig. 8).

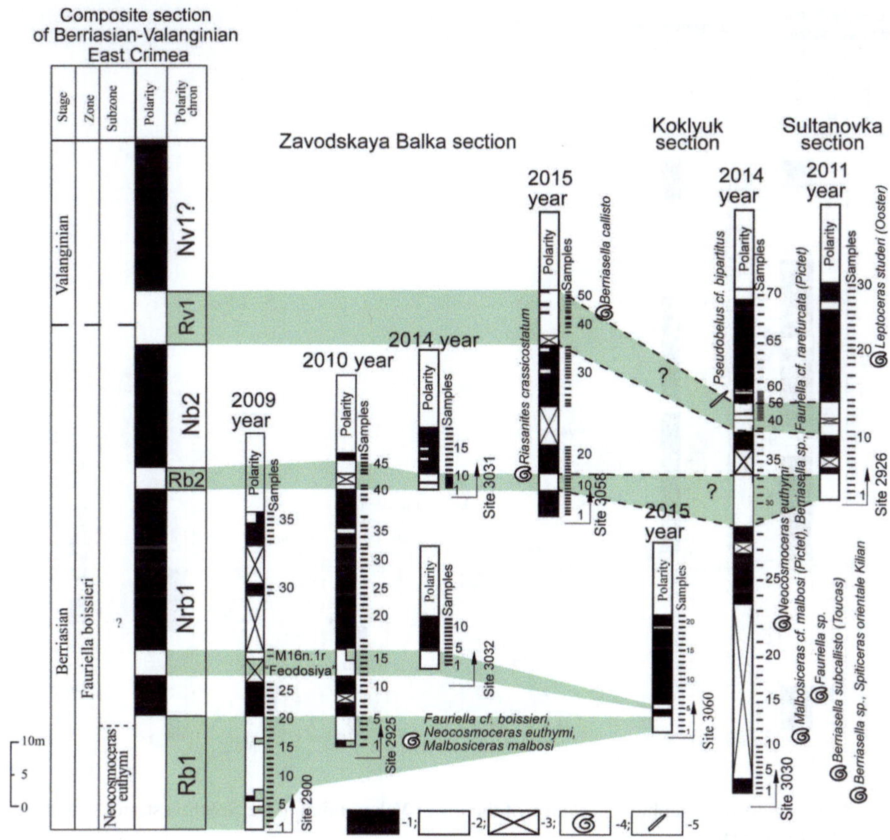

Fig. 7 Magnetostratigraphic correlation of the Berriasian-Valanginian boundary interval in the East Crimea. *Legend* 1, 2—normal and reverse geomagnetic polarity, respectively (in half of the column thickness—tentative determination of the polarity sign); 3—no paleomagnetic data available; 4, 5—finds of ammonites and belemnites

3. The main carrier of magnetization in the section is probably detrital magnetite, which is detected by low fields of magnetic saturation, DTMA data (Fig. 5). Note that formation of secondary magnetite in the shallow-water, oxic environment appears rather unlikely.
4. The large variance of paleomagnetic vectors, typical for rudaceous sediments.
5. Comparability of obtained paleomagnetic results with paleontological data and known data on the regime of geomagnetic field in the Valanginian [33].

In this way, the obtained data reveal that paleomagnetic columns of all studied sections (Dlinnaya, Sheludivaya, Patil, Bolshoy Kermen, Selbukhra) are compatible and can be viewed as an alternation of seven subzones (intervals) of different sign (Rb2, Nb2, Rv1, Nv1, Rv2, Nrv, Rnv1). Magnetostratigraphic data on each single section may reasonably cast doubt, because of numerous anomalous paleomagnetic

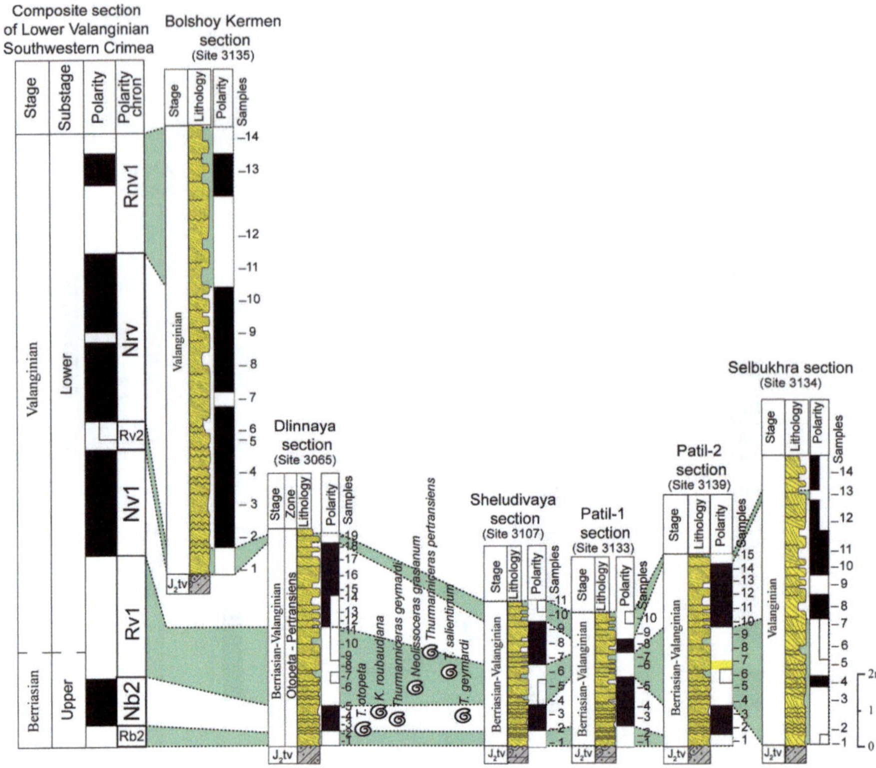

Fig. 8 Magnetostratigraphic correlation of the lower Valanginian in the Southwestern Crimea. See Fig. 7 for the legend

directions and abundant sedimentation breaks. However, the similar paleomagnetic structure can be traced in as many as 5 sections giving some confidence to the obtained data (Fig. 8).

4.2.3 The Rezanaya Mountain Section

The reference section of the Upper Valanginian in Southwest Crimea, located near the town of Rezanaya, apprears more favorable in the paleomagnetic sense.

The analysis of Zijderveld diagrams has shown that the major part of Valanginian samples from the Rezanaya section are characterized by two-component $\mathbf{J_n}$ vectors, with low coercivity (low-temperature) component probably of viscous origin demagnetized at 18–25 mT (200–350 °C), and high coercivity (high-temperature) extending up to 80–100 mT or 550 °C (Fig. 6 Ig, h). Interbed grouping is rather poor, and large number of anomalous directions are observed; data from 12 stratigraphic levels were excluded from further consideration. Still, the remaining data appear suitable for magnetic polarity determination and magnetostratigraphic analysis (Fig. 6 IIh).

Paleomagnetic column of the upper Valanginian substage in the Rezanaya section is subdivided into seven subzones (Fig. 9). Four are characterized by exclusively normal polarity (Nv, the probable analogue of the upper part of magnetozone Nrv from the rest of Southwest Crimea, Nv2, Nv3 and Nv4), and three subzones show mainly reverse polarity (Rnv1, Rnv2, Rnv3).

5 Discussion

5.1 East Crimea

In East Crimea, the reliable magnetostratigraphic characteristic for the Berriasian-Valanginian boundary interval is obtained for the Zavodskaya balka section [8]. Micropaleontologic exploration of the section revealed here the layers containing Lingulina trilobitomorpha, Haplophragmoides vonontianus. The range of detected micropaleontologic layers is comparable with the range of ammonite zones T. Otopeta—S. Verrucosum (upper Berriasian—Valanginian) [36]. The find of Berriasian ammonite Berriassella callisto [8] within the magnetozone of reverse polarity makes it possible to correlate the latter with the chron M14r (Fig. 7). In the similar way, biostratigraphic data allow to correlate a rather complex paleomagnetic zonality of the Zavodskaya balka section with magnetic chrons M16, M15 and M14. Considering an extreme rarity of the ammonite findings in the Berriasian-Valanginian boundary interval, the base of M14r chron should be used in East Crimea as a criteria for substantiating the base of the Valanginian by analogy with the Geomagnetic Polarity Timescale (Fig. 9). In general, the base of the Valanginian may be traced in the Koklyuk and Sultanovka sections because there seems to be possible to establish the succession of magnetic chrons M16, M15 and M14 in both sections, despite a poor quality of paleomagnetic data.

5.2 Southwest Crimea

Micropaleontologic studies of the Dlinnaya, Patil-1, 2 and Rezanaya sections have detected the presence of foraminifers complex with *Lenticulina espitaliei*, *Haplophragmium* ex gr. *aequale*; *Conorboides* cf. *hofkeri*; *Lenticulina espitaliei*, *Lenticulina* ex gr. *ouachensis*, implying the Berriasian-Valanginian or Valanginian age of the host sediments. The most part of recognized ostracod species are known from Berriasian sediments of Crimea and the Caucasus, but they are identified by us in the open nomenclature; perhaps they are descendant forms (Berriasian forms may be their ancestors) and are on the same evolutionary line of development.

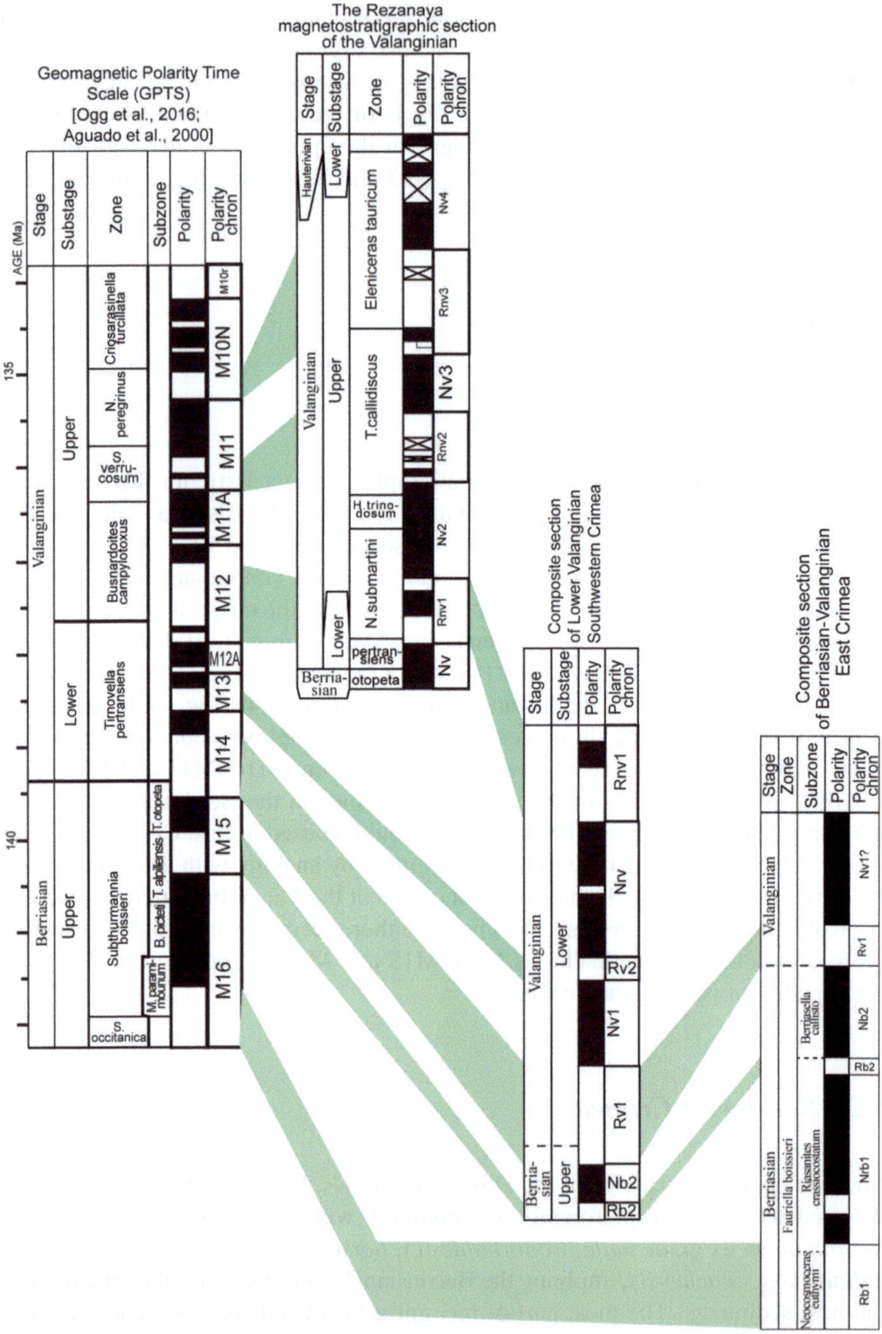

Fig. 9 General magnetostratigraphic scheme of the Valanginian in the Crimean Mountains

The cooccurrence of Berriasian-Valanginian species *Protocythere praetriplicata* and Valanginian-Barremian species *P. Triplicata* is noted for the Valanginian [2, 10]. In the upper part of the Rezanaya section, a transitional form *P.* ex gr. *Triplicata* was found. It is important to note that the representatives of *P. Triplicata* were found in the Hauterivian-Barremian of the Rezanaya section [38], that also does not contradict the hypothesis about the Valanginian age of the studied strata. In such manner, considering the available paleontologic and paleomagnetic data, established in the sections Dlinnaya, Sheludivaya, Patil-1, 2, Selbukhra and Bolshoi Kermen, the analogues of chrons M15—M12 may be identified in the composite magnetostratigraphic section of the Berriasian-Valanginian transition in the Crimean Mountains (Fig. 9).

The results of magnetostratigraphic data on the Rezanaya section contradict the biostratigraphic data on ammonites [12, 33]. The detected zone of mostly reverse polarity Rnv2, located within T. callidiscus zone, matches with magnetic chron M11r which is considered coeval with the ammonite zone S. verrucosum. However, considering $\delta^{13}C$ isotope data obtained from the Rezanaya section [18, 45] where isotope peak occurs within the ammonite zone T. callidiscus and $\delta^{13}C$ isotope data on the sections of Northern Mediterranean [18] where a similar peak is detected in the lower part of the ammonite zone N. Peregrinus, it is important to note a good correlation obtained with non-biostratigraphic methods. Assuming the proposed correlation, the overlying zone of mostly reverse polarity (Rnv3) is to be correlated with magnetic chron M10N, and the underlying subzone Rnv1 with magnetic chron M12. In this way, the analysis of obtained data allows to establish the following chrons: M12, M11A, M11, M10N in the section of mtn. Rezanaya (Fig. 9).

6 Conclusions

The microfaunistic analysis conducted for four sections of Southwestern Crimea allowed:

1. To establish four foraminiferal assemblages: "shorefase" complex with *Haplophragmium* ex gr. *Aequale* (middle Tithonian—lower Valanginian); complex with *Lenticulina espitaliei*, *Haplophragmium* ex gr. *Aequale* (uppermost Berriasian—lower Valanginian); resedimented (?) complex with *Conorboides* cf. *Hofkeri* (uppermost Berriasian); complex with *Lenticulina espitaliei*, *Lenticulina* cf. *ouachensis* (Valanginian);
2. To trace the complex with *Lenticulina espitaliei*, *Haplophragmium* ex gr. *Aequale* in three sections;
3. To compare the complex with *Lenticulina espitaliei*, *Lenticulina* cf. *ouachensis*, established in the Rezanaya section in Southwestern Crimea, with the upper part of the Zavodskaya balka section and the middle-upper parts of the Koklyuk section of East Crimea using several common species;

4. The most part of the studied species of ostracods is known from the Berriasian of Crimea and the Caucasus, but they are reported mainly in open nomenclature. In the upper part of the Rezanaya section (complex with *Lenticulina espitaliei*, *Lenticulina* cf. *ouachensis*), an intermediate form between the species *Protocythere praetriplicata* (Berriasian—Valanginian) and *P. triplicata* (Valanginian—Barremian) was found, namely *P.* ex gr. *triplicata*. This form most likely indicates the Valanginian stage.

The paleomagnetic analysis allowed:

1. In the Zavodskaya Balka section, the Upper Berriasian biostratigraphic subdivisions have been for the first time recognized in a continuous succession: the Euthymi, Crassicostatum and Callisto subzones and magnetic chron analogues from M16n to M14r (Fig. 9).
2. Established in the sections Dlinnaya, Sheludivaya, Patil-1, 2, Selbukhra and Bolshoi Kermen, the analogues of chrons M15—M12 may be identified in the composite magnetostratigraphic section of the Berriasian-Valanginian transition in the Crimean Mountains (Fig. 9).
3. To establish next following chrons: M12, M11A, M11, M10N in the section of mtn. Rezanaya (Fig. 9).

By the results of correlation, the following conclusions were obtained:

1. A composite magnetostratigraphic section of the Valanginian for the Crimean Mountains was constructed and correlated with Geomagnetic Polarity Timescale (Fig. 9).
2. The sections located in East Crimea represent the fullest stratigraphic succession of the Berriasian-Valanginian boundary interval in the Crimean Mountains. In these sections, the presence of an age analogues of the T. otopeta zone has been substantiated and the base of the Valanginian stage has been traced by the analogy with West European sections, using the base of the M14r magnetic chron as a tie-point.
3. In Southwest Crimea the analogues of the majority of Valanginian magnetic chrons were detected. Despite a hypothetical character of the identification of the lower Valanginian magnetozones and their comparison with chrons M15-M12A, the obtained data allow to provide a detailed correlation of condensed lower Valanginian sections in Southwest Crimea.

Acknowledgements This study has been supported by RFBR (projects 18-35-00134 mol-a and 17-05-00716 a). Authors convey thanks to A. Yu. Guzhikov (Saratov State University) for the help in paleomagnetic data interpretation; to L. A. Kartseva (BIN RAS), to E. S. Platonov (Geologorazvedka) for taking photos of foraminifers and ostracods; to E. S. Ochkasova (Geologorazvedka) for the preparation of microfaunistic samples and technical help. We are very grateful to R. V. Veselovskiy and O. S. Dzyuba for reviewing our article.

References

1. Aguado, R., Company, M., Tavera, J.M.: The Berriasian/Valanginian boundary in the Mediterranean region: new data from the Caravaca and Cehegin sections, SE Spain. Cretac. Res. **21**, 1–21 (2000)
2. Andreev, Yu.N.: The ostracodes of Central Asia Cretaceous (composition, distribution, development, geological meaning). Diss. Dr. Geol.-Min. Science, Moscow State University, Moscow, vol. 2, 404 pp. (1986) (in Russian)
3. Andreev, Yu.N., Oertli, H.J.: Quelques ostracodes cretaces d'Asie centrale et formes proches d'Europe. In: Rauser-Chernousova, D.M. (ed.) Problems of Micropaleontology, Iss. 13, pp. 95–121. Nauka, Moscow (1970) (in Russian with French abstract)
4. Arkadiev, V.V., Bagaeva, M.I., Guzhikov, A.Yu., Manikin, A.G., Yampolskaya, O.B.: Bio- and magnetostratigraphy characteristic of the Upper Berriasian section "Zavodzkaya balka" (East Crimea, Feodosia). Bull. Saint Petersburg St. Univ. (Geology, Geography), series 7(2), 3–16 (2010) (in Russian with English abstract)
5. Arkadiev, V.V., Bogdanova, T.N., Guzhikov, A.Yu., Lobacheva, S.V., Myshkina, N.V., Platonov, E.S., Savelyeva, Yu.N., Shurekova, O.V., Yanin, B.T.: Berriasian of Mountain Crimea. LEMA, Saint Petersburg, 472 pp. (2012) (in Russian)
6. Arkadiev, V.V., Baraboshkin, E.Yu., Bagaeva, M.I., Bogdanova, T.N., Guzhikov, A.Yu., Manikin, A.G., Piskunov, V.K., Platonov, E.S., Savel'eva, Yu.N., Feodorova, A.A., Shurekova, O.V.: New data on Berriasian biostratigraphy, magnetostratigraphy, and sedimentology in the Belogorsk area (Central Crimea). Stratigr. Geol. Correl. **23**(2), 155–191 (2015a)
7. Arkadiev, V.V., Guzhikov, A.Yu., Savelieva, J.N., Feodorova, A.A., Shurekova, O.V., Bagayeva, M.I., Grishchenko, V.A., Manikin, A.G.: New data on bio- and magnetostratigraphy of Upper Berriasian section "Zavodskaya balka" (Eastern Crimea, Feodosiya). Bull. Saint Petersburg St. Univ. (Geology, Geography), series 7(4), 4–36 (2015b) (in Russian with English abstract)
8. Arkadiev, V., Guzhikov, A., Baraboshkin, E., Savelieva, Yu., Feodorova, A., Shurekova, O., Manikin, A.: Biostratigraphy and magnetostratigraphy of the upper Tithonian–Berriasian of the Crimean Mountains. Cretac. Res. **87**, 5–41 (2018)
9. Astahova, T.V., Gorak, S.V., Kraeva, E.Y., et al.: Geology of Shelf of the USSR. Stratigraphy (Shelf and Black Sea Coast). Naukova Dumka, Kiev, 184 pp. (1984) (in Russian)
10. Babinot, J.F., Damotte, R., Donze, P., Grosdidier, E., Oertli, H.J., Scarenzi-Carbini, G.: Crétacé inférieur. In: Oertli, H. (ed.) Atlas des Ostracodes de France. Bulletin des Centres de Recherche-Exploration-Production elf Aquitaine, Mémoire, vol. 9, pp. 163–209 (1985)
11. Baraboshkin, E.Yu.: Boreal-Tethyan correlation of lower Cretaceous Ammonite scales. Mosc. Univ. Geol. Bull. **59**(6), 9–20 (2004)
12. Baraboshkin, E.Yu., Yanin, B.T.: The correlation of the Valanginian of the South-Western and Central Crimea. In: Milanovsky, E.E. (ed.) Geology Essays on the Crimea. Proceedings of the Crimean Geological Scientific and Educational Center Named After Professor A.A. Bogdanov. Moscow State University, Geological Faculty Press, Moscow, pp. 4–26 (1997) (in Russian)
13. Bartenstein, H.: Feinstratigraphisch wichtige Ostracoden aus dem nordwestdeutschen Valendis. Paläontol. Z. **33**(4), 224–242 (1959)
14. Bartenstein, H., Oertli, H.J.: Index ostracodes in the lower Cretaceous of Heligoland. Bull. Cent. Rech. Pau (SNPA) **91**, 5–25 (1975)
15. Grishchenko, V.A., Guzhikov, A.Yu., Baraboshkin, E.Yu.: First magnetostratigraphic data on the lower Valanginian of the mtn. Dlinnaya section (SW Crimea) Cretaceous system of Russia and near abroad: problems of stratigraphy and paleogeography. In: Materials of Eight All-Russian Conference, Crimean Republic, 26 Sept–3 Oct 2016. Chernomorpress, Simferopol, pp 31–33 (2016) (in Russian)
16. Grishchenko, V.A.: The reflection of construction and forming conditions of lower Valanginian of Southwestern Crimea in the magnetism of rocks. In: Trofimuk Readings—2017: Materials of All-Russian Conference of Youth with Participation of Foreign Scientist. Trofimuk Institute of Petroleum Geology and Geophysics, Siberian Branch Russian Academy of Sciences, Novosibirsk, pp 15–18 (2017) (in Russian)

17. Grishchenko, V.A., Guzhikov, A.Y.: Paleomagnetic directions distortion caused by viscous-plastic deformations estimated from anisotropy of magnetic susceptibility (case study of Berriasian clays from East Crimea). In: Nurgaliev, D., Shcherbakov, V., Kosterov, A., Spassov, S. (eds.) Recent Advances in Rock Magnetism, Environmental Magnetism and Paleomagnetism, pp. 25–35. Springer, Cham (2019)
18. Groecke, D.R., Price, G.D., Robinson, S.A., Baraboshkin, E.J., Mutterlose, J., Ruffel, A.H.: The Upper Valanginian (Early Cretaceous) positive carbon-isotope event recorded in terrestrial plants. Earth Planet. Sci. Lett. **240**, 495–509 (2005)
19. Guzhikov, A.Yu.: Solving unsolvable problems in stratigraphy (Comments to the paper "New data on the magnetostratigraphy of the Jurassic–Cretaceous boundary interval, Nordvik Peninsula (northern East Siberia)" by V.Yu. Bragin, O.S. Dzyuba, A.Yu. Kazansky, and B.N. Shurygin). Russ. Geol. Geophys. **54**, 349–354 (2013)
20. Guzhikov, A., Bagayeva, M., Arkadiev, V.: Magnetostratigraphy of the Upper Berriasian "Zavodskaya Balka" section (East Crimea, Feodosiya). Volumina Jurassica **XII**(1), 175–184 (2014)
21. Hoedemaeker, P.J., Reboulet, S., Aguirre-Urreta, M.B., et al.: Report on the 1st International Workshop of the IUGS lower Cretaceous Ammonite Working Group, the "Kilian Group" (Lyon, 11 July 2002). Cretac. Res. **24**, 89–94 (2003)
22. Kolpenskaya, N.N.: Ostracods/Berriasian of Northern Caucasus (Urukh Section)/Biochronology and Correlation of Fanerozoic of Oil and Gas Bearing Basins of Russia, № 2, Kirichkova, A.I. (ed.). VNIGRI, Saint Petersburg, pp. 115–129 (2000) (in Russian)
23. Kupriyanova, N.V.: Lower Cretaceous Ostracods of the Kolguev Island (Well 140)/Stratigraphy and Fauna of Paleozoic and Mesozoic of Arctic. VNIIOkeanologia, Saint Petersburg, pp. 92–98 (2000) (in Russian)
24. Kuznetsova, K.I., Gorbachik, T.N.: Stratigraphy and Foraminifers of Jurassic and Lower Cretaceous of Crimea. Nauka, Moscow, 133 p. (1985) (in Russian)
25. Lev, O.M.: Early Cretaceous Ostracods of Boyarka River. Paleontologic Substantiation of the Partition of Paleozoic and Mesozoic of Arctic Regions of USSR. Collection of Scientific Papers. PGO Sevmorgeologiya, Leningrad, pp. 103–117 (1983) (in Russian)
26. Lyubimova, P.S.: Ostracods of lower Cretaceous sediments of Precaspian depression. L. VNIGRI Trans. **244**, 199 pp. (1965) (in Russian)
27. McFadden, P.L.: A new fold test for palaeomagnetic studies. Geophys. J. Int. **103**, 163–169 (1990)
28. McFadden, P.L., McElhinny, M.W.: Classification of the reversal test in paleomagnetism. Geophys. J. Int. **103**, 725–729 (1990)
29. Molostovskiy, E.A., Khramov, A.N.: Magnetostratigraphy and Its Meaning in Geology. Saratov University Publishing, 180 pp. (1997) (in Russian)
30. Neale, J.W.: The Cretaceous. In: Bate, R., Robinson, E. (eds.) A Stratigraphical Index of British Ostracoda. Geological Journal Special Issue No. 8. Seel House Press, Liverpool, pp. 325–384 (1978)
31. Nil, J.V.: Ostracods from lower Valanginian of Central Crimea. Paleontol. J. **1**, 87–100 (1966) (in Russian)
32. Ogg, J.G., Hinnov, L.A.: Cretaceous. In: Gradstein, F., Ogg, J.G., Schmitz, M.D., Ogg, G.M. (eds.) The Geologic Time Scale. Elsevier, pp. 793–853 (2012)
33. Ogg, J.G., Ogg, G.M., Gradstein, F.M.: A Concise Geologic Time Scale. Elsevier, 242 pp. (2016)
34. Opdyke, N.D., Channell, J.E.T.: Magnetic Stratigraphy, p. 344. Academic Press, New York (1996)
35. Savelieva, J.N.: Paleoecological analysis of Berriasian Ostracods of the Central Crimea. Volumina Jurassica **XII**(1), 163–174 (2014)
36. Savelieva, Y.N., Shurekova, O.V., Feodorova, A.A., Arkadiev, V.V., Grishchenko, V.A., Guzhikov, A.Yu., Manikin, A.G.: Microbiostratigraphy of the Berriasian–Valanginian boundary in East Crimea: foraminifers, ostracods, organic-walled dinoflagellate cysts. Geol. Carpath. **68**(6), 517–529 (2017)

37. Savelieva, J.N., Feodorova, A.A., Shurekova, O.V., Arkadiev, V.V.: Integrated palaeontological characteristics (ammonites, ostracods, foraminifers, and dinocysts) of the Berriasian deposits of central Crimea. Volumina Jurassica **XII**(1), 129–162 (2014)
38. Shurekova, O.V., Savelieva, Yu.N.: Micropaleontologic characteristic (ostracods, dynocysts) of Hauterivian-Aptian sediments of the Verhorechye section (Southwestern Crimea). In: Baraboshkin, E.Yu. (ed.) Proceedings of Conference on Cretaceous. Chernomorpress, Simferopol, pp. 289–291 (2016) (in Russian)
39. Slipper, I.J.: Marine lower Cretaceous. In: Whittaker, J.E., Hart, M.B. (eds.) Ostracods in British Stratigraphy. The Micropalaeontological Society, Special Publications, The Geological Society, London, pp. 309–343 (2009)
40. Supplement to the Stratigraphic Code of Russia: VSEGEI, Saint-Petersburg, 112 pp. (2000) (in Russian)
41. Tesakova, E.M., Rachenskaya, L.P.: New ostracodes (Crustacea, Ostracoda) of the genus Costacythere Griindel from the Berriasian of Central Crimea. Paleontol. J. **30**(4), 428–435 (1996a)
42. Tesakova, E.M., Rachenskaya, L.P.: New ostracodes (Crustacea, Ostracoda) of the genera bairdia M'Coy, neocythere mertens, macrodentina martin, hechticythere griindel, and cypridea bosquet from the Berriasian of Central Crimea. Paleontol. J. **30**(5), 542–550 (1996b)
43. Triebel, E.: Ostracoden-Untersuchungen. 1. Protocythere und Exophthalmocythere, zwei neue Ostracoden-Gattungen aus der deutschen Kreide. Senckenbergiana **20**, 179–200 (1938)
44. Van der Voo, R.: Paleomagnetism of the Atlantic, Tethys, and Iapetus oceans, p. 412. Cambridge University Press, Cambridge (1993)
45. Yampolskaya, O.B., Baraboshkin, E.Yu., Guzhikov, A.Yu., Pimenov, M.V., Nikul'shin, A.S.: Lower Cretaceous paleomagnetic section in the southwest Crimea. Mosc. Univ. Geol. Bull. **61**(1), 1–16 (2006)
46. Yanin, B.T.: Correlation between the general and local stratigraphic units recognized in the lower Cretaceous of the southwestern Crimea (Kacha-Bodrak Interfluve). Mosc. Univ. Geol. Bull. **52**(3), 28–34 (1997)

Magnetostratigraphy of the Maastrichtian from Volga Right Bank Area Near Saratov (Lower Volga Region)

A. Yu. Guzhikov, A. A. Guzhikova, A. G. Manikin and V. A. Grishchenko

Abstract We present here a synthesis of magnetostratigraphic studies of the Upper Campanian-Maastrichtian carried out for six sections in Volga Right bank area near Saratov, including new results from three sections: Vishnevoye, Lokh, and Pudovkino. On the basis of paleomagnetic correlations, ages of local stratigraphic units have been determined more precisely, and diachroneity of their boundaries has been documented. The uppermost layers of the Lokh formation have been reassigned to Upper Maastrichtian in Lokh and Pudovkino sections, contrary to previous assignment of these strata to Lower Maastrichtian. Tracing from the north to the south, the age of the top of the Nalitovo formation changes from the Early Maastrichtian (C31r magnetic chron) to the Late Campanian (C32r), and the age of the bottom of the Lokh formation from the Early Maastrichtian (from C31r) to the Late Maastrichtian (to C31n or C30n). Interrelation has been established between anisotropy of magnetic susceptibility and dynamics of slope processes.

Keywords Magnetostratigraphy · Geomagnetic polarity · Rock magnetism · Anisotropy of magnetic susceptibility · Campanian · Maastrichtian · Russian plate · Lower Volga

A. Yu. Guzhikov (✉) · A. A. Guzhikova · A. G. Manikin · V. A. Grishchenko
Saratov State University, Astrakhanskaya, 83, 410012 Saratov, Russia
e-mail: aguzhikov@yandex.ru

A. A. Guzhikova
e-mail: blackhole3@yandex.ru

A. G. Manikin
e-mail: agmanikin@mail.ru

V. A. Grishchenko
e-mail: grishenko-vladimir@bk.ru

T. B. Yanovskaya et al. (eds.), *Problems of Geocosmos—2018*, Springer Proceedings in Earth and Environmental Sciences, https://doi.org/10.1007/978-3-030-21788-4_7

1 Introduction

The authors have been investigating the magnetostratigraphy of Maastrichtian in the Volga Right bank area near Saratov since 2012. This is a part of a larger project aimed at elaborating the stratigraphic chart of the Upper Cretaceous strata in the East European Platform within Russia [23] by means of integrated study of Upper Cretaceous reference sections from the Southeast of the Russian Plate. Magnetostratigraphic data contribute to producing a more refined and exact chart, allowing to individualize formations and reveal the most accurate interrelations between the local stratigraphic units and the units of the General Stratigraphic Scale.

So far, magnetostratigraphy results for Upper Campanian-Maastrichtian strata have been acquired from six sections (Fig. 1). Results from the three of these have been published previously: the Bolshevik and the Kommunar quarries near the town of Volsk [7] and the Syrt Mountain section neat the village of Nizhnyaya Bannovka [6]. Results from the other three sections located near the villages of Lokh, Vishnevoye, and Pudovkino, respectively, are presented here.

The area of the Volga Right bank near Saratov lies at the junction of the Ulyanovsk-Saratov Trough (the northern part of the area) and the Volga Monocline (the southern part) [23] (Fig. 1).

According to the subregional stratigraphic chart of the Upper Cretaceous from the Middle and the Lower Volga Regions [23], the Maastrichtian from the eastern part of the Ulyanovsk-Saratov Trough is represented exclusively by carbonate rocks, mostly by chalky marls and chalk, and is subdivided into two formations: the Karsun (Lower Maastrichtian) and the Radishchevo (Upper Maastrichtian). In the vicinity of Volsk, in the area where Campanian-Maastrichtian was characterized paleomagnetically [7] the Karsun formation overlies discontinuously the carbonate marls of Lower Campanian Sengilej formation.

In the western part of the Ulyanovsk-Saratov Trough, the Maastrichtian is composed of carbonate-terrigenous beds—marls of the Lokh formation (Lower Maastrichtian), and aleuritic and sandy marls of the Nikolayevka formation (Upper Maastrichtian). The Lokh formation is underlain by the Nalitovo formation (Upper Campanian) composed of alternating marls, carbonate and siliceous clays.

The Maastrichtian from the Volga Monocline within the southern part of the Saratov Volga Region is represented by the carbonate-terrigenous Lokh formation (Lower Maastrichtian) and the terrigenous Karamysh formation (Upper Maastrichtian). The Lokh formation is composed of carbonate clays, with aleurite and sand contents growing upwards in the section. The Karamysh formation is composed mostly of highly clayey, calcareous sands. In the south of the Volga Right bank area near Saratov the Lokh formation overlies the Nalitovo formation (Upper Campanian) represented by siliceous clays. The magnetostratigraphic characteristics of the Nalitovo and the Lokh formations and probably of the lowermost of the Karamysh formation have been studied previously at the Nizhnyaya Bannovka reference section (Syrt Mountain) [6].

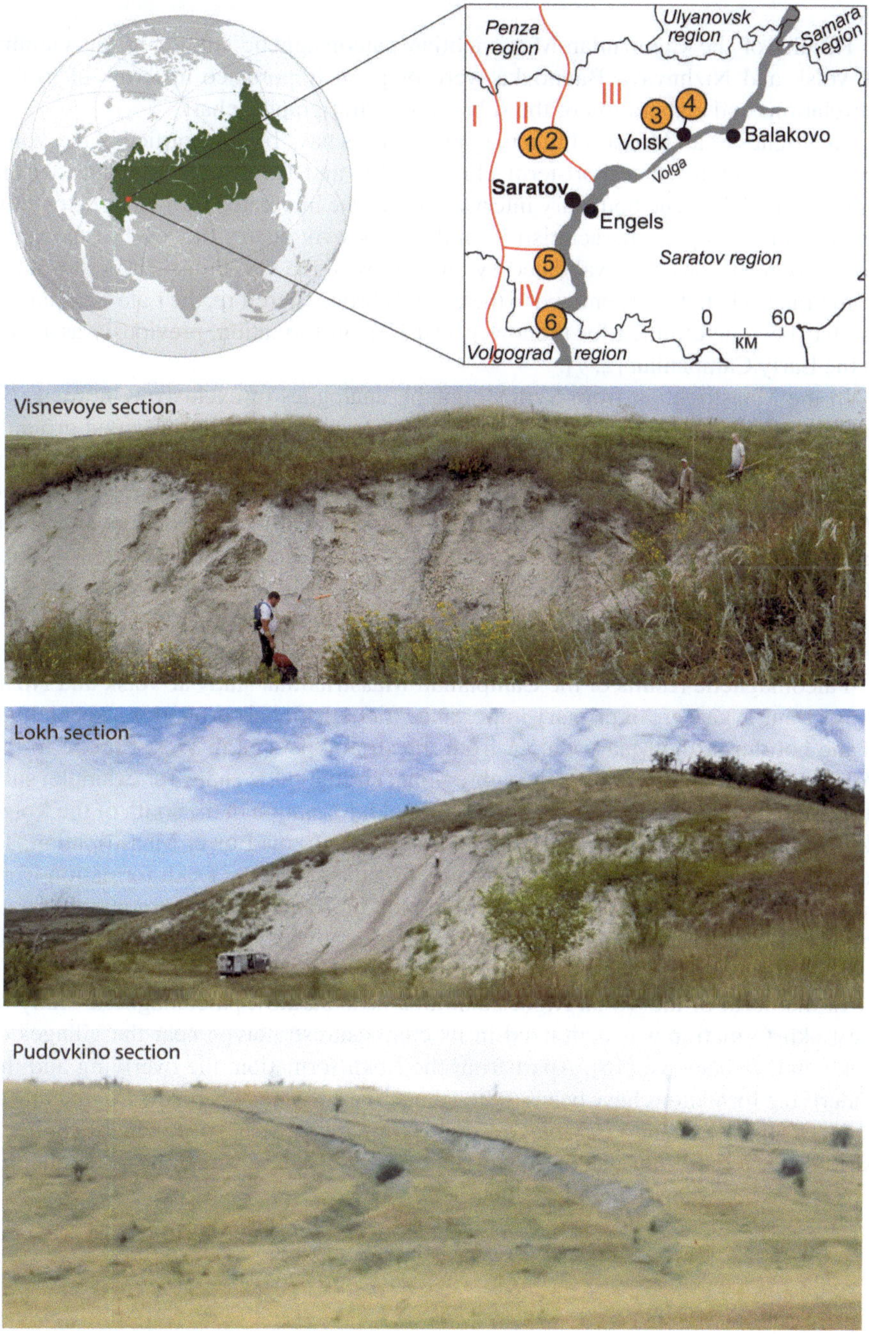

Fig. 1 Schematic location of the Upper Campanian-Maastrichtian sections (red lines on the map—boundaries of the geostructures) and photos of the studied sections. Sections: 1—Visnevoye, 2—Lokh, 3—Bolshevik, 4—Kommunar, 5—Pudovkino, 6—Syrt Mountain (Nizhnyaya Bannovka). Geostructures: I—Murom-Lomov Trough, II—western part of the Ulyanovsk-Saratov Trough, III—eastern part of the Ulyanovsk-Saratov Trough, IV—Volga Monocline

Results of the Campanian-Maastrichtian paleomagnetic studies in the vicinity of Volsk and Nizhnyaya Bannovka were of great importance in terms of global correlations and elaboration of the subregional stratigraphic chart.

Analogues of all Maastrichtian magnetic chrons have been identified in the Volsk section, except for the short-term chrons C30r and C29r, corresponding to the Cretaceous-Paleogene boundary interval. Due to the integrated (bio-, magnetostratigraphic and isotopic) characteristics, in the Bolshevik quarry the level of the stage lower boundary has been validated by analogy with the GSSP of the Maastrichtian from France [14]; that is confined to the C32n2 chron. Therefore, the Late Campanian age has been determined for the lowermost Karsun formation, previously assigned to the Early Campanian [2, 7].

In the Maastrichtian from Syrt Mountain, analogues of exclusively normal (N) polarity chrons C31n and (or) C32n have been recognized. This provided grounds for regarding the age of the Lokh formation in that section as the Late Maastrichtian, instead of the previously determined Early Maastrichtian [23]. Later on, the Late Maastrichtian age of the deposits has been substantiated with micropaleontological data (foraminifers, radiolarians, nannoplankton and calcareous dinocysts) [6]. Substantial disparity in determinations of the stratigraphic age has argued for the priority of paleomagnetic study of the Lokh formation among other Upper Cretaceous beds from the Lower Volga Region.

Paleomagnetic results of the Campanian-Maastrichtian study at Volsk and Nizhnyaya Bannovka were of primary importance for both global correlation, i.e. tracking of the bottom of the Maastrichtian from the stratotype area in West Europe to the Volga Region, and for refining the subregional stratigraphic chart, for example, substantiating the late Maastrichtian age of the Lokh formation in the south of the Volga Right bank area near Saratov instead of assigning it to the Lower Maastrichtian. To settle the contradictions between substantiating the age of the Lokh formation from biostratigraphic data in the north and complex bio- and magnetostratigraphic data in the south of the Volga Right bank area near Saratov, paleomagnetic study of this formation was carried out in other sections as well.

In the north of the Volga Right bank area near Saratov, paleomagnetic study of the Lokh formation was continued in its composite stratotype near the villages of Lokh and Vishnevoye [16]. Apart from the Lokh formation, the overlying and the underlying formations have been studied there: the Nikolayevka (in the Lokh section) and the Nalitovo (in the Vishnevoye section) formations, respectively. In the southern part of the Volga Right bank area near Saratov, the Lokh formation was studied in the section near the village of Pudovkino.

2 Sampling and Methods

2.1 Field Works and Characteristics of the Sections

Pioneer tools (pick, shovel, excavation knife) were used for paleomagnetic sampling. Samples were taken from the beds without any signs of supergene changes, generally in 0.3–0.5 m from the day surface (pits down to 1 m deep were dug if necessary). Oriented rock specimens were taken from every stratigraphic level. To be used for laboratory measurements, the specimens were sawn into three or four $2 \times 2 \times 2$ cm cubes. When it was impossible to acquire a monolith specimen of required dimensions, several oriented specimens were taken from one level (specimen minimum size was determined by the possibility to saw at least one 2-cm cube). The specimens were oriented by means of a magnetic compass. Sampling intervals were ~90 cm in Lokh and Vishnevoye sections and ~50 cm in the Pudovkino section.

The section of the Lokh formation near the village of Lokh lies in an abandoned quarry on the SW slope of the hill facing the NE fringe of the village. The section coordinates: 52° 08′ 24.9″ W, 45° 51′ 25.8″ E. The beds lie subhorizontally though no precise determination of bedding elements is possible due to the lack of pronounced bedding planes. Currently, an interval of apparent thickness of ~16 m is accessible for examination; of these, ~10 m belong to the upper part of the Lokh formation and ~6 m to the lowermost of the Nikolayevka formation. Oriented specimens were taken from the Lokh and the Nikolayevka formations, from 11 and 7 levels, respectively (Fig. 2a). In late 1990s, intervals of the Lokh (the LokhA member) and the Nikolayevka (the LokhB and LokhC members) formations were studied there; their thicknesses were at that time 11.5 and 16.5 m, respectively [1]. The bottom of the Lokh formation and the top of the Nikolayevka formation, however, were not examined. It should be noted, that the outward appearances of the marls from the Lokh and the Nikolayevka formations are similar (the authors of the first description of the Maastrichtian from the vicinity of Lokh [1] did not differentiate the rock names in the uppermost and the lowermost of the section). But from the data of grain-size analysis the bottom of the Nikolayevka formation is marked by a layer with high (up to 20%) content of the >0.05 mm fraction [1], clearly represented in the magnetic susceptibility plot by the increase up to 15×10^{-5} SI units (Fig. 2a).

Initially, based on the results of micropaleontologic examination the Lokh formation (the LokhA member) was referred to the subzones CC23a [22] according to the nannoplankton, and BF10a according to benthic foraminifers. These zones were compared to the belemnite zone B. licharewi [1], i.e. to the uppermost of the Campanian stage in modern view. Later on, the authors have reconsidered their ideas of the LokhA member age according to nannoplankton and of the interrelation between the belemnite and the foraminifer units [1, 16]. Thus, the Lokh formation was parallelized with the lower zone of the Maastrichtian—B. lanceolata.

The study of the lowermost of the Lokh formation, alongside with the Nalitovo formation, was continued in the section near the village of Vishnevoye (Petrovsk District), about 20 km to the west of Lokh (Fig. 1). The section (coordinates:

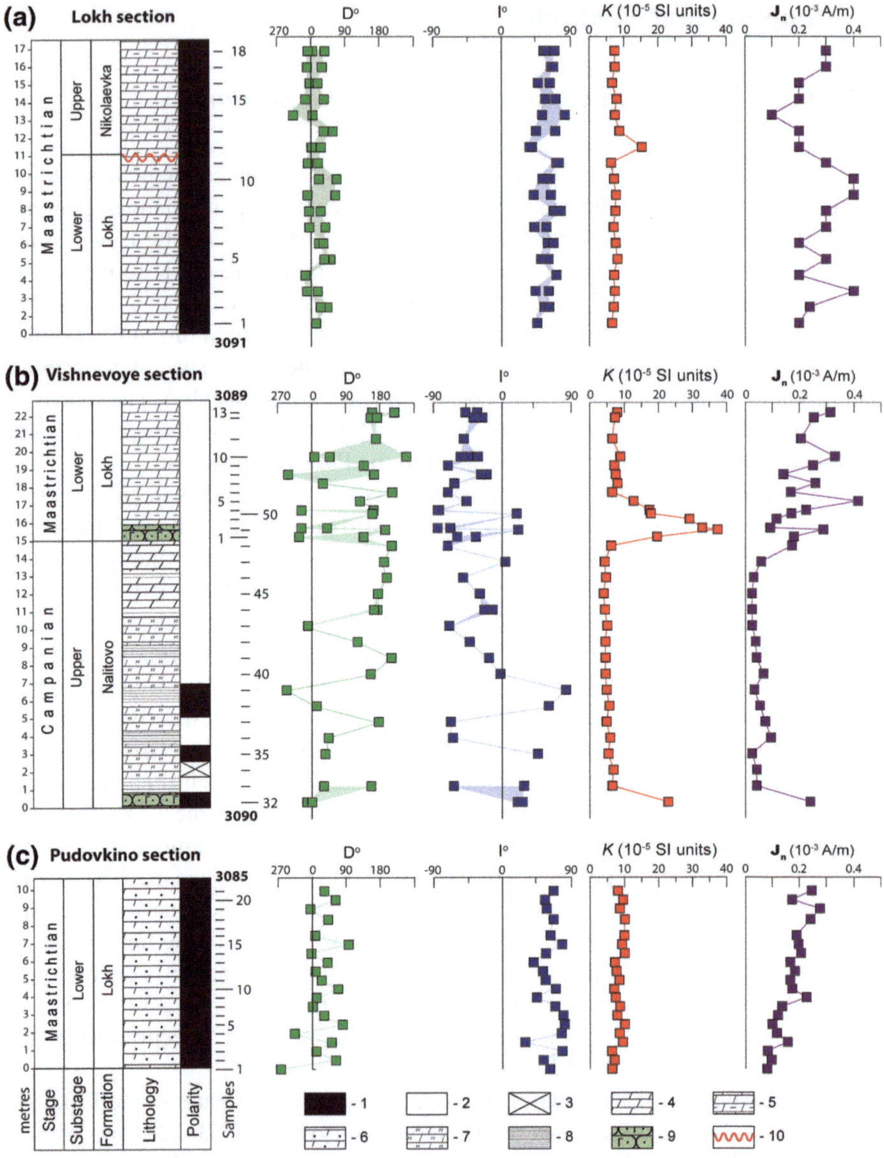

Fig. 2 Paleomagnetic and rock-magnetic characteristics of the studied sections: Lokh (**a**), Vishnevoye (**b**), Pudovkino (**c**). Geomagnetic polarity: 1—normal (N), 2—reversed (R), 3—missing polarity data. Lithology: 4—marls, 5—clay marls, 6—carbonate clays with aleurite and sand, 7—siliceous marls, 8—clays, 9—glauconitic sandstones, 10—biostratigraphic hiatus

52° 05′ 17.4″ N, 45° 33′ 50.6″ E) outcrops in the head of a gully incised in the interfluve of the Medveditsa and Chardym rivers and opening into the Chardym valley near Vishnevoye. According to the complex of macro- and microfauna data, the ages of the Nalitovo and the Lokh formations there are substantiated as the Late Campanian and the Early Maastrichtian, respectively [17, 18]. The Lokh formation, with the apparent thickness of ~8 m, is represented by marls with glauconitic sandstones in the base. The 15-m thick Nalitovo formation is represented by marls and siliceous clays, with glauconitic sandstone in the base as well. Oriented samples were taken from 14 and 18 levels of the Lokh and the Nalitovo formations, respectively. All the layers dip to the north and to the northwest. Bedding elements were measured from the two most pronounced solid surfaces of glauconitic sandstones. The dip azimuth of 18° and the dip angle of 18°, determined at the level of the sample 3090/32, were extrapolated to all the overlying levels through sample 3089/4. The dip azimuth of 345° and the dip angle of 13°, measured at the level of sample 3089-5, were extrapolated to all the overlaying beds (Fig. 2b).

The Lokh formation section near Pudovkino, Saratov District, ~65 km to the north of Syrt Mountain (Nizhnyaya Bannovka), was examined in major washouts on a steep slope of the Volga Upland, about 1.5 km to the NNW of the northern margin of the village on the bank of the Volga. The section coordinates: 51° 17′ 50.3″ N, 45° 45′ 42.0″ E. The layers lie subhorizontally. Notwithstanding the extensive erosional pattern, plastic clays of the Lokh formation are extremely poorly exposed. We managed to study an interval in the upper part of the Lokh formation, with the apparent thickness of ~10 m (with the actual thickness of the formation in that area of at least 30 m), and to collect oriented samples from 21 levels (Fig. 2c).

2.2 *Magnetic Mineralogy Analysis*

Petromagnetic and magnetomineralogical examinations involved studies of initial magnetic susceptibility (K) and its anisotropy (AMS), natural remanent magnetization ($\mathbf{J_n}$) measurements, acquisition of saturation isothermal remanent magnetization ($\mathbf{J_{rs}}$) and coercivity of remanence ($\mathbf{H_{cr}}$) determination, and thermomagnetic analysis (TMA). TMA was performed measuring magnetization ($\mathbf{J}$) temperature dependence. Samples were heated up to 700 °C in a 650 mT DC magnetic field (first heating). Magnetic minerals were diagnosed from their respective Curie temperatures and/or phase transition temperatures recorded in $\mathbf{J}(T)$ curves as characteristic bends. Samples were then reheated (second heating) in order to detect newly formed alteration products that may be used to judge on the type of the initial mineral.

Susceptibility measurements were made with an MFK1-FB kappabridge, and the Anisoft 4.2. program was used for AMS analysis. $\mathbf{J_n}$ and $\mathbf{J_{rs}}$ were measured with a JR-6 spinner magnetometer. Isothermal remanent magnetization acquisition and backfield demagnetization were performed by means of an adjustable electromagnet providing a magnetic field up to 700 mT. The value of remanent magnetization acquired in a 700 mT field was taken as the $\mathbf{J_{rs}}$ value. A fraction thermoanalyzer TAF-2 (magnetic balance) was used for TMA.

Paleomagnetic examination was made by the standard procedure [4, 13], involving measurements of the sample $\mathbf{J_n}$ with a JR-6 spinner magnetometer after a series of sequential demagnetizations by either alternating field (mostly up to 50–100 mT in increments of 5 mT) using a LDA-3 AF demagnetizer, or temperature (from 100 to 550 °C in 50 °C increments) in a home-built furnace constructed by V. P. Aparin. Demagnetization was stopped when $\mathbf{J_n}$ values became comparable to the instrument threshold sensitivity or a magnetic bias in a laboratory ambient field was detected. Mineral phase transformations upon heating were controlled by measuring the sample K values after each thermal demagnetization. Paleomagnetic examinations of the samples from most levels were duplicated using a 2G Enterprises cryogenic (SQUID) magnetometer at IPE RAS (Moscow). The Remasoft 3.0 program was used for component analysis.

Because of the horizontal bedding of layers near the villages of Lokh and Pudovkino, the data on the $\mathbf{J}_n$ components and on the axes of AMS ellipsoids from the homonymous sections was analyzed in the geographical coordinate system (geo-coordinates in the respective figures). The same data from the Vishnevoye section, where the layer dips were recorded, were considered in the paleogeographical coordinate system (tilt-coordinates in the respective figures).

3 Results

3.1 Rock Magnetic and Magnetomineralogical Investigations

In all sampled sections, including the Syrt Mountain section studied previously [5], the beds from the Lokh and the Nalitovo formations are weakly magnetic. Magnetic susceptibility (K) varies from 4 to 15 $\times$ 10^{-5} SI units, $\mathbf{J_n}$ from 0.02 to 0.04 $\times$ 10^{-3} A/m respectively. Increased K values (as high as 15–33 $\times$ 10^{-5} SI units) are indicative only of the layers containing glauconite, which is known to show a high paramagnetic susceptibility [10] (Fig. 2a–c).

According to the magnetic saturation results, a magnetically soft phase is detected in all studied samples: over 90% of $\mathbf{J_{rs}}$ is acquired in fields below 100 mT, and is demagnetized at 30–40 mT (Fig. 3a). This is characteristic of fine-grained magnetite.

TMA curves of both the first and the second heating show only a single, poorly pronounced, bend about 580 °C, Curie temperature of magnetite. Negative magnetization values above 530 °C may be caused by extremely low concentration of Fe_3O_4; this leads to domination of the sample holder diamagnetic effect in the vicinity of the mineral Curie point. It should be noted that magnetization decrease continues up to 700 °C and above (Fig. 3b); alongside with the sample magnetic softness this is suggestive of native iron presence [21].

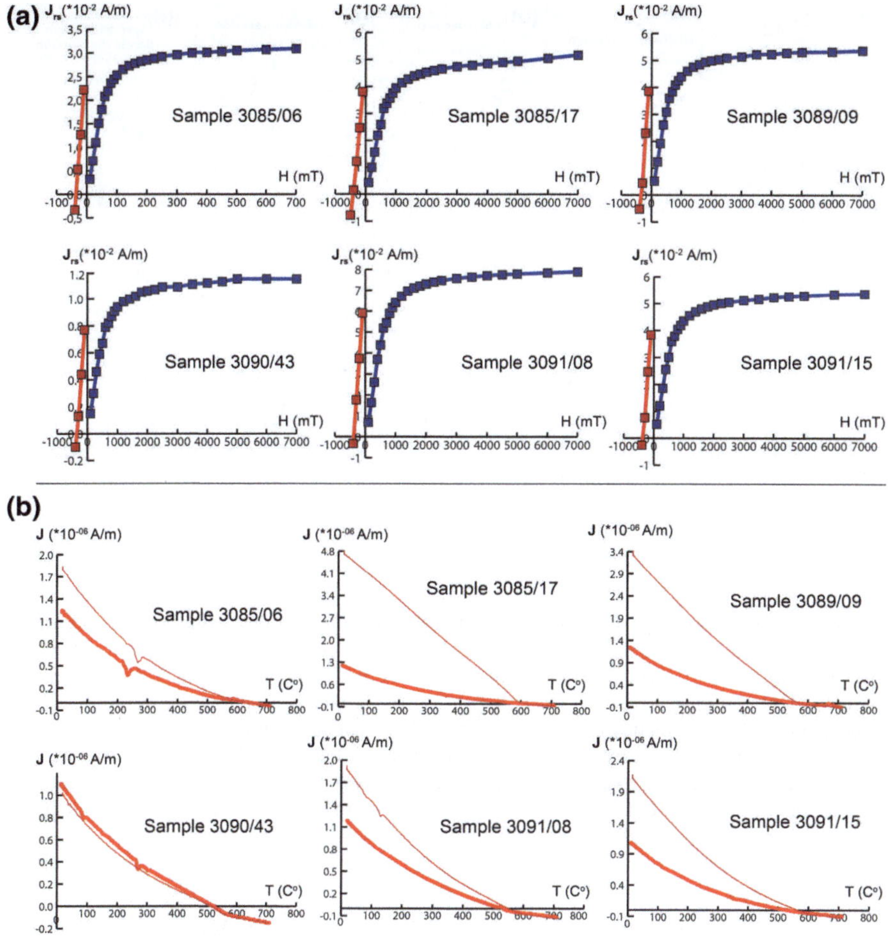

Fig. 3 Results of magnetomineralogical analysis. **a** Curves of magnetic saturation and destruction, **b** TMA curves: first (thick line) and second (thin line) heating

According to the data of magnetic saturation, TMA and microprobe analysis, space-borne ferriferous and magnetite particles appear to be principal magnetization carriers in the Nalitovo, Lokh and probably Karamysh formations from the Campanian-Maastrichtian section of Syrt Mountain (Nizhnyaya Bannovka) [6], as well as in the Karsun and Radishchevo formations from the Maastrichtian section at Bolshevik and Kommunar quarries (the town of Volsk) [7]. Experimental results of magnetic saturation experiments and thermomagnetic analyses of the samples from Syrt Mountain are identical to those from the Lokh, Vishnevoye and Pudovkino sections. Therefore, by analogy with the Syrt section, dominantly cosmogenic origin of the ferromagnetic fraction in the coeval beds from the study sections may be regarded as the most probable one. An enrichment of the Maastrichtian beds with

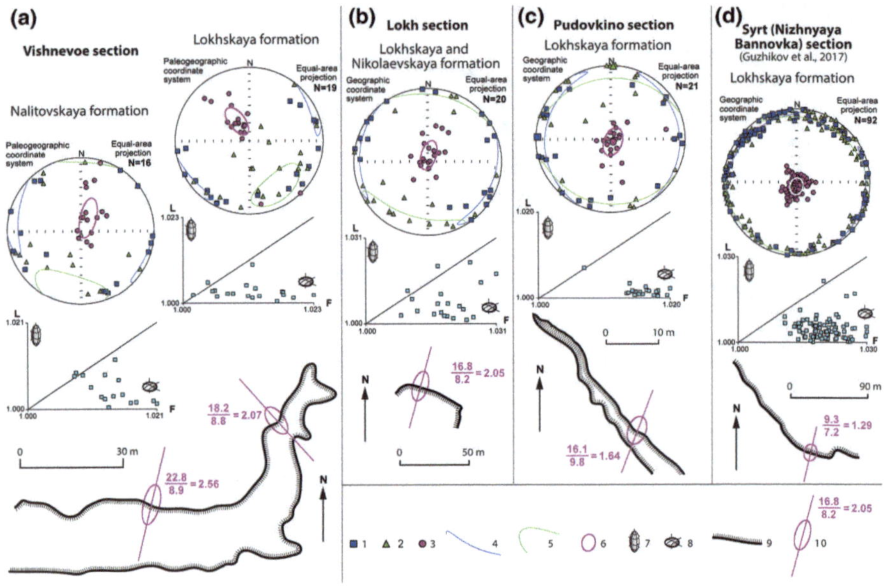

Fig. 4 Data on anisotropy of magnetic susceptibility for the sections Vishnevoe (**a**), Lokh (**b**), Pudovkino (**c**) and Syrt (Nizhnyaya Bannovka) [6] (**d**): from top to bottom—stereograms of the AMS axes, F–L diagrams [5], schematic image of the objects studied in field (in plan). Axes directions: 1—long (K1), 2—mean (K2) 3—short (K3); 4, 5, 6—confidence ellipses for K1, K2 and K3, respectively; 7, 8—schematic images of elongate and flat magnetic particles, respectively; 9—slope direction in the ravine (**a**), career wall (**b**), breakthrough (**c**) and landslide cliff (**d**); 10—strike of the confidence ellipse long axis for K3 and relation between the long half-axis (in numerator) of the confidence ellipse to the short half-axis (in denominator). Confidence ellipses are situated in the places where the oriented samples were taken

meteoric dust has been previously observed in the north of the Volga Right bank area near Saratov, in the sections near the villages of Teplovka in Novyje Burasy District, Klyuchi and Belogrodnya in the Volsk District [12].

Diagrams of magnetic anisotropy parameters F–L record mostly flat-shaped magnetic particles [5] characteristic of clayey rocks with submicron ferromagnetic particles (probably, cosmic dust in this particular case) aggregated on clay mineral flakes (Fig. 4a–d).

Some highly interesting patterns are observed in the distribution of the axes of magnetic susceptibility ellipsoids in the studied sections. Universally, the primary magnetic texture of sediments formed in calm hydrodynamic settings (with subvertical short-axes projections (K3), and projections of the medium (K2) and long (K1) axes uniformly distributed along the equator) is distorted due to elongated distributions of short axes (Fig. 4a–d). Coincidence of orientations of the long axes of K3 confidence ovals and the directions of gully or hill slopes in the sampling sites (Fig. 4a–d) indicate that in all cases, a downslope displacement of plastic rocks is the principal reason of the K3 elongated distributions.

Slope viscous-plastic deformations of the deposits are manifested stronger in the sections from the north of the Volga Right bank area near Saratov. Maximum elongation of K3 projections is recorded in the Vishnevoye section sampled in the gully sides (Fig. 4a). K3 distribution is less elongated in the Lokh section sampled it the quarry vertical wall on the hill slope (Fig. 4b). In the Pudovkino section, studied along the washout sides on the side-hill, the elongation index is still lower (Fig. 4c), and there are practically no deviations from the sediment primary magnetic texture in the Lokh formation at Syrt Mountain [6], sampled in a landslide cliff (Fig. 4d). Since the calcite contents decreases regularly from the north to the south of the Volga Right bank area near Saratov (marls in the Vishnevoye and the Lokh are replaced with carbonate clays and aleurite-sandy carbonate clays in Pudovkino and Nizhnyaya Bannovka, respectively), it is obvious that karst processes contribute substantially to the intensity of the slope deformation.

3.2 *Paleomagnetic Investigations*

There is good agreement between the results of paleomagnetic investigations, irrespective of demagnetization procedure or the type of measuring instruments (Fig. 5).

In the studied sections, we managed to distinguish characteristic remanent magnetizations $\mathbf{J_n}$ (**ChRM**), with maximum deviation angles less than 10° for 136 samples from 70 sampled stratigraphic levels (Figs. 5, 6 and 7). Paleomagnetic vector projections grouped in the northern bearings of the lower hemisphere, were interpreted as corresponding to the geomagnetic field of normal polarity (N). The **ChRM** projections on the upper hemisphere would then correspond to the reversed polarity (R) (Fig. 8a–c). In some samples, low-coercivity and low-temperature components are present. These samples are demagnetized at weak fields (<5–10 mT) and low temperature (<100–150 °C), respectively. Their magnetization directions are distributed chaotically, and are most probably associated with viscous magnetization acquired during sample transportation and pretreatment.

Paleomagnetic columns from the Lokh and the Pudovkino sections are represented by polarity magnetozones of exclusively normal polarity (Figs. 2a, c and 7a, c). The paleomagnetic column from the Vishnevoye section is formed by a long reversed polarity magnetozone (Figs. 2b and 7b), with three relatively short normal polarity intervals seen below the latter (Fig. 2b). Theoretically, these normal intervals may represent independent polarity zones, but, since they have been proved with samples from one or two levels (while a magnetozone should be confirmed with samples from at least three levels [9]), such conclusions are premature at least.

Results of statistical analysis of **ChRM** distributions are presented in Table 1. To acquire statistical values for rocks belonging to a single type, scarce data on glauconitic sandstones has been omitted. In the Vishnevoye section, only reversed ChRM's were analyzed. With glauconitic sandstones excluded, normal polarity in that section has been recorded only in three levels, which is insufficient for statistical calculations. Rocks in the Lokh and Pudovkino sections exhibit acceptable paleo-

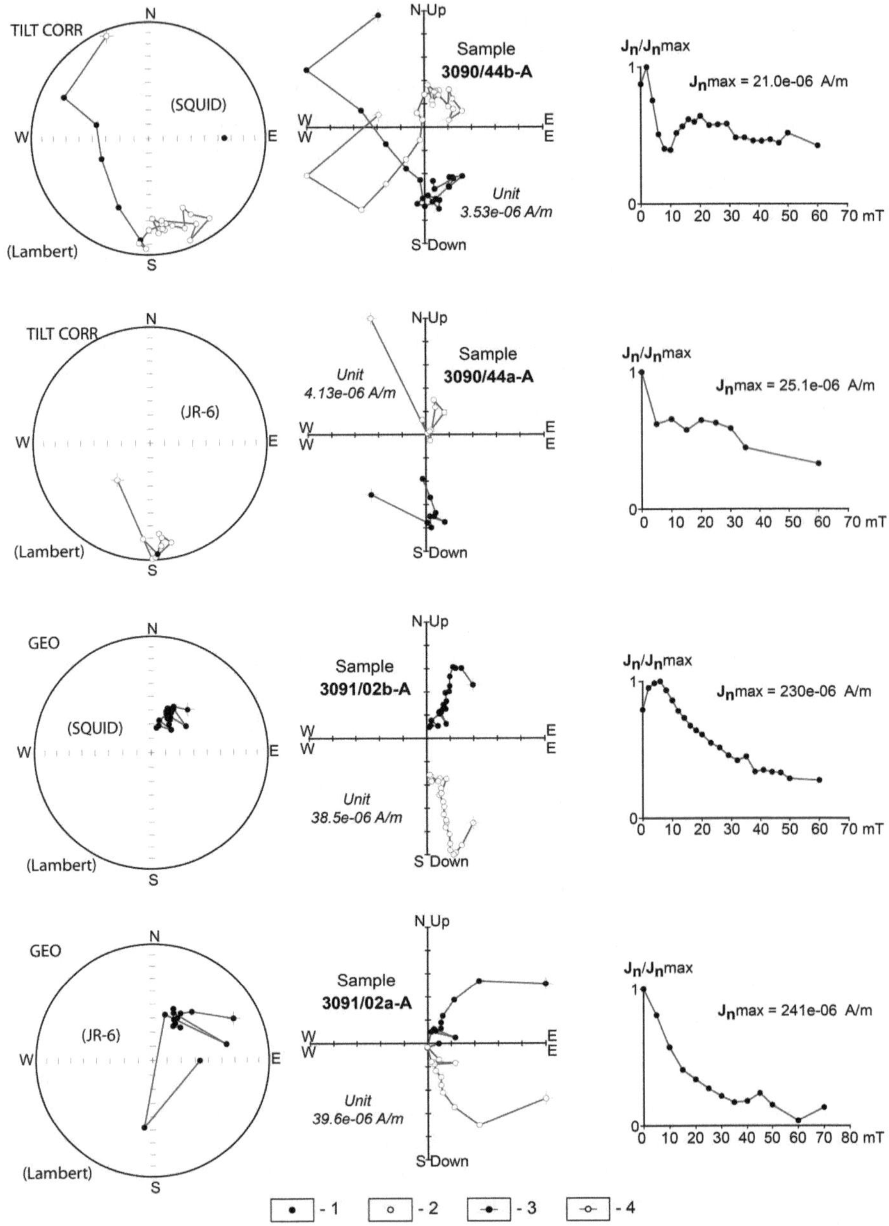

Fig. 5 A comparison of results of paleomagnetic studies on a spinner magnetometer JR-6 and on a 2G Enterprises cryogenic magnetometer. The results of component analysis (left to right: stereographic presentation of $\mathbf{J_n}$ changes in the process of demagnetization, Zijderveld diagrams, sample demagnetization plots). 1, 2—Projection of $\mathbf{J_n}$ on the lower hemisphere and upper hemisphere, respectively; 3, 4—Projection of $\mathbf{J_n}$ on the horizontal and vertical plane, respectively

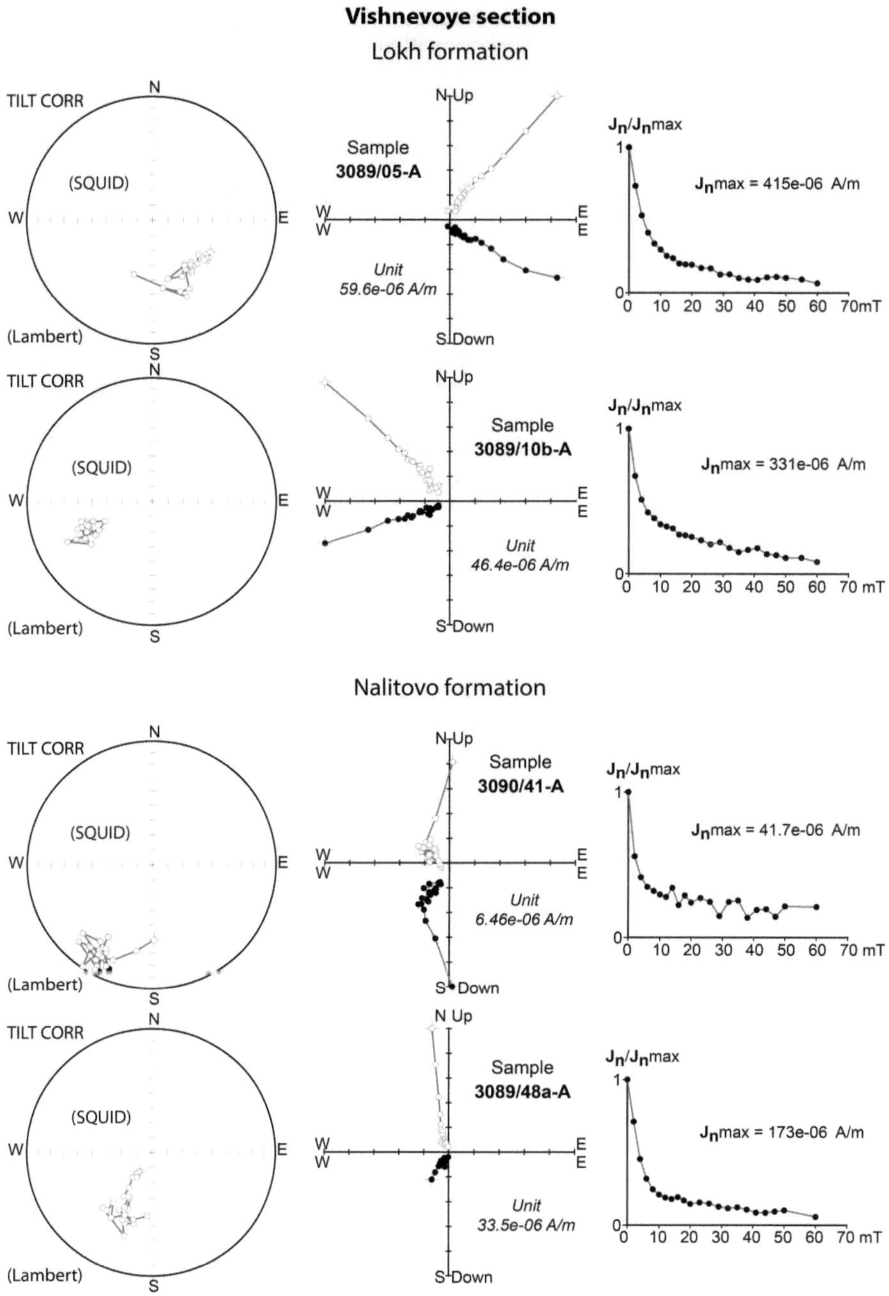

Fig. 6 The results of component analysis of samples from the Vishnevoye section. For explanations see Fig. 5

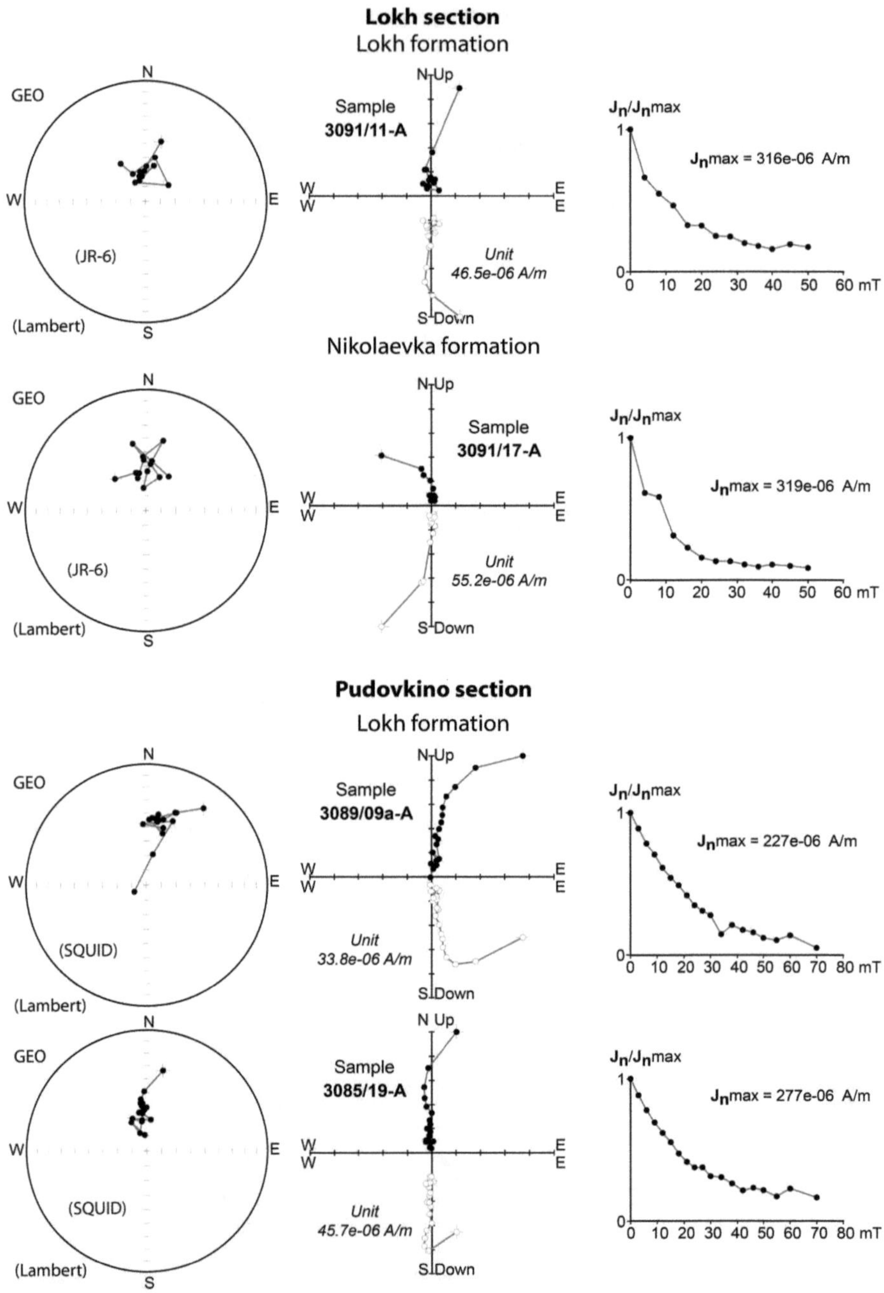

Fig. 7 The results of component analysis of samples from the Lokh and Pudovkino sections. For explanations see Fig. 5

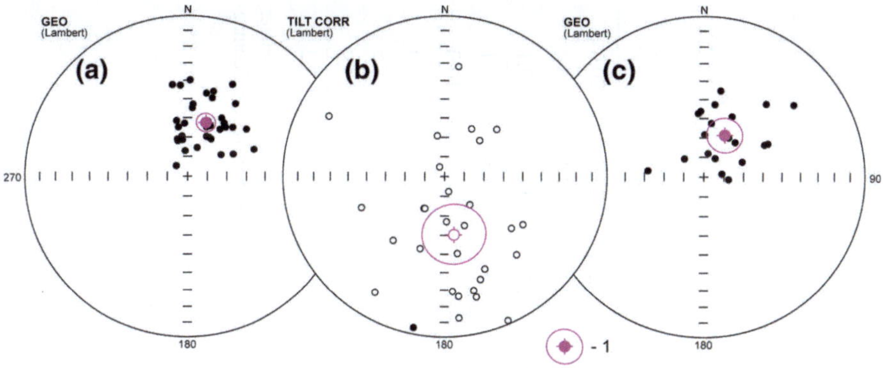

Fig. 8 Stereographic projections of paleomagnetic direction (**ChRM**) of the studied sections: **a** Lokh, **b** Vishnevoye (without a directions of normal polarity and a directions in glauconitic sandstones), **c** Pudovkino. 1—average and confidence circle. For other explanations see Fig. 5

magnetic quality, with the interbed paleomagnetic precision parameters of the order of twenty and the radii of confidence circles less than 10°. Low precision parameter value and large confidence circle in the Vishnevoye section are determined by a combination of several factors (Fig. 8b; Table 1). Alongside with intense viscous-plastic deformations of the rocks, the abnormally high dispersion of paleomagnetic vectors may be caused by an uncomplete erasure of the viscous remanence, directed along the present geomagnetic field. **ChRM** would then be a vector sum of two components: the primary one, corresponding to reversed polarity, and the undemagnetized viscous magnetization. Note that this effect would have practically no impact on the paleomagnetic statistics for normally magnetized samples, because the directions of ancient and present normal polarity geomagnetic fields are in our case quite close. Nevertheless, the reversal test of the samples from the Vishnevoye and the Lokh sections is positive at the C level according to McFadden and McElchinny classification [11] (Table 1).

Positive result of the reversal test makes an important argument for the primary character of magnetization. Moreover, there are some other indications of ancient age of $\mathbf{J_n}$:

1. All sections are peculiar for low (fraction of unity) Koenigsberger parameter values (factor Q) and interbed paleomagnetic precision parameters (15–25, not considering extremely low values in the Vishnevoye section which are due to intense slope deformations, Table 1). This is not typical for NRM of chemical origin [13]. The most probable magnetization carrier (meteoritic dust) in this section is of allothogenic origin.
2. Polarity sign does not exhibit any dependence on lithological composition. R-magnetozone in the Vishnevoye section comprises clays and marls, as well as glauconitic sandstone in the base of the Lokh formation. At the same time, glauconitic sandstone in the lowermost of the Nalitovo formation demonstrates nor-

Table 1 Statistic paleomagnetic characteristics of the Upper Campanian-Maastrichtian from the studied sections

Section	n	D°	I°	k	α_{95}°	γ°	γ_c°	Φ°	Λ°	dp°	dm°	A_{95}°
Vishnevoye	30	170.5	−59.9	3.8	15.7	13.9	14.5	77.0	259.9	17.9	23.7	20.6
Lokh	36	18.9	61.0	25.8	4.8			73.7	166.7	5.6	7.3	6.4
Pudovkino	20	26.3	66.2	14.6	14.6			73.0	134.5	11.9	14.5	13.1
Syrt [6]	59	349.9	65.1	20.9	20.9			82.5	291.2	6.8	5.5	6.1
Average paleomagnetic pole for the studied sections								83.0	188.5			16.0
65 MY (averaged over 5 MY) [3]								79.8	209.5			3.6
70 MY (averaged over 5 MY) [3]								80.0	213.2			4.8
70 MY (averaged over 10 MY) [3]								81.4	206.1			5.9
75 MY (averaged over 5 MY) [3]								81.3	188.6			7.2

n—number of samples in the selection; D, I—average paleomagnetic declination and inclination, respectively; k—paleomagnetic precision parameter; α_{95}—radius of 95% confidence circle; Φ, Λ, dp, dm, A_{95}—latitude, longitude, semiaxes of error oval and confidence circle radius for average paleomagnetic poles, respectively. Results of reversal test [11]: γ—angle between paleomagnetic vectors, γ_c—critical angle

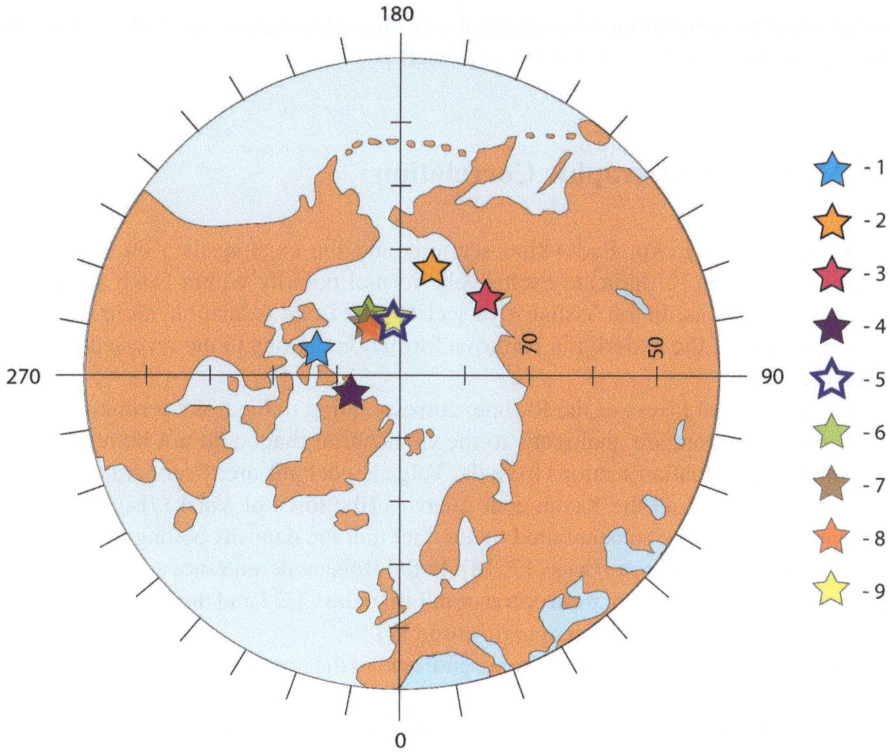

Fig. 9 Positions of paleomagnetic poles with confidence circles (A_{95}) determined for the sections of Upper Campanian-Maastrichtian (1–4): 1—Vishnevoye, 2—Lokh, 3—Pudovkino, 4—Syrt (Nizhnyaya Bannovka) [6]; 5—average paleomagnetic pole for the studied sections; 6 to 9—standard paleomagnetic poles for stable Europe [3]

mal polarity. Marls from the Lokh section correspond to the N-magnetozone, while similar marls from the Vishnevoye section to the R-magnetozone (Fig. 2).

3. Coordinates of the average paleomagnetic pole in all examined sections, including the Lokh formation section from Syrt Mountain [6], coincide statistically with the positions of 65–75 Ma standard poles for stable Europe [3] (Fig. 9). Meanwhile, the poles, calculated for each section individually, statistically differ from the standard pole. Averaging obviously tends to level down the random distortions caused by landslide deformations.

Each of the above features taken separately merely does not contradict a hypothesis of the ancient nature of $\mathbf{J_n}$, but together they allow to state that the polarity record found in the present work does reflect the geomagnetic field regime at the end of the Cretaceous period. The acquired data meet seven criteria out of eight accepted for evaluating the reliability of magnetostratigraphic records [8], and six out of seven of Van der Voo criteria for evaluating the quality of paleomagnetic data [24]. This is sufficient to consider the magnetic polarity characteristics of the study sections to be reliable from the formal standpoint as well.

4 Magnetostratigraphic Correlation

New data on the Lokh, Pudovkino sections and the existing data on Nizhnyaya Bannovka section [6] indicate exclusively normal polarity within Lokh and Nikolayevka formations. In the Vishnevoye section, the Lokh formation, alongside with the upper half of the underlying Nalitovo formation, belong to the reversed polarity zone (Fig. 10).

The uppermost layers of the R-zone, corresponding to the Lokh formation in the Vishnevoye section, are analogous to the C31r chron that could not be revealed in any other Maastrichtian sections from the Volga Right bank area near Saratov, except for the Bolshevik and the Kommunar quarries (the town of Volsk) (Fig. 10). This interpretation may be substantiated by the fact that the deposits belong to the CC23 and CC24 nannoplankton zones [17, 18]. In the Bolshevik reference section (Volsk), the part of the Karsun formation corresponding to the CC23 and the CC24 zones [19, 20] correlates with the C31r magnetic chron [7].

The lowermost layers of the R-zone, confined to the upper part of the Nalitovo formation, should probably also be identified as C31r. The assumption is based on that the Campanian-Maastrichtian boundary interval in the Geological Time Scale (GTS) [15] lacks extended reversed polarity chrons other than C31r comparable to a thick (~18 m) fragment of the R-zone, characterizing the upper half of the Nalitovo formation (Fig. 10). A further support to this hypothesis is brought by that the uppermost layers of the Nalitovo formation characterized by the reversed polarity, belong to the CC23 subzone according to nannoplankton [17, 18]. If our magnetostratigraphic correlation is correct, the upper layers of the Nalitovo formation, traditionally dated to the Late Campanian [17, 18], should be instead assigned to the Early Maastrichtian, by analogy with West European sections (Fig. 10).

Since the lower part of the Lokh formation outcrops near Vishnevoye, while its upper part outcrops near Lokh, the normal polarity magnetozone in the Lokh section comprising both the uppermost of the Lokh formation and the Nikolayevka formation may be analogous only to the chrons that are younger than C31r: C31n and/or C30 (Fig. 10).

In Nizhnyaya Bannovka, correspondence of this N-magnetozone to the Late Maastrichtian C31n or C30n chron, or to a superposition of the C31 and C30n chrons is confirmed by determinations of benthic and planktonic foraminifers, radiolarians and nannoplankton indicative of the Late Maastrichtian age of the deposits [6]. The N-magnetozone in Pudovkino is also identified with the C31n or the C30n or with C31n + C30n, since the beds studied in that section represent complete lithological and petromagnetic analogues of the Lokh formation from Syrt Mountain (Fig. 10).

In the Lokh section, the thickness of the unexamined interval in the lowermost of the Lokh formation does not exceed a few meters [1], while in the Vishnevoye

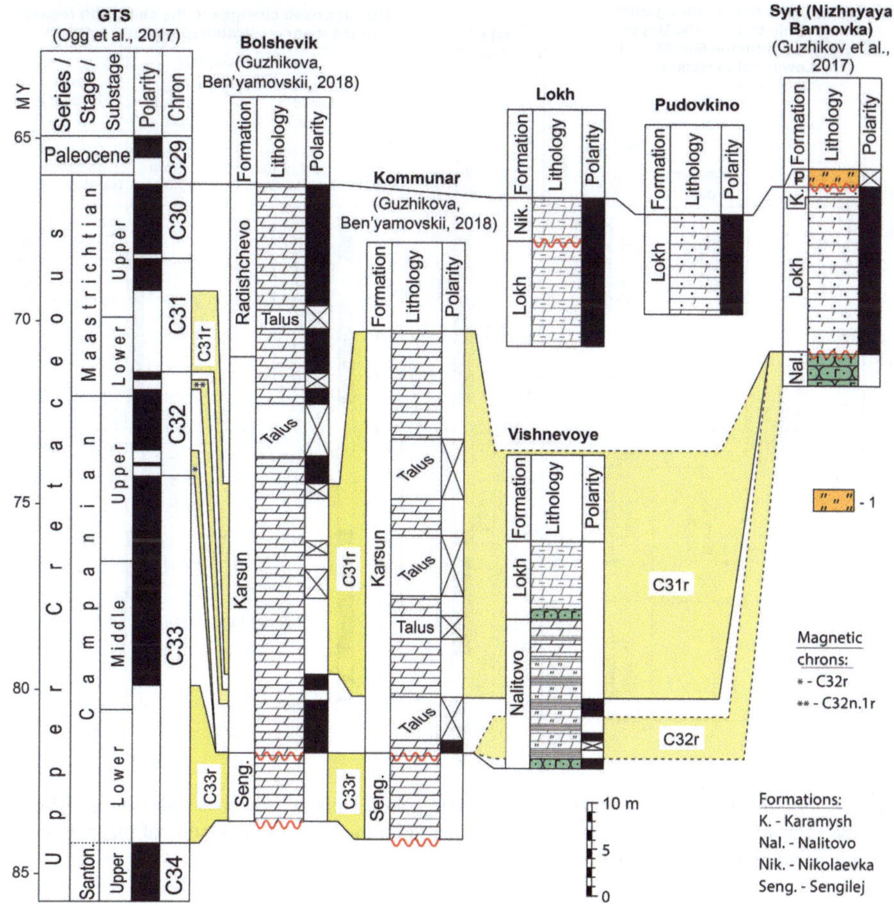

Fig. 10 Magnetostratigraphy correlation of sections of the Upper Campanian-Maastrichtian of Volga Right bank area near Saratov and their comparison with the geological time scale (GTS). 1—opoka (siliceous rocks). For other explanations see Fig. 2

section, at least an 8-m thick interval in the lowermost part of the same formation is peculiar for reversed polarity. This is the reason to conclude that the base of the Lokh formation is of older age in the Vishnevoye than in the Lokh. In other words, the acquired data documents diachroneity of the Lokh formation bottom over the distance of about 20 km. The issue of exact relationship between the Vishnevoye and the Lokh sections of the Lokh formation remains open; to solve the problem, paleomagnetic characteristics of the uppermost Lokh formation from the Vishnevoye are required. Regretfully, these strata are not exposed. Anyway, it is still untimely to recommend the Lokh section as the stratotype of the Lokh formation [16] and even as a composite stratotype, in combination with the Vishnevoye section.

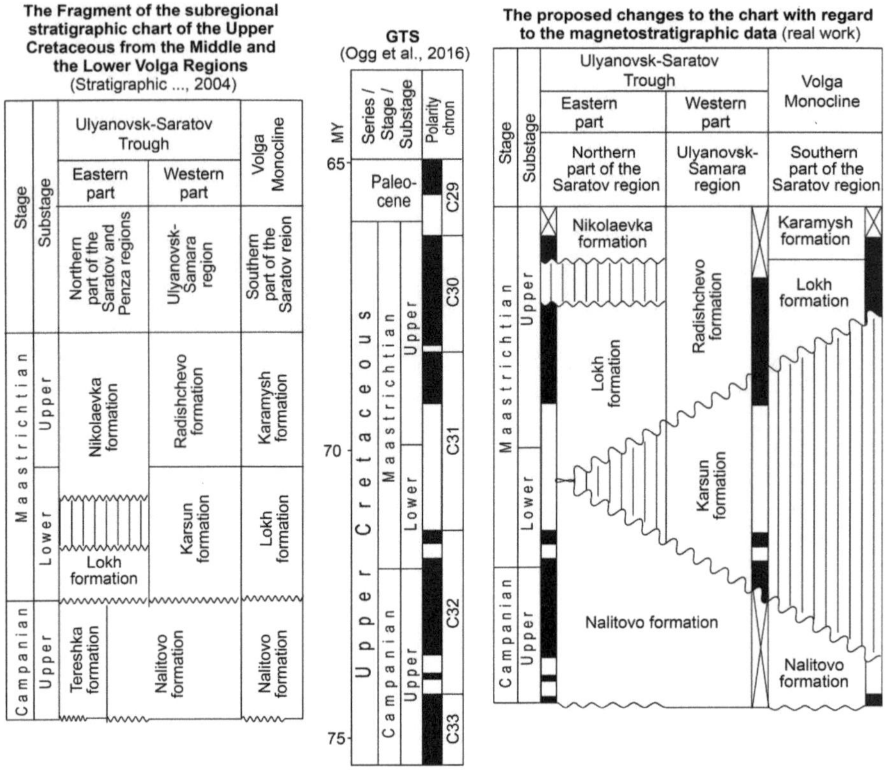

Fig. 11 Fragment of a subregional stratigraphic chart of the Upper Cretaceous beds from the Middle and the Lower Volga regions after changes, proposed on the basis of the results of magnetostratigraphic correlation (see Fig. 10)

With regard for the magnetostratigraphic data on the Upper Campanian-Maastrichtian from the Volga Right bank area near Saratov, the following changes (Fig. 11) should be introduced into a subregional stratigraphic chart of the Upper Cretaceous beds from the Middle and the Lower Volga Regions [23]:

- the Lokh formation in the northern part of the Saratov region (the Vishnevoye and Lokh sections) corresponds to both of the Maastrichtian substages (the Early Maastrichtian C31r chron and the Late Maastrichtian C31n chron);
- the Lokh formation in the southern part of the Saratov region (the Pudovkino and Syrt sections) probably corresponds to the Upper Maastrichtian substage only (the Late Maastrichtian C31n or C30n chrons or to their superposition);
- the uppermost of the Nalitovo formation in the northern part of the Saratov region (the Vishnevoye section) is dated to Early Maastrichtian (the C31r chron).

While tracking from the north to the south of the Volga Right bank area near Saratov (over the distance of about 200 km), the age of the top of the Nalitovo formation changes from the Early Maastrichtian (C31r magnetic chron) in the Vishnevoye section to the Late Campanian (C32r) in the Syrt section, and the age of the bottom of the Lokh formation decreases from the Early Maastrichtian (C31r) in the Vishnevoye section to the Late Maastrichtian (C31n or C30n) in the Pudovkino and the Syrt sections. Diachroneity of formation boundaries estimated by comparison with the Geological Time Scale reaches 2–3 million years (Fig. 11).

It is reasonable to indicate the boundary diachroneity in the local stratigraphic units relative to the boundaries of the general stratigraphic units (stages, substages) (Fig. 11).

5 Conclusions

1. In the Vishnevoye section, the analogues of magnetic chron C31r were established previously found in the Volga region only in the Bolshevik and the Kommunar quarries (the town of Volsk).
2. The deposits studied in the Lokh and Pudovkino sections correspond to chron C31n or C30n, or their superposition.
3. By comparison of new magnetostratigraphic data with the Geologic Time Scale, ages of the local stratigraphic units have been determined more precisely:
   - in the Vishnevoye section, the uppermost of Nalitovo formation has been assigned to the Early Maastrichtian, while previously referred to as Late Campanian;
   - in the Lokh and Pudovkino sections, the uppermost of Lokh formation has been assigned to the Late Maastrichtian, while previously referred to as Early Maastrichtian.

   These changes are recommended to be included to the Subregional stratigraphic chart of Upper Cretaceous for Middle and Lower Volga region.
4. The results of the paleomagnetic correlation from the north to the south of the Volga Right bank area near Saratov (the distance of about 200 km) show that the age of the top of the Nalitovo formation becomes more ancient, and the age of the base of the Lokh formation becomes younger. Diachroneity of formation boundaries is estimated as 2–3 million years and should be mentioned in the Subregional stratigraphic chart.
5. The Lokh and Vishnevoye sections cannot serve as a composite stratotype of Lokh formation because of a gap of unknown thickness between them.
6. Interrelation between magnetic textures and plastic rock deformations resulting from slope and karst processes deserves special interest both from the methodological point of view in paleomagnetic studies, and in practical engineering-geological work.

Acknowledgements We are grateful to V. A. Fomin (SSU) for contributing to the field studies of the described sections, and to V. E. Pavlov, R. V. Veselovskiy and other staff members at the Laboratory of the Main Geomagnetic Field and Petromagnetism, IPE RAS (Moscow) for affording an opportunity of measurements at a cryogenic magnetometer.

References

1. Alekseev, A.S., Kopaevich, L.F., Ovechkina, M.N., Olferiev, A.G.: Maastrichtian and lower Paleocene of northern Saratov region (Russian platform, Volga river): foraminifera and calcareous nannoplankton. Bull. Inst. R. Sci. Nat. Belg. Sci. Terre **69**, 15–45 (1999)
2. Baraboshkin, E.Y., Benyamovski, V.N., Guzhikov, A., Aleksandrova, S., Pervushov, E., Selzer, V.B., Ovechkina, M.N., Kaljakin, E.A., Kopaevich, L., Vishnevskaya, V.S., Guzhikova, A., Pokrovsky, B., Baraboshkin, E.E., Iakovishina, E.V.: Integrated study of campanian/Maastrichtian boundary interval at Volga region (Russia) and Aktolagay plateau (West Kazakhstan) of the Russian platform. In: 10th International Symposium on the Cretaceous—Abstracts, Vienna, 21–26 Aug 2017, p. 22. Berichte der Geologischen Bundesanstalt **120** (2017)
3. Besse, J., Courtillot, V.: Apparent and true polar wander and the geometry of the geomagnetic field over the last 200 Myr. J. Geophys. Res. **107**(B11), 2300 (2002). https://doi.org/10.1029/2000JB000050
4. Collinson, D.W.: Methods in Rock Magnetism and Palaeomagnetism. Techniques and Instrumentation. Chapman & Hall, London (1983)
5. Flinn, D.: On the symmetry principle and the deformation ellipsoid. Geol. Mag. **102**(1), 36–45 (1965)
6. Guzhikov, A.Y., Baraboshkin, E.Y., Beniamovsky, V.N., Vishnevskaya, V.S., Kopaevich, L.F., Pervushov, E.M., Guzhikova, A.A.: New bio- and magnetostratigraphic data on Campanian-Maastrichtian deposits of the classical Nizhnyaya Bannovka section (Volga river right bank, southern Saratov region). Stratigr. Geol. Correl. **25**(1), 39–75 (2017)
7. Guzhikova, A.A., Ben'yamovskii, V.N.: The Campanian–Maastrichtian magnetostratigraphy of the Volga region (vicinity of Volsk town). Russ. Geol. Geophys. **59**(3), 276–284 (2018)
8. Khramov, A.N., Shkatova, V.K.: General magnetostratigraphic scale of polarity of Phanerozoic. In: Zhamoida, A.I. (ed.) Supplements to the Stratigraphic Code of Russia, pp. 24–44. Izd. VSEGEI, St. Petersburg (2000) (in Russian)
9. Khramov, A.N., Sholpo, L.E.: Palaeomagnetism. Nedra, Leningrad (1967) (in Russian)
10. Lurcock, P.C., Wilson, G.S.: The palaeomagnetism of glauconitic sediments. Global Planet. Change **110**, 278–288 (2013)
11. McFadden, P.L., McElhinny, M.W.: Classification of the reversal test in palaeomagnetism. Geophys. J. Int. **103**, 725–729 (1990)
12. Molostovsky, E.A., Fomin, V.A.: Sedimentogenesis in Maastrichtian-Danian basins of the Russian plate and adjacent areas in the context of plume geodynamics. Russ. J. Earth Sci. **8**, 1–13 (2006)
13. Molostovskii, E.A., Khramov, A.N.: Magnetostratigraphy and Its Significance for Geology. Saratov. Gos. Univ, Saratov (1997) (in Russian)
14. Odin, G.S., Lamaurelle, M.A.: The global Campanian-Maastrichtian stage boundary. Episodes **24**(4), 229–238 (2001)
15. Ogg, J.G., Ogg, G.M., Gradstein, F.M.: A Concise Geologic Time Scale. Elsevier (2016)
16. Olfer'ev, A.G., Alekseev, A.S.: Stratigraphic Chart of the Upper Cretaceous Deposits of the East European Platform. Explanatory Note. Paleontol. Inst. Ross. Akad. Nauk, Moscow (2005) (in Russian)

17. Olfer'ev, A.G., Beniamovsky, V.N., Vishnevskaya, V.S., Ivanov, A.V., Kopaevich, L.F., Pervushov, E.M., Seltzer, B.V., Tesakova, E.M., Kharitonov, V.M., Shcherbinina, E.A.: Upper Cretaceous deposits in the northwest of Saratov Oblast. Part 1: litho-and biostratigraphic analysis of the Vishnevoe Section. Stratigr. Geol. Correl. **15**(6), 610–655 (2007)
18. Olfer'ev, A.G., Beniamovsky, V.N., Vishnevskaya, V.S., Ivanov, A.V., Kopaevich, L.F., Ovechkina, M.N., Pervushov, E.M., Seltzer, B.V., Tesakova, E.M., Kharitonov, V.M., Shcherbinina, E.A.: Upper Cretaceous deposits in the northwest of Saratov region, part 2: problems of chronostratigraphy and regional geological history. Stratigr. Geol. Correl. **16**(3), 267–294 (2008)
19. Olfer'ev, A.G., Beniamovsky, V.N., Ivanov, A.V., Ovechkina, M.N., Seltzer, B.V., Kharitonov, V.M.: Upper Cretaceous deposits in the north of the Saratov Region. Part 1: the Bol'shevik Quarry section in the environs of the town of Vol'sk. Bull. Mosk O-va Ispyt. Prir. Ser. Geol. **84**(2), 5–22 (2009) (in Russian)
20. Olfer'ev, A.G., Beniamovsky, V.N., Ivanov, A.V., Ovechkina, M.N., Seltzer, B.V., Kharitonov, V.M.: Upper Cretaceous deposits in the northwest of Saratov Region. Part 2: biostratigraphic subdivision of the Bol'shevik Quarry section in the environs of the town of Vol'sk. Bull. Mosk O-va Ispyt. Prir. Ser. Geol. **84**(4), 29–46 (2009) (in Russian)
21. Pechersky, D.M., Nurgaliev, D.K., Fomin, V.A., Sharonova, Z.V., Gilmanova, D.M.: Extraterrestrial iron in the Cretaceous-Danian sediments. Izv. Phys. Solid Earth **47**(5), 379–401 (2011)
22. Perch-Nielsen, K.: Mesozoic calcareous nannofossils. In: Bolli, H.M., Saunders, J.B., Perch-Nielsen, K. (eds.) Plankton Stratigraphy, pp. 329–426. Cambridge Univ. Press, Cambridge (1985)
23. Stratigraphic Scheme of Upper Cretaceous Deposits in the East European Platform. Kart. Fabr. VSEGEI, St. Petersburg (2004) (in Russian)
24. Van der Voo, R.: Palaeomagnetism of the Atlantic, Tethys, and Iapetus Oceans. Cambridge Univ. Press, Cambridge (1993)

The First Paleomagnetic Data on Lower Carboniferous Volcanics of the Central Magnitogorsk Zone in the South Urals

I. V. Golovanova, K. N. Danukalov and R. Yu. Sal'manova

Abstract This paper gives new paleomagnetic research results concerning Lower Carboniferous volcanics of the Central Magnitogorsk Zone in the South Urals, with the aim to independently assess geological ideas about the development history of the Magnitogorsk island arc. We studied 9 sections of effusive rocks with primarily basic composition in the Grekhovka and Berezovka Formations located in the meridional flow of the Ural River from the village of Kizilskoe on the north to the village of Ershovka on the south. Laboratory paleomagnetic investigations were conducted according to currently accepted methods, including thermal demagnetization and component analysis of the isolated magnetization directions. The mean direction of the high-temperature magnetization component obtained on 28 sites and the paleolatitude calculated therefrom virtually coincide with global data on the eastern margin of the paleocontinent Baltica. Consequently, according to paleomagnetic data the Central Magnitogorsk Zone was a part of the continent in the Early Carboniferous. A combined analysis of the new results on Lower Carboniferous rocks and our previous data on Devonian rocks suggest that paleomagnetic data and modern ideas about the evolution of the Uralian Orogen associated with the collision between the Magnitogorsk island arc and the passive continental margin do not contradict each other.

Keywords Paleomagnetism · High-temperature magnetization component · Paleomagnetic pole · Lower Carboniferous · South Urals · Magnitogorsk island arc · Baltica

The Magnitogorsk Megazone of the South Urals represents the preserved fragments of an island-arc system developed actively during the Devonian. According to

I. V. Golovanova (✉) · K. N. Danukalov · R. Yu. Sal'manova
Institute of Geology, Ufa Federal Research Centre, Russian Academy of Sciences, Ufa, Russia
e-mail: golovanova@ufaras.ru

K. N. Danukalov
e-mail: danukalov@mail.ru

R. Yu. Sal'manova
e-mail: vrushana@mail.ru

T. B. Yanovskaya et al. (eds.), *Problems of Geocosmos—2018*, Springer Proceedings in Earth and Environmental Sciences, https://doi.org/10.1007/978-3-030-21788-4_8

modern ideas ([12, 13] and references therein), the Magnitogorsk island arc above the subduction zone dipping eastward (in modern coordinates) was formed during the Devonian in the Ural paleo-ocean opposite the south-Uralian part of the Baltic passive continental margin. A further "soft" arc–continent collision took place in the Late Devonian resulting in the accretion of the Magnitogorsk island-arc system onto the East European continental margin. The process of the early-stage accretion along the continental margin had terminated by the end of the Famennian time. The boundary of the emerging continental margin was displaced into the present-day Turgay Trough. Here, a new subduction zone emerged in the Early Carboniferous dipping toward the continent to form the latter's active margin. Within the backland domain of the active margin, a rift zone developed associated with the accumulation of subalkaline contrast volcanics in the Berezovka and Grekhovka Formations of the Central Magnitogorsk Zone. Paleomagnetic data in the indicated age range can serve as an additional independent test for these geological ideas. By the beginning of our works, no reliable paleomagnetic determinations of the Lower Carboniferous age have been published for the region under investigation. It should also be noted that there are very few paleomagnetic data on the Lower Carboniferous rocks of paleocontinent Baltica (and its folded margin) and published results are not always unambiguous [8, 11]. The aim of this work is to acquire new reliable paleomagnetic data from Lower Carboniferous volcanics in the central part of the Magnitogorsk-Bogdanovka graben that occupies the middle and eastern areas of the Central Magnitogorsk Zone. It is also necessary to compare these data to paleomagnetic reference directions for the paleocontinent Baltica.

In the Central Magnitogorsk Zone [14] the Berezovka, Grekhovka and Kizil Formations are established that cover the time interval from the Late Tournaisian to the Early Bashkirian. The stratigraphic subdivision of these volcanogenic and sedimentary units overlaying and partially replacing volcanics is difficult and often inconsistent. The Berezovka Formation is composed of tuffaceous sandstones, conglomerates, and aleurolites, and of volcanomictic sandstones, basalts, spilites, and rare rhyolites. The formation contains limestone interbeds and is dated by foraminifera within the Late Tournaisian to the beginning of the Early Viséan; however, in some particular sections both the upper and the lower boundaries of the formation have a diachronous pattern. The upper-lying Grekhovka Formation is composed mainly of basalts, less frequent andesitic basalts, trachyrhyolites, trachyrhyodacites, with subordinate interbeds of tuffaceous rocks and foraminiferal limestones. The age of its upper and lower boundaries is also variable, in a range between the end of the Early Viséan and the entire Late Viséan. The Kizil Formation is represented by limestones containing foraminifera, conodonts and brachiopods its age ranging from the end of the Early Viséan to the earliest Bashkirian; the bottom of the formation is facially replaced by volcanics in some of the sections. The diachronous boundary pattern of the formations [12] is associated with spatial changes in the age of the volcanic processes typical for rift and hot spot volcanism.

We studied 9 sections of Lower Carboniferous volcanics in the Grekhovka and Berezovka Formations along the meridional flow of the Ural River (see Fig. 1). The collected samples were grouped into sites. Each site contains 8–10 samples and characterizes a separate lava flow. The interpretable results were obtained for 8 sections.

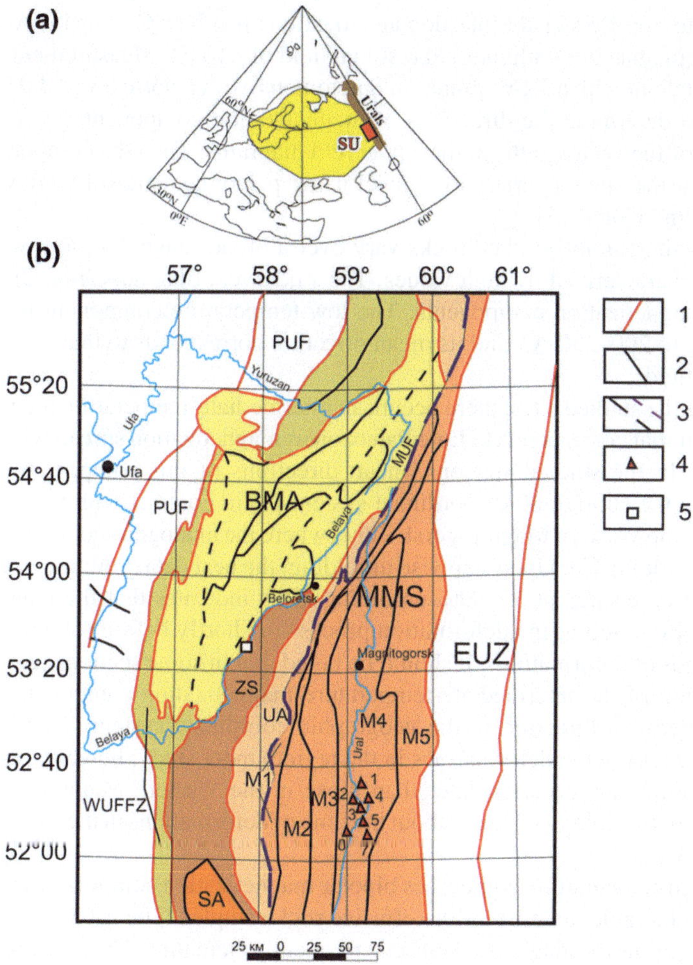

Fig. 1 **a** The location of the Baltica block with the Precambrian basement (highlighted in yellow), the Urals (brown band), and the study area in the South Urals (SU, red rectangle). **b** Schematic map of the South Urals simplified after Kozlov [10]. Legend: (1) Boundary between the first-order structures; (2) Boundary between the second-order structures; (3) The largest tectonic faults, thick blue line is the Main Uralian Fault. Tectonic structures: PUF—Pre-Uralian foredeep; WUFFZ—West Uralian frontal folding zone; BMA—Bashkir Meganticlinorium; ZS—Zilair Synclinorium; UA—Uraltau Anticlinorium; MUF—Main Uralian Fault; MMS—Magnitogorsk Megasynclinorium; M1—Voznesensk-Prisakmarskaya Subzone; M2—Irendyk Subzone; M3—Kizil-Utrazym Subzone; M4—Central Magnitogorsk Zone; M5—East Magnitogorsk Zone; EUZ—East Uralian Zone. Sampling sites: (4) 1—Grekhovka; 2—Sokolki-1; 3—Sokolki-2 and Ravine; 4—Ural village; 5—Left bank; 6—Kamenka; 7—Karaganka. (5) Position of Kaga section

Laboratory experiments were conducted according to currently accepted methods. In order to determine the natural remanent magnetization, NRM, components thermal demagnetization procedure was used to resolve the NRM vector into components according to their temperature stability. One specimen of each large sample was subjected to step-by-step thermal demagnetization up to 700 °C using the ASC TD-48 thermal demagnetizer with internal residual field of <10 nT. Measurements of NRM were carried out with a JR-6 spinner magnetometer with a noise level of 0.05 mA/m. In order to determine the directions of magnetization components we used linear segments of the vector paths in the Zijderveld diagrams. Then the components were analyzed at the site (separate flow) level using paleomagnetic software written by Enkin [3] and Cogné [2].

NRM values in the studied rocks vary over a broad range, but, as a whole, lava flows are characterized by high values. The Zijderveld diagrams (Fig. 2) show one to three magnetization components. The low-temperature component is destroyed by heating to 200–250 °C, and its mean direction corresponds to that of the modern magnetic field.

Among the studied sites, there occurs an intermediate-temperature magnetization component that corresponds to Late Paleozoic remagnetization spread widely across the Urals. In the studied sections, mean directions of this component are better grouped in the stratigraphic coordinate system [9] in contrast with the majority of sections in the Western Magnitogorsk Zone, where the analogous component is more often post-folding [5]. Upon stepwise unfolding, the best data grouping (=dispersion minimum) is reached at 55–80% unfolding. This indicates that the magnetization occurred either during the deformation process or shortly before folding and at the initial stages of deformations. It should be noted that in some sections (e.g. Sokolki) the direction of the intermediate-temperature magnetization component coincides with the Permian direction in the geographic coordinate system. This implies an intricate pattern of the deformations in the region under study; however, as a whole, these deformations occurred later than those in the Western Magnitogorsk Zone, consistent with geologists' ideas about the time of deformations in the Magnitogorsk-Bogdanovka graben.

The high-temperature, sometimes bipolar magnetization component was isolated in the considerable majority of the effusive rock samples. The carrier for this component is primarily magnetite and less frequently hematite. The dispersion in the directions of high-temperature components is rather large. Sometimes in the individual volcanogenic units (e.g. in the section Sokolki-2) one can find distinct directions of the high-temperature component with the north-eastern declination and steep negative inclination, this being in marked contrast to those in the majority of other flows and hardly explainable by geomagnetic secular variations. It is not yet possible to draw a definite conclusion about the nature of this component; perhaps, it represents an unresolved sum of other components. The final paleomagnetic direction is derived from 28 sites of effusive rocks, more often of basic, and sometimes of alkaline composition (Fig. 3; Table 1).

Table 1 High-temperature component directions in the Lower Carboniferous volcanics of the Central-Magnitogorsk zone (South Urals)

Data	N/n	S/D	In situ				Tilt-corrected			
			D,°	I,°	k	α_{95},°	D,°	I,°	k	α_{95},°
Grekhovka (N: 52.75°, E: 58.95°)										
1-1	7/6	159/41	282.4	19.7	30.0	12.4	282.2	−18.5	30.0	12.4
1-2	7/6	159/41	261.3	11.1	51.3	9.4	261.4	−29.9	51.3	9.4
1-3	8/6	184/52	77.1	−49.2	46.8	9.9	87.2	−0.6	46.8	9.9
1-4	9/8	184/52	68.1	−45.4	30.2	10.2	80.1	0.2	30.2	10.2
1-5	9/5	148/41	263.1	27.5	15.6	20.0	261.8	−12.5	15.6	20.0
1-6	9/5	200/62	251.5	42.9	6.9	31.3	259.8	−2.3	19.1	18.0
1-7	8/6	159/30	281.9	23.9	18.8	15.8	280.0	−4.2	18.8	15.8
1-8	8/4	188/29	82.1	−25.7	20.4	20.9	83.4	7.7	17.5	22.6
1-9	15/6	167/32	249.5	40.5	16.6	16.9	254.7	2.9	16.1	17.2
1-10	8/7	174/53	280.3	61.9	38.3	9.9	270.2	4.4	42.6	9.4
Sokolki-1 (N: 52.62°, E: 58.97°)										
2-1	7/7	207/40	241.3	45.5	24.4	12.5	264.2	21.6	24.4	12.5
2-2	7/7	207/40	265.2	3.2	29.4	11.3	259.4	−25.3	29.4	11.3
2-3	8/8	207/40	252.7	−7.1	34.0	9.6	241.4	−27.3	34.0	9.6
2-4	9/6	187/48	252.2	14.6	8.4	24.5	249.5	−24.5	8.4	24.5
2-5	8/6	194/41	252.0	2.5	32.6	11.9	245.4	−26.5	32.6	11.9
Sokolki-2 (N: 52.60°, E: 58.97°)										
3-1	9/9	182/41	74.4	−10.5	69.3	6.2	72.4	27.4	84.9	5.6
3-2	7/7	192/40	118.6	−34.1	61.2	7.8	117.7	5.8	61.2	7.8
3-3	7/6	209/31	245.8	18.7	27.2	13.1	249.7	−1.0	27.2	13.1
3-4	8/6	190/50	74.4	−73.5	15.5	17.6	92.2	−24.9	15.5	17.6
Ravine (N: 52.58°, E: 58.95°)										
3-5	8/6	192/35	257.8	32.0	9.8	22.6	263.7	2.1	9.8	22.6
Ural village (52.60°, E: 59.00°)										
4-1	9/6	197/33	256.2	26.8	20.7	15.1	258.9	−3.6	16.6	16.9
4-2	8/5	59/12	235.4	−14.2	15.5	20.0	238.5	−14.6	15.5	20.0
4-3	12/5	140/58	237.1	39.3	18.9	18.1	242.4	−11.8	21.4	16.9
Left bank (N: 52.55°, E: 59.00°)										
5-1	8/4	195/27	237.5	28.9	34.8	15.8	244.1	9.3	34.8	15.8
Kamenka (N: 52.53°, E: 59.00°)										
6-1	14/4	177/29	65.8	0.8	14.6	24.9	62.9	27.6	14.6	24.9
6-2	8/4	189/29	96.3	−2.2	13.4	26.0	96.0	26.8	13.4	26.0
6-3	8/3	211/41	244.7	11.5	49.3	17.7	245.1	−9.7	49.3	17.7

(continued)

Table 1 (continued)

Data	N/n	S/D	In situ				Tilt-corrected			
			D,°	I,°	k	α_{95},°	D,°	I,°	k	α_{95},°
Karaganka (N: 52.48°, E: 59.13°)										
7-1	9/9	340/25	237.7	−29.3	83.2	5.7	239.3	−7.5	66.0	6.4
All	(28/28)		256.2	23.6	9.4	9.4	258.7	−9.2	15.8	7.1
	110.3%						258.5	−12.6	15.9	7.1

Notes N/n, number of samples (sites) studied/used. S/D, strike and dip angle. D, declination, I, inclination, k, concentration parameter, α_{95}, radius of 95% confidence circle. Localities are labeled as in Fig. 1b; ALL is the overall mean for all sites listed in the table

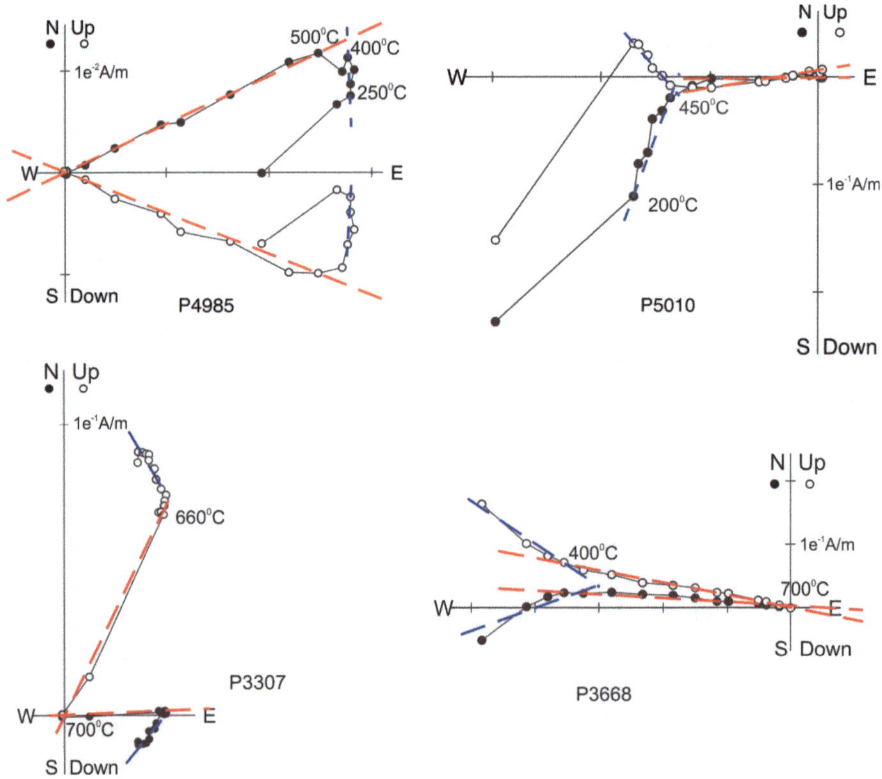

Fig. 2 Representative thermal demagnetization plots of Lower Carboniferous volcanics from the Central-Magnitogorsk zone of the South Urals, in stratigraphic coordinates. Isolated intermediate-temperature (blue) and high-temperature (red) components are denoted by thick dashed lines. Full (open) circles represent vector endpoints projected onto the horizontal (vertical) plane

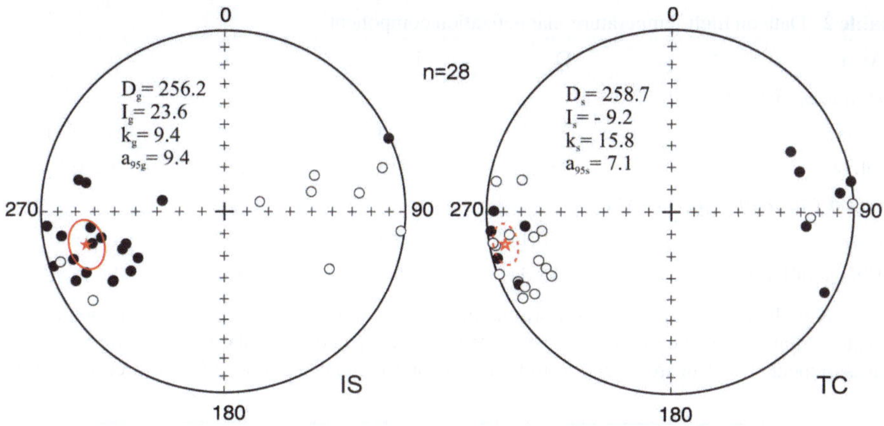

Fig. 3 Stereoplots of the high-temperature component (HTC) for the studied Lower Carboniferous sections in situ and after tilt correction. Full (open) symbols are projected onto the lower (upper) hemisphere

The mean direction calculated for 28 sites is much better grouped in the stratigraphic coordinate system and comprises D = 258.7°, I = −9.2°, k = 15.8, α_{95} = 7.1° (see Table 1). The reversals test for the high-temperature component is positive, the class is C, γ/γ_{cr} = 11.0/15.6. The fold test performed in the Enkin's "direction–correction" modification [4] gives a positive result. We therefore consider the isolated magnetization component to be primary.

The overall mean paleomagnetic pole for the Lower Carboniferous data falls at Λ = 335.2°, Φ = −10.5°, A95 = 6.7°. This result is in a good agreement with the data for Baltica [15]. The expected pole coordinates for Baltica 340 million years ago are Λ = 333.1°, Φ = −9.3°, A95 = 8.3°, just 2.4° from our pole, which is clearly within the limits of error.

Previously, we have performed the paleomagnetic study of Lower and Middle Devonian volcanics and jasperoids in the West Magnitogorsk Zone [6, 7]. The mean direction of the high-temperature magnetization component measured in the stratigraphic coordinate system on 27 sites is D = 253.5°, I = 0.3°, k = 19.4, α_{95} = 6.5°, in good agreement with the published results of paleomagnetic research on Devonian rocks in the Magnitogorsk Zone, Sibay point [1]. A combined analysis of new results on Lower Carboniferous rocks and the data obtained previously on Devonian rocks makes it possible to compare coeval paleolatitudes of the Magnitogorsk island arc and the eastern margin of the East European continent and to determine mutual displacements of these two tectonic structures (see Table 2; Fig. 4).

No primary magnetization has been preserved in the coeval rock sections studied along the eastern margin of the East European continent. Therefore, in order to assess the relative position between the continent and the island arc we used the data on the location of the pole for the Baltic paleocontinent from the review paper [15] recalculated into paleomagnetic directions for the continental margin (Kaga point).

Table 2 Data on high-temperature magnetization component

Area	N	D	I	α_{95}	φ
Devonian, 390 Ma					
Kaga		278.3	2.4	11.0	1.2 ± 11.0
Sibay	27	253.5	0.3	6.5	0.2 ± 5.7
Lower Carboniferous, 340 Ma					
Kaga		260.1	−8.5	8.2	4.3 ± 8.2
Village of Ural	28	258.7	−9.2	7.1	4.6 ± 7.2

Notes N is the number of sites, D, I are the declination and inclination in the stratigraphic coordinate system, α_{95} is the radius of the confidence circle, φ is the paleolatitude and the error of its measurement. The data are given for midpoints in the Devonian and Lower Carboniferous sections

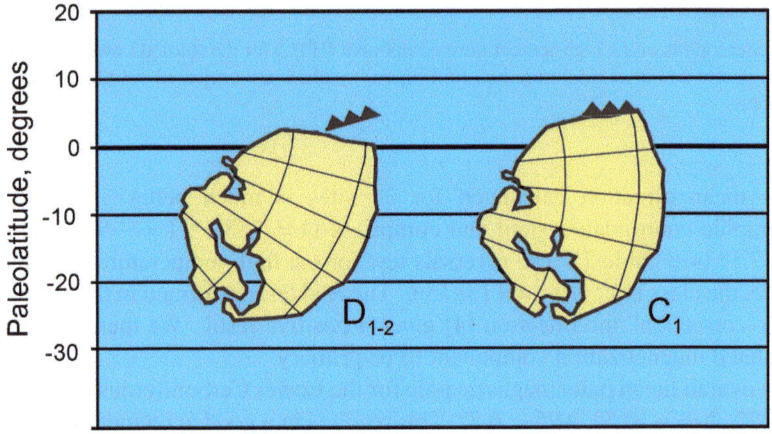

Fig. 4 The Baltica–Magnitogorsk Arc relationship at different times according to paleomagnetic poles listed in Table 2

The results of paleomagnetic research (see Table 2; Fig. 4) allow us to conclude the following:

1. In the Lower and Middle Devonian the passive margin of the East European continent and the adjacent Magnitogorsk island arc occupied a near-equatorial position, which is also confirmed by paleoclimatological investigations (existence of barrier reefs, etc.). The paleodeclination values obtained in this and previous [6, 7] studies suggest that the orientation of the continental margin and the island arc was close to sublatitudinal.
2. The latitudinal position of the Magnitogorsk island arc in the Lower and Middle Devonian differed from that of the continent by about 1°, and even if we take into account rather large measurement errors in the determination of the latitude, the distance between the continental margin and the island arc probably exceeded 200 km, but it was unlikely to be more than 1000 km (most likely, less).

3. The research results show that the Magnitogorsk island arc in the Devonian was at some distance from the passive continental margin, and the latter, during their convergence, rotated counterclockwise relative to the island arc by approximately 25°. The complete similarity of inclinations and paleolatitudes of the continental margin and island arc sites indicates that, as a result of subduction and collision, the island arc and the passive continental margin were brought together, and the passive margin was thus built up in the Carboniferous by accretion of the island arc. The oceanic "gap" between the continent and the island arc was transformed into the suture zone of the Main Ural Fault (see Fig. 1). Thus, the paleomagnetic data support modern ideas about the evolution of the Uralian Orogen concerning, in particular, the collision process between the Magnitogorsk island arc and the passive margin of Baltica.
4. Since the studied object in the post-Devonian time was already structurally part of the platform, our results yield a new Lower Carboniferous paleomagnetic pole for Baltica.

Acknowledgements The research was performed in the framework of government contract no. 0252-2017-0013 and supported in part by the Russian Foundation for Basic Research (grant no. 15–05–03036).

References

1. Burtman, V.S., Gurarii, G.Z., Dvorova, A.V., Kuznetsov, N.B., Shipunov, S.V.: The Uralian Paleoocean in the Devonian (according to paleomagnetic data). Geotectonics **34**(5), 397–406 (2000)
2. Cogné, J.P.: PaleoMac: A Macintosh application for treating paleomagnetic data and making plate reconstructions. Geochem. Geophys. Geosyst. **4**(1), article no. 1007 (2003). http://dx.doi.org/10.1029/2001GC000227
3. Enkin, R.I.: A Computer Program Package for Analysis and Presentation of Paleomagnetic Data, Sidney. Pacific Geoscience Centre, Geological Survey of Canada, Canada (1994)
4. Enkin, R.J.: The direction-correction tilt test an all purpose tilt/fold test for paleomagnetic studies. Earth Planet. Sci. Lett. **212**, 151–166 (2003). https://doi.org/10.1016/S0012-821X(03)00238-3
5. Golovanova, I.V., Danukalov, K.N., Kosarev, A.M., Khidiyatov, M.M., Bazhenov, M.L.: Paleomagnetizm Magnitogorskoy zony Yuzhnogo Urala. Predvaritelnye rezultaty (Paleomagnetism of the Magnitogorsk zone in the South Urals. Preliminary results). In: Materialy mezhdunarodnoy shkoly-seminara "Problemy paleomagnetizma i magnetizma gornykh porod" (Proceedings of the International Workshop-School "Problems of Rock Paleomagnetism and Magnetism"), pp. 29–35. Shmidt Institute of Physics of the Earth, St. Petersburg State University. Filigran, Yaroslavl (2016) (in Russian)
6. Golovanova, I.V., Danukalov, K.N., Puchkov, V.N., Kosarev, A.M., Sal'manova, R.Y.: Polozheniye Magnitogorskoy dugi v devone po paleomagnitnym dannym/Geologiya, poleznyye iskopayemyye i problemy geoekologii Bashkortostana, Urala i sopredel'nykh territoriy (The position of the Magnitogorsk arc in the Devonian according to paleomagnetic data/geology, minerals and geoecology problems of Bashkortostan, the Urals and adjacent territories). In: 12-ya Mezhregional'naya nauchno-prakticheskaya konferentsiya (12th Interregional Scientific and Practical Conference), pp. 85–93. Own Publishing House, St. Petersburg (2018) (in Russian)

7. Golovanova, I.V., Danukalov, K.N., Puchkov, V.N., Kosarev, A.M., Sal'manova, R.Y.: Paleomagnetic study of Devonian and Carboniferous rocks of the Southern Urals: an independent test of collision between the Magnitogorsk Island arc and the passive margin of the continent of Laurussia. Doklady Earth Sci. **482**(1), 1134–1137 (2018). https://doi.org/10.1134/S1028334X18090052
8. Iosifidi, A.G., Mikhailova, V.A.: Paleomagnetism of carboniferous formations of the Russian platform: reinvestigation of old collections Izvestiya. Phys. Solid Earth **53**(3), 461–476 (2017)
9. Khidiyatov, M.M., Golovanova, I.V., Danukalov, K.N., Kadyrov, A.F., Levashova, N.M., Bazhenov, M.L.: Paleomagnetizm nizhnekamennougolnykh vulkanogennykh i osadochnykh tolshch Magnitogorsko-Bogdanovskogo grabena na Yuzhnom Urale. Predvaritelnye rezultaty (Paleomagnetism of Lower Carboniferous volcanogenic and sedimentary rocks in the Magnitogorsk-Bogdanovka Graben of the South Urals. Preliminary results). In: Paleomagnetizm i magnetism gornykh porod: teoriya, praktika, eksperiment (Paleomagnetism and Rock Magnetism: Theory, Practice, Experiment), pp. 229–234. "Borok" Geophysical Laboratory, Division of the Shmidt Institute of Physics of the Earth. Filigran, Yaroslavl (2015) (in Russian)
10. Kozlov, V.I. (ed.): Geological Map of the Russian Federation and Adjacent Territory of the Republic of Kazakhstan. Scale: 1:1 000000 (a New Series). Sheet N-40(41)-Ufa. VSEGEI, St. Petersburg (2002) (in Russian)
11. Piper, J.D.A., Atkinson, D., Norris, S., Thomas, S.: Palaeomagnetic study of the Derbishire lavas and intrusions, central England: definition of carboniferous apparent polar wander. Phys. Sci. Planet. Inter. **69**, 37–55 (1991)
12. Puchkov, V.N.: The Uralides and Timanides: their structural relationship and position in the geologic history of the Ural-Mongolian fold belt. Russ. Geol. Geophys. **44**(1–2), 28–39 (2003). https://doi.org/10.1134/S1028334X18090052
13. Puchkov, V.N.: Geologiya Urala i Priuraliya [aktualniye voprosy stratigrafiyi, tektoniki, geodinamiki i metallogeniyi] (Geology of the Urals and Cis-Urals [actual problems of stratigraphy, tectonics, geodynamics and metallogeny]), 280 p. DesignPoligraphService, Ufa (2010) (in Russian)
14. Salikhov, D.N., Yarkova, A.V.: Nizhnekamennougolny vulkanizm Magnitogorskogo sinklinoriya (Lower Carboniferous Volcanism of the Magnitogorsk Synclinorium), 138 p. BSC UrB RAS, Ufa (1992) (in Russian)
15. Torsvik, T.H., Van der Voo, R., Preeden, U., MacNiocaill, C., Steinberger, B., Doubrovine, P.V., van Hinsbergen, D.J.J., Domeir, M., Gaina, C., Tohver, E., Meert, J.G., McCausland, P.J.A., Cocks, R.M.: Phanerozoic polar wander, palaeogeography and dynamics. Earth Sci. Rev. **114**, 325–368 (2012). https://doi.org/10.1016/j.earscirev.2012.06.007

Paleomagnetism and Petromagnetism of Sedimentary Rocks of the Zhuravlevka-Amur Terrane (Junction Zone Between the Sikhote-Alin and Mongol-Okhotsk Orogenic Belts)

A. Yu. Peskov, M. V. Arkhipov, A. S. Karetnikov, A. V. Kudymov and A. N. Didenko

Abstract This paper presents paleomagnetic data on the Berriasian-Valanginian sandstones of the Gorinskaya and Pionerskaya Formations of the Komsomolskaya Group from the coastal outcrops of the Sakhalin Bay (54.09° N, 140.05° E). Structurally, these rocks belong to the Zhuravlevka-Amur (ZhA) terrane of the Sikhote-Alin orogenic belt. Paleomagnetic studies indicate that these rocks were formed at a latitude of 10°–18° N, which is comparable to paleolatitudes (10° N–4° S) obtained previously from the Berriasian-Valanginian basalts of the Rozhdestvenskaya Formation (Sakhalin Island). The obtained paleolatitude means that after 140 Ma the Zhuravlevka-Amur terrane drifted northward by over 4000 km along the continental margin of Eurasia at an average rate of about 10 cm/year.

Keywords Continental margin · Paleomagnetism · Paleolatitude · Sikhote-Alin orogenic belt

A. Yu. Peskov (✉) · M. V. Arkhipov · A. S. Karetnikov · A. V. Kudymov · A. N. Didenko
Kosygin Institute of Tectonics and Geophysics, Far Eastern Branch, Russian Academy of Sciences, Khabarovsk, Russia
e-mail: peskov@itig.as.khb.ru

M. V. Arkhipov
e-mail: arkhipov_mv@mail.ru

A. S. Karetnikov
e-mail: natali@itig.as.khb.ru

A. V. Kudymov
e-mail: kadi77760@mail.ru

A. N. Didenko
e-mail: alexey_didenko@mail.ru

T. B. Yanovskaya et al. (eds.), *Problems of Geocosmos—2018*, Springer Proceedings in Earth and Environmental Sciences, https://doi.org/10.1007/978-3-030-21788-4_9

1 Introduction

The junction zone between two orogenic belts under consideration is represented by the Upper Triassic, Lower-Middle Jurassic turbidites of the Ulban terrane of the Mongol-Okhotsk orogenic belt (MOOB) and the Cretaceous rocks of the ZhA terrane of the Sikhote-Alin orogenic belt (SAOB) (Fig. 1). Sandstones of the Gorinskaya and Pionerskaya Formations of the Komsomolskaya Group that were sampled in coastal outcrops of the Sakhalin Bay (54.09° N, 140.05° E) form the focus of this study. Structurally, these rocks belong to the Zhuravlevka-Amur turbidite basin [4] or the Amur accretionary prism [9]. The Gorinskaya Formation with a thickness of 1100–1300 m is dominated by sandstones; less common are siltstones and their flyschoid interlayering. It is assigned a Berriasian age based on buchiids [6]. The overlying Pionerskaya Formation (900–1430 m) is divided into two sub-formations and represented by siltstone, sandstone, rarely sedimentary breccias, gritstone, and mudstone. Buchiids indicate its Berriasian-Valanginian age [6].

Despite the growing interest in tectonics and geodynamics of East Asia, the junction zone between the Mongol-Okhotsk and Sikhote-Alin orogenic belts remains poorly understood. The lack of paleomagnetic data for this area prevents the

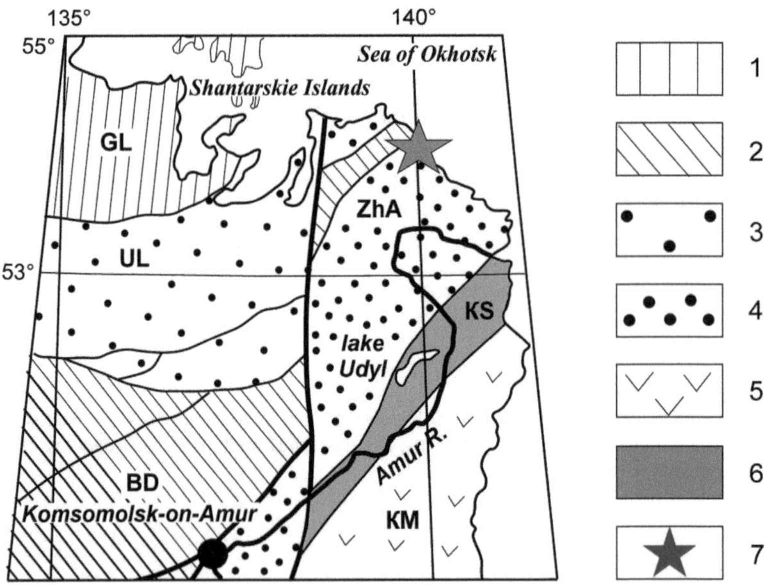

Fig. 1 Schematic map of terranes of the Sikhote-Alin orogenic belt after [4, 5] with additions. 1—Paleozoic terranes (GL, Galam); 2, 3—Jurassic terranes: 2—fragments of accretionary prisms (BD, Badzhal), 3—near-continental turbidite basin (UL, Ulban); 4 to 6—Cretaceous, essentially Early Cretaceous, terranes: 4—near-continental pull-apart turbidite basin (ZhA, Zhuravlevka-Amur), 5—Barremian-Albian island-arc system (KM, Kema), 6—Albian accretionary prism (KS, Kyselevka-Manoma); 7—region of paleomagnetic studies

paleo-latitudinal position and kinematic parameters of the Zhuravlevka-Amur terrane being quantified. The objective of the study is to obtain the first petromagnetic and paleomagnetic data on Berriasian-Valanginian sandstones of the Zhuravlevka-Amur terrane.

2 Methods

To calculate the anisotropy of initial magnetic susceptibility (AMS), the magnetic susceptibility ($\varkappa$) was measured by the MFK-1FA Kappabridge in 64 positions (AGICO, Czech Republic). The obtained data were statistically processed by the Anisoft software package (AGICO, Czech Republic). Paleomagnetic studies included: (a) thermomagnetic demagnetization in 100–5 °C steps to 680 °C involving a step-by-step measuring of magnetization on a JR-6A spinner magnetometer (AGICO, Czech Republic) inside the Helmholtz coils; and (b) cleaning with alternating field (AF) demagnetization up to 90 mT using a SQUID magnetometer (2G Enterprises, USA). Natural remanent magnetization (NRM) components were isolated by principal component analysis using R. Enkin's PMGSC (version 4.2) software [3]. The fold test was conducted on the components of magnetization to constrain their age and to determine whether the magnetization is primary or secondary. Coordinates of the paleomagnetic pole and the paleolatitude at which the studied rocks formed were calculated using R. Enkin's PMGSC (version 4.2) software [3].

A total of 95 samples were studied (51 samples from the Pionerskaya Formation and 44 samples from the Gorinskaya Formation).

3 Petromagnetic Results

The study has found that the NRM mean values of the studied specimens from the Gorinskaya and Pionerskaya Formations differ by more than two orders of magnitude: 1.4 mA/m (Pionerskaya Fm) and 331 mA/m (Gorinskaya Fm). Mean susceptibility values ($\varkappa$) are respectively 1.37E−04 SI units (Pionerskaya Fm) and 2.98E−04 SI units (Gorinskaya Fm).

The AMS of the studied specimens of rocks from the Pionerskaya Formation is both linear (predominantly) and planar (Fig. 2a–c). Rocks belonging to turbidite flows with high sedimentation rates are most probably responsible for the linear type anisotropy [4]. The average degree of anisotropy (P) is 3.5% ranging from 2 to 5%. Specimens of the Gorinskaya Formation show predominantly the planar type anisotropy (Fig. 2d, e). The degree of anisotropy for specimens of the Gorinskaya Formation is appreciably higher than in the previous case ranging from 2.2 to 22.8% with an average of 6.5%. Such high P values are most probably the result of secondary processes that affected the magnetic fabric of the material. It should be noted

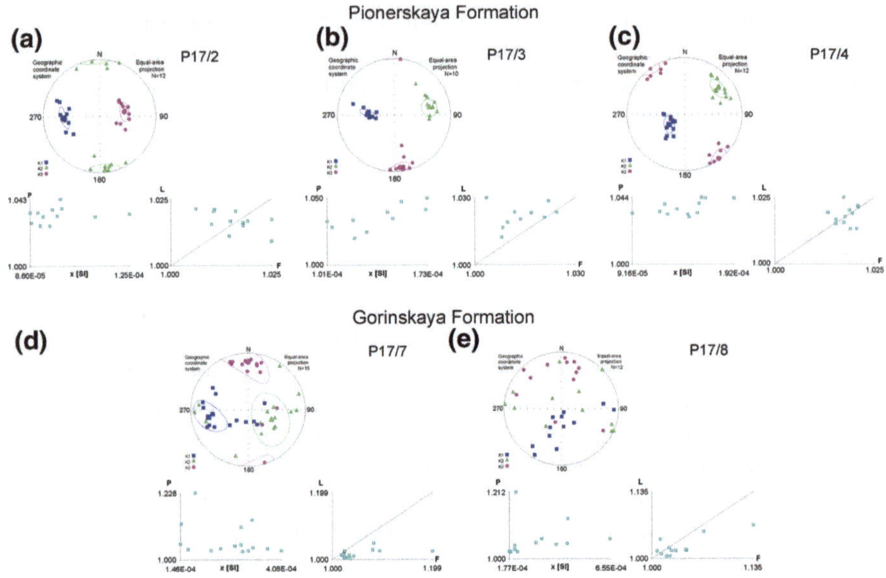

Fig. 2 Results of the initial magnetic susceptibility anisotropy analysis of specimens from the Pionerskaya (**a–c**) and Gorinskaya (**d, e**) formations. Symbols: P—degree of anisotropy; L—linear anisotropy; F—planar anisotropy; K1, K2, K3—maximum, average, and minimum axes of the anisotropy ellipsoid

that specimens from the Gorinskaya and Pionerskaya Formations, when heated to 400–420 °C, do not show changes of AMS parameters.

The *Königsberger ratio* (Qn) was calculated for the studied specimens and rock-magnetic properties were plotted on the log-log diagram (Fig. 3). Qn values of the majority of rock specimens from the Pionerskaya Formation range from 0.1 to 1, whereas for the rock specimens of the Gorinskaya Formation the Königsberger ratio is >1 and for the bulk of the studied specimens >10.

One can infer that magnetic properties of sandstones from the Pionerskaya and Gorinskaya Formations have essential differences. The log-log NRM versus magnetic susceptibility plot show clear difference in the magnetic hardness of studied specimens (Fig. 3). The appreciable difference in the rock magnetic properties of specimens from the two formations we interpret as due to the presence of different carriers of magnetic properties of rocks, as well as the different degree of secondary alteration.

High values of Qn in rock specimens of the Gorinskaya Formation (>10), as well as the behaviour of specimens during temperature demagnetization (specimens were completely demagnetized at 330–400 °C) indicate the presence of pyrrhotite as a carrier of magnetization. High coefficients of anisotropy indicate secondary (chemical) alteration in the studied rocks, which can be explained by the formation of secondary pyrrhotite.

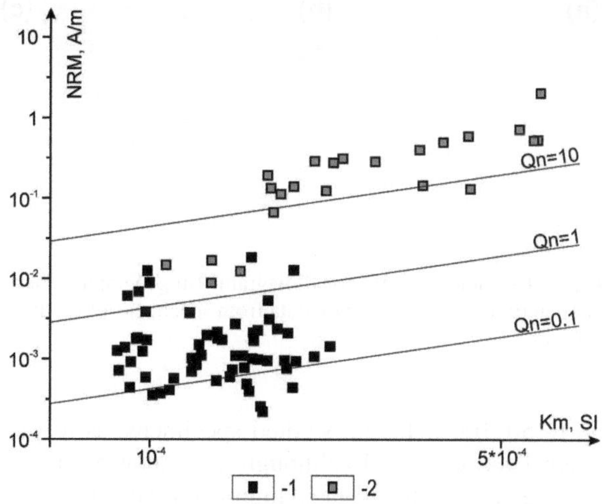

Fig. 3 Log-log plot showing distribution of rock-magnetic parameters for specimens from Pionerskaya (1) and Gorinskaya (2) formations. NRM—natural remanent magnetization, A/m; Km—magnetic susceptibility, SI units

The anisotropy coefficient in the studied specimens of the Pionerskaya Formation is not high suggesting that secondary alteration is either absent in the rock or weakly developed. Considering the obtained rock magnetic data, as well as the behaviour of specimens during temperature demagnetization (the demagnetization curve has a peak between 330 and 400 °C, while the specimens were demagnetized between 550 and 580 °C), magnetite and, to a lesser extent, pyrrhotite can be assumed to be carriers of magnetization in the studied rocks of the Pionerskaya Formation.

4 Paleomagnetic Results

4.1 Gorinskaya Formation

Detailed thermal demagnetization of rock specimens from the Gorinskaya Formation isolated one component of magnetization. All the specimens were demagnetized at temperatures between 330 and 400 °C (Fig. 4a). The isolated components of magnetization in the Gorinskaya specimens are distributed chaotically on the stereogram and shall not be used for further analysis.

4.2 Pionerskaya Formation

Detailed thermal demagnetization of specimens from the Pionerskaya Formation was carried out to 600 °C (in some cases to 680 °C) and alternating-field demagnetization

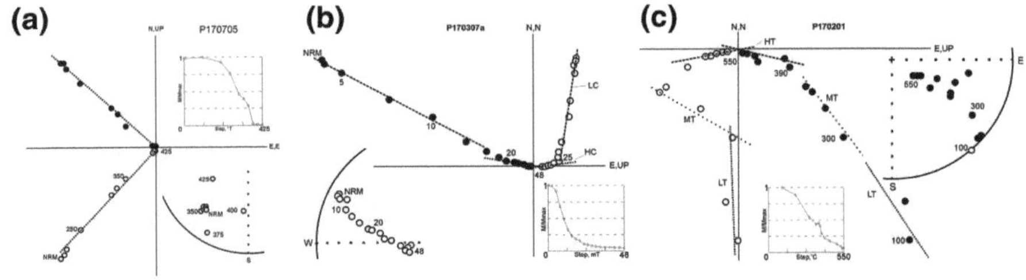

Fig. 4 Examples of orthogonal diagrams (in geographic coordinates) of alternating-field (**b**) and thermal (**a**, **c**) demagnetization data from specimens of Gorinskaya (**a**) and Pionerskaya (**b**, **c**) formations

up to 50–100 mT. The studied specimens, as a rule, display two components of magnetization (Fig. 4b) although some show as many as three components (Fig. 4c).

Directions of the isolated high-coercivity (HC) and high-temperature (HT) components are similar suggesting that they may represent primary magnetization [8]. In the Pionerskaya Formation, HT component is isolated between 380–430 °C and 540–580 °C and decays to the origin. As an example, an orthogonal diagram of thermomagnetic demagnetization is shown for sample P170201 (Fig. 4c). The NRM of this sample comprises three components, and the HT component of magnetization is isolated at temperatures ranging from 400 to 550 °C (Fig. 4c). During the Af-demagnetization, HC component is isolated between 20–25 mT and 40–60 mT and, like HT component, trends toward the origin (with an exception of a few specimens which were excluded from the statistics). As an example, an orthogonal diagram of AF demagnetization is shown for sample P170307a whose HCC is isolated in the 25–48 mT interval (Fig. 4b). These components (HT and HC) are considered to represent the characteristic magnetization (ChRM) for sedimentary rocks of the Pionerskaya Formation of the ZhA terrane.

In stratigraphic coordinates, clustering of the ChRM directions increases more than six-fold at the sample (Ks/Kg = 6.2) and by more than 100 times at the site level (Ks/Kg = 134.7) (Table 1). The best grouping is attained at about 100% (97%) untilting suggesting that the ChRM was acquired before the folding event (Fig. 5b).

Sample and site mean ChRM directions were calculated (Table 1). Coordinates of the paleomagnetic pole, as well as a paleolatitude of 14.2 ± 4° N at which the rocks under consideration formed, were calculated from the sample-mean ChRM direction (Table 1).

Table 1 Results of paleomagnetic studies

Sites	φ, °	λ, °	Rock type	N_1	N_2	Dip azim., °	Dip angle, °	Dec_g, °	Inc_g, °	K_g	a_{95}, °	Dec_s, °	Inc_s, °	K_s	a_{95}, °
P17/2	54.09	140.05	Sandstone	15	14	319	40	294.2	62.3	14.8	10.7	306.7	24.0	14.8	10.7
P17/3	54.10	140.03	Sandstone	11	5	137	110	248.1	−67.5	20.1	15.3	293.5	26.4	20.1	15.3
P17/4	54.10	140.02	Sandstone	25	15	317	87	162.6	60.3	11.1	12.0	302.8	29.4	11.1	12.0
Sample-mean (ChRM) N = 35					Kmax = 96.6%			229.3	68.0	2.2	21.8	302.7	26.8	13.7	6.8
Site-mean (ChRM) N = 3					Kmax = 97.3%			236.0	47.5	1.1	180	301.0	26.7	148.2	10.2
Paleomagnetic pole coordinates (for sample-mean)								Plat = 30.4°; Plong = 31.1°; dm = 7.4; dp = 4.0							

Note φ, λ—geographical latitude and longitude of sampling site, respectively; N1—number of collected samples; N2—number of samples used in calculating the mean; Dec, Inc—paleomagnetic declination and inclination, respectively, in geographic (g) and stratigraphic (s) coordinates; K—grouping; a95—radius of the confidence ellipse around the mean; Plat, Plong—latitude and longitude of the paleomagnetic pole, respectively; dp; dm—semi-axes of the confidence ellipse around the pole

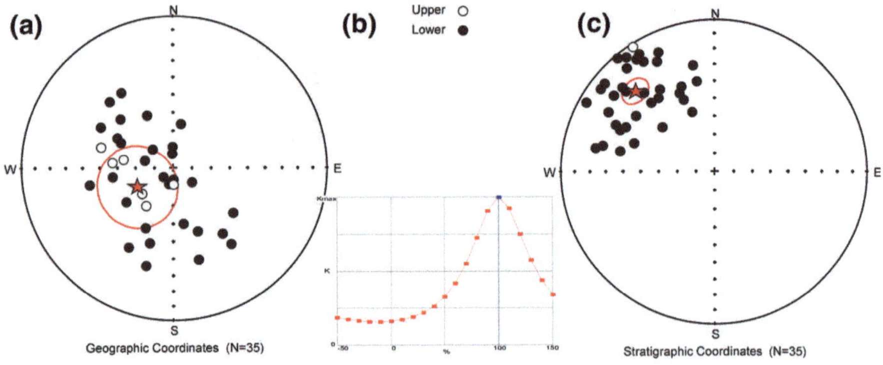

Fig. 5 Stereograms of ChRM distribution for samples of Pionerskaya formation in the geographic (**a**) and stratigraphic (**c**) coordinate systems. **b** Fold test (the relationship between the degree of clustering (K) and the degree of untilting)

5 Conclusion

Rock magnetic studies have shown that magnetic properties of rocks of the Pionerskaya and Gorinskaya Formations differ significantly. High values of the degree of anisotropy (P), and the Königsberger ratio for rocks of the Gorinskaya Formation suggest secondary alteration in these rocks attributed to the presence of secondary pyrrhotite as a carrier of magnetization, which impaired the results of paleomagnetic cleaning (demagnetization), during which attempts to isolate a primary paleomagnetic signal have not met with success. Magnetite and, to a lesser extent, pyrrhotite are carriers of magnetization in the studied rocks of the Pionerskaya Formation.

Paleomagnetic studies of rock specimens from the Pionerskaya Formation (thermomagnetic and AF demagnetization) yielded the characteristic magnetization component of pre-folding origin. Based on the direction of the characteristic component, the paleomagnetic pole was calculated for Berriasian-Valanginian rocks of the Pionerskaya Formation (Zhuravlevka-Amur terrane): Plat = 30.4°; Plong = 31.1°; dm = 7.4; dp = 4.0.

The new data are consistent with the published paleomagnetic data [2]. These data lend further support to the terrane model for the formation of the Sikhote-Alin orogenic belt and allow the tectonomagnetic model of its formation to be complemented.

The Berriasian-Valanginian sandstones of the Pionerskaya Formation formed in tropical latitudes of between 10° and 18° N, which is comparable to the paleolatitudes (10° N–4° S) earlier obtained from the Berriasian-Valanginian basalts of the Rozhdestvenskaya Formation (Sakhalin Island) [1]. The resulting paleolatitudes indicate that later on rocks of the Pionerskaya Formation moved along the Eurasian continental margin for more than 4000 km (≈4500 km). The available paleomagnetic data on Cenomanian-Turonian volcanic rocks of the Zhuravlevka-Amur terrane of the Sikhote-Alin orogenic belt [7] showing evidence of tectonic coherence of the continental paleo-margin and the above-mentioned terrane, allow the kinematic parameters of the Amur terrane to be determined. According to our estimates, the Amur terrane drifted in a northerly direction at an average rate of ≈10 cm/year.

Acknowledgements The authors express their sincere thanks and gratitude to A. A. Kosterov and A. Yu. Kazansky. The article benefited greatly from their comments and suggestions. The authors would also like to thank T. L. Koryakina for translating the article into English.
Rock-magnetic studies were funded by RFBR according to the research project № 18-35-00048. Paleomagnetic studies were supported by the Russian Science Foundation (project no. 16-17-00015).

References

1. Abrajevitch, A.V., Zyabrev, S.V., Didenko, A.N., Kodama, K.: Paleomagnetism of the West Sakhalin basin: evidence for northward displacement during the Cretaceous. Geophys. J. Intern. **190**, 1439–1454 (2012)
2. Didenko, A.N., Peskov, A.Yu., Kudymov, A.V., Voinova, I.P., Tikhomirova, A.I., Arkhipov, M.V.: Paleomagnetism and accretionary tectonics of northern Sikhote Alin. Izv. Phys. Solid Earth **53**(5), 733–749 (2017)
3. Enkin, R.J.: A Computer Program Package for Analysis and Presentation of Paleomagnetic Data, 16 p. Pacific Geoscience Centre, Geological Survey of Canada (1994)
4. Khanchuk, A.I. (ed.): Geodynamics, Magmatism and Metallogeny of the Russian East, 981 p. + colour map. In 2 books. Dal'nauka, Vladivostok (2006) (in Russian)
5. Golozybov, V.V.: Tectonics of Jurassic and Lower Cretaceous Complexes in North-Western Margins of Pacific Ocean, 239 p. Dal'nauka, Vladivostok (2006) (in Russian)
6. Kaidalov, V.A., Anoikin, V.I., Belomestnova, T.D.: State Geological Map of the Russian Federation. Scale 1: 200 000. Sheet M-54-I. Explanatory Notes, 185 p. VSEGEI, St. Petersburg (2009) (in Russian)
7. Kudymov, A.V., Voinova, I.P., Tikhomirova, A.I., Didenko, A.N.: Geology, geochemistry, and paleomagnetism of rocks of the Utitsa Formation, north Sikhote Alin. Russ. J. Pac. Geol. **9**(5), 323–337 (2015)
8. Metelkin, D.V., Kazansky, A.Y.: The Essentials of Magnetotectonics: Tutorial, 27 p. Novosibirsk University, Novosibirsk (2014) (in Russian)
9. Zyabrev, S.V.: Oceanic deposits of the Amur terrane: their ages and tectonic significance. Russ. J. Pac. Geol. **5**(2), 155–163 (2011)

References

Analysis of Strong-Field Hysteresis in High Coercivity Magnetic Minerals

A. Kosterov, E. S. Sergienko, A. G. Iosifidi, P. V. Kharitonskii and S. Yu. Yanson

Abstract To evaluate the effect of undersaturation of magnetic hysteresis loops measured in moderate (<2 T) fields in magnetically hard minerals such as goethite or hematite, we measured room temperature hysteresis loops in a 7 T field and DC backfield demagnetization curves in fields up to 3 T using an MPMS 3 instrument. Sediments from different regions of the East European platform, mostly of Carboniferous age were used for this study. Similar experiments were also carried out for a small collection of archaeological ceramics (bricks) apparently containing a High Coercivity Low unblocking Temperature (HCLT) magnetic phase (ε-Fe_2O_3?). Hysteresis measurements were complemented by thermomagnetic analysis at low and high temperatures, microscopic observations, and X-ray diffraction studies. High-field magnetic hysteresis loops alone appear insufficient to definitively discriminate goethite from hematite, though there is, expectedly, a tendency that increasing goethite content leads to magnetic hardening, with coercive force reaching 1 T and coercivity of remanence 1.7 T. At the same time, ε-Fe_2O_3 can seemingly be distinguished from either hematite or goethite due to its high saturation magnetization. However, combining hysteresis measurements with low- and high-temperature thermomagnetic analysis provides a much better insight into the magnetic mineralogy of

A. Kosterov (✉) · E. S. Sergienko · P. V. Kharitonskii · S. Yu. Yanson
St. Petersburg State University, St. Petersburg, Russia
e-mail: a.kosterov@spbu.ru

E. S. Sergienko
e-mail: e.sergienko@spbu.ru

P. V. Kharitonskii
e-mail: peterkh@yandex.ru

S. Yu. Yanson
e-mail: jansn.sv@gmail.com

A. G. Iosifidi
All-Russian Research Institute of Petroleum Research (VNIGRI), St. Petersburg, Russia
e-mail: iosag@mail.ru

Institute of Terrestrial Magnetism, Ionosphere and Radio Wave Propagation, St. Petersburg, Russia

P. V. Kharitonskii
St. Petersburg Electrotechnical University "LETI", St. Petersburg, Russia

T. B. Yanovskaya et al. (eds.), *Problems of Geocosmos—2018*, Springer Proceedings in Earth and Environmental Sciences, https://doi.org/10.1007/978-3-030-21788-4_10

samples. Still, acquiring the reference data on well characterized hematite, goethite, and ε-Fe_2O_3 samples is highly desirable.

Keywords Magnetic hysteresis · Day-Dunlop plot · Hematite · Goethite · HCLT phase

1 Introduction

High coercivity magnetic minerals such as hematite or goethite are an important constituent of sedimentary rocks, particularly redbeds, where they carry a significant fraction of rock natural remanent magnetization (NRM) [1, 2]. In addition, a novel high coercivity magnetic mineral referred to as the High Coercivity Low unblocking Temperature (HCLT) phase was recently discovered [3, 4] in many samples of archaeological ceramics which is a prime target for archaeomagnetic studies. This phase was provisionally identified [5] with a peculiar allotropic form of iron oxide (III) ε-Fe_2O_3 with orthorhombic crystal lattice [6, 7].

The analysis of magnetic hysteresis loops in such samples is strongly hampered by not saturating their magnetization in moderate (<2 T) fields available in most laboratories. The appearance of the corresponding minor hysteresis loops can be but a qualitative indicator of the presence of high coercive minerals in the sample. Moreover, minor loops are of limited use in the construction of a Day-Dunlop diagram [8–10], which, despite the recent critique [11] may still serve as a useful tool for analyzing rock magnetic properties.

To contribute to the analysis of hysteresis loops in high coercivity magnetic minerals, we investigated: (1) samples of Carboniferous sedimentary rocks from the East European Platform, and (2) bricks from different regions of European Russia, with ages from modern to early 18th century.

2 Samples and Experimental Techniques

Sedimentary rock samples used in the present study originate from two collections. The first one consisting predominantly of redbeds that belong to the Shatsk Formation, Vereisky horizon, was sampled in the Ryazan region by late R. A. Komissarova in 1965. This collection was recently reinvestigated using modern paleomagnetic methods [12]. The second collection including clays, carbonate rocks and quartzites of late Visean age was sampled in the Leningrad region at Lininka and Ragusha rivers in 2017 [13].

Ceramics samples are bricks from different regions of European Russia, from modern to about 300 years old. Historical samples were collected (1) on the islands of northern Ladoga from churches and related buildings (early 20th century) and (2) in the Yaroslavl region from ruined churches (early to mid-18th century). Modern bricks appear heterogeneous, the center of the brick having a pronounced gray to black color which gradually changes to red in the outer zone. This may indicate that the central zone was heated to a lower temperature than the outer one. Samples from the central and the outer zones were thus studied separately. The historical bricks do not show any change in coloration (red) between the central and outer zones.

Hysteresis loops were measured in a maximum field of 7 T at room temperature (295 K) using an MPMS 3 instrument (Quantum Design, USA) in the vibrating sample mode with a logarithmic field increment, 400 points per loop. Coercivity of remanence (H_{cr}) values were determined from backfield demagnetization curves of isothermal remanent magnetization (IRM) acquired in a 5 T field, measured in fields up to 1 T, and in exceptional cases up to 3 T, using the same instrument. For comparison, hysteresis loops and backfield IRM (1 T) demagnetization curves were also measured, for selected samples, in a maximum field of 1.8 T using a LakeShore 7410 vibrating sample magnetometer.

To further constrain the magnetic mineralogy of the samples, temperature dependence of the initial magnetic susceptibility was measured in air up to 700 °C using an MFK-1FA Kappabridge and a CS4 furnace. Low-temperature magnetic measurements were carried out using the MPMS 3 instrument. Isothermal remanent magnetization acquired in a 5 T field at 1.8 K after cooling in zero (zero field cooling, ZFC) and in a strong (5 T, field cooling, FC) magnetic field, respectively, was measured during warming to 300 K in zero field. In addition, IRM (5 T), acquired at 300 K, was measured during the cooling-warming cycle between 300 and 1.8 K in a zero field. For selected samples, this experiment was repeated for a maximum temperature of 400 K.

Samples were also characterized using scanning electron microscopic (SEM) observations and X-ray structural analysis. SEM studies were performed using the following equipment: (i) a Hitachi TM3000 electron microscope (Japan) with an Oxford ED3000 microanalyzer (Oxford, UK); and (ii) a system with focused electron and ion probes QUANTA 200 3D (FEI, the Netherlands) with the Pegasus 4000 analytical complex (EDAX, USA). Accelerating voltages were 15 kV for TM3000, and 20 kV for QUANTA 200 3D microscopes, respectively. X-ray studies were performed on powder samples using a Bruker D2 Phaser benchtop powder diffractometer with copper and cobalt anodes.

Samples for SEM studies were either fresh chips or prepared as follows. The rock was ground into powder, and washed in water and alcohol in order to increase the concentration of the ore phase in the sample without destroying individual grains. In both cases, it is impossible to determine grains exact chemical composition, since the probe size (1–2 μm) and the depth of secondary X-ray generation often exceed the size of the studied mineral grains. This is the case, in particular, when a mineral phase grows as a thin film on a silicate matrix grain. Energy-dispersive spectra (EDS) analysis then yields but an integral picture, where the determined composition unavoidably includes the contribution from the non-magnetic matrix. However, when combined with morphological and structural features, microanalysis data often provide an unambiguous determination of mineral phases. Microprobe analysis was performed at 5–10 points of the given mineral phase, with generally good convergence of results.

3 Results and Discussion

3.1 Microscopy and X-Ray Studies

SEM studies show that in Shatsk samples the neoformed hematite, often with Ti and Al impurities, with grain sizes from <1 to 10 μm is widely present. It is characterized by diverse morphologies, including separate oolitic grains with a scaly structure, skeletal or shell-like exsolutions over mineral aggregates of the non-magnetic matrix, hematite roses (Fig. 1a, b). In all, Shatsk rocks show signs of a high-degree alteration of primary material, mostly in reducing conditions [14]. This is confirmed by the results of X-ray powder diffraction showing significant, up to >50% quartz content.

According to X-ray powder diffraction data, samples from Lininka and Ragusha outcrops come from carbonate and clayey rocks. Hematite and goethite are present in a relatively small amount. Both minerals are poorly crystallized, their diffraction maxima overlap each other and the diffraction maxima of layered minerals, making difficult to correctly evaluate the unit cell parameters. In some samples, spherical formations are present, consisting of goethite, hematite, kaolinite and mica. With SEM, a significant amount of ore minerals, predominantly hematite and goethite, of various morphology and composition is observed in both clays and carbonates (Fig. 1c, d). These occur as films, separate small submicron grains, scaly exsolutions on clay minerals, and apparently belong to several generations of secondary minerals.

SEM studies of bricks were carried out on freshly prepared chips without sputtering, in low vacuum. This method allows to evaluate the morphology and location in the rock matrix of iron-bearing grains but only qualitatively assesses their composition. In all studied samples SEM revealed very similar picture: several iron-bearing phases of different genesis, composition and morphology. Nests of 5–20 μm are frequently observed (Fig. 1e), formed with crystals of magnetite-like habitus and, probably, composition; they contain a small, generally <10%, amount of titanium, the size of an individual crystal being about 1 μm or less. Individual grains are found with the traces of resorption and of characteristic irregular shape which could be maghemite, about 20 μm in size. Presumably, these are chemically altered agglomerates of the nests described above; their iron content is higher and no titanium impurity is present. Finally, numerous dust-like precipitates of the smallest ($\ll$1 μm) grains, of indefinite shape with Ti impurity are scattered throughout the non-magnetic alumosilicate matrix (Fig. 1f).

3.2 Magnetic Mineralogy

High-temperature susceptibility curves (Fig. 2) do not always determine the magnetic mineralogy of the studied samples unambiguously. In some simple cases (Fig. 2a, b), samples are reasonably stable during heating, and yield Curie temperatures of 622 °C (sample L61) and 607 °C (sample L62) respectively, which can correspond

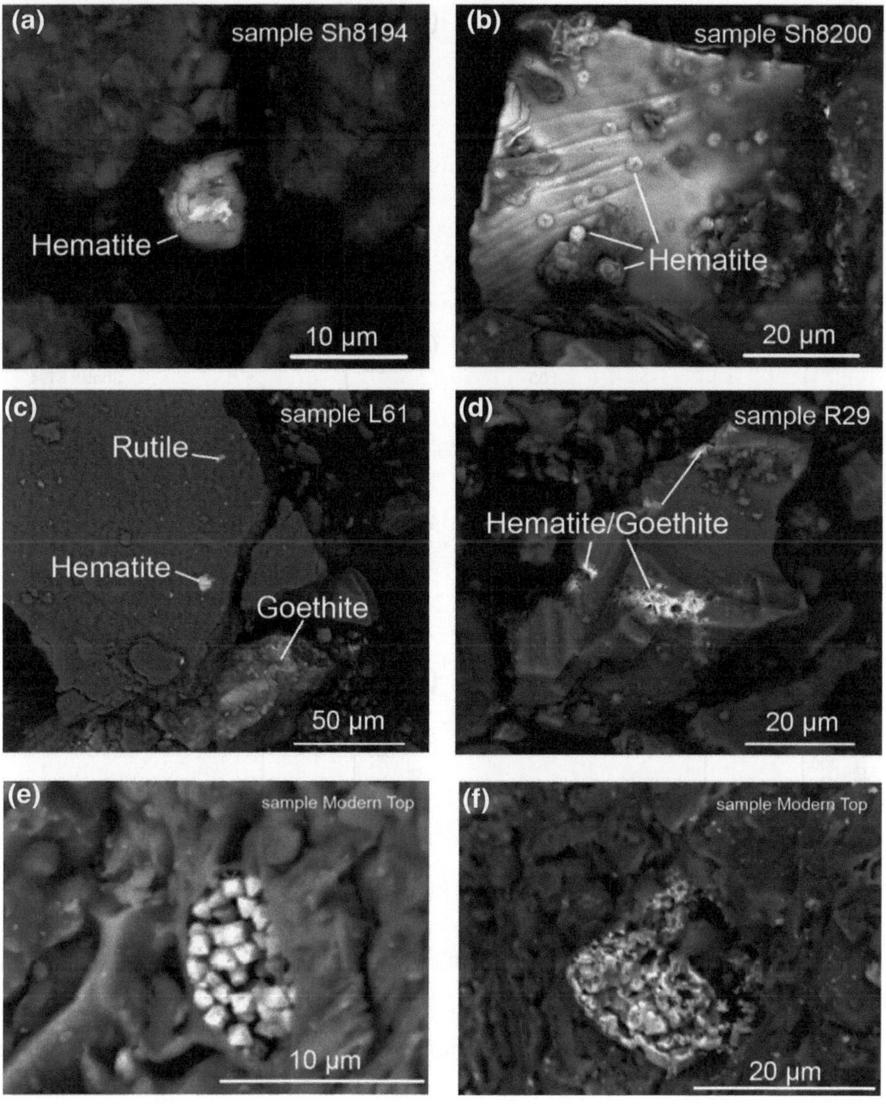

Fig. 1 SEM images. **a**, **b**—clays of the Shatsk suite; **c**—clay from the Lininka outcrop; **d**—calcareous dolomite from the Ragusha outcrop; **e**, **f**—recent brick sample

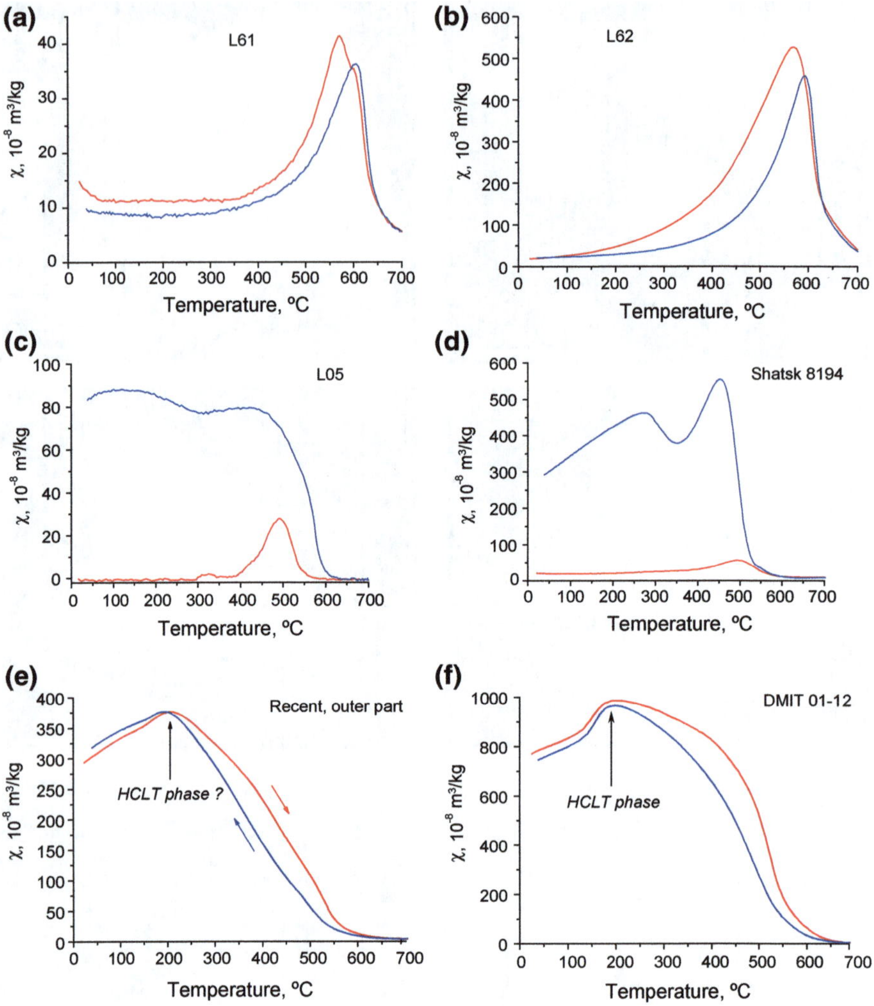

Fig. 2 Temperature dependences of the initial magnetic susceptibility. Red curves indicate heating, blue—cooling

to hematite with ~10% of impurity atoms. At the same time, in all Shatsk and in a major part of Lininka and Ragusha samples a strongly magnetic phase starts forming at ~450 °C (Fig. 2c, d), making it impossible to detect the presence of hematite from thermomagnetic curves. Neoformation of a strongly magnetic mineral also prevents detecting magnetite that may be present in virgin samples.

A phase with Curie temperatures of 500–600 °C is present in all samples of both modern and historical bricks (Fig. 2e, f). It could be identified with strongly oxidized magnetite or maghemite, possibly cation-substituted, in agreement with SEM studies. Curie temperatures corresponding to hematite were not detected. At the same time, most samples display inflection points at about 200 °C, characteristic of the HCLT phase [4, 5].

Further information on the samples magnetic mineralogy can be obtained from the analysis of low-temperature thermomagnetic curves. In some sediment samples, SIRM (5 T), acquired at 1.8 K after ZFC and FC from 300 K, decreases monotonously up to this temperature, without showing any peculiar features (Fig. 3a, b). On the other hand, if cooling is performed from, and subsequent warming to, 400 K, SIRM (T) curves flatten out, and respective FC and ZFC curves converge, at ca. 360 K for sample R29 and at ca. 300 K for L61. It is then natural to assume that these temperatures are the Curie (Néel) points of a particular magnetic mineral. The fact that SIRM (5 T), acquired at 400 K, is significantly less than that acquired at 300 K (Fig. 3c, d), confirms this hypothesis. On the other hand, a noticeable excess of FC-SIRM (1.8 K) over ZFC-SIRM (1.8 K) indicates that this mineral is magnetically hard. Considering the results of microscopy and X-ray diffraction, goethite seems the most likely candidate. However, it should be noted that these Curie temperatures are noticeably lower than the Curie temperature of pure goethite (120 ± 2 °C) [15]. Thus goethite in these samples, similarly to hematite, likely contains a noticeable amount of impurities. Alternatively, the temperature at which FC and ZFC curves converge can be viewed as a maximum unblocking temperature of goethite phase [16]. It would then require a very fine, possibly <1 μm, grain size. We note however that Guyodo et al.'s [16] 350-nm goethite particles while having the maximum unblocking temperature of ~300 K as in our sample L61, showed also a much larger, about a factor of ten, contrast between FC and ZFC remanences at 5 K. In our case, FC/ZFC remanence ratio is about 1.5 for L61 and about 3 for R29, respectively. Therefore, the latter explanation of FC/ZFC convergence temperature may hold for R29 implying a grainsize of several μm, but clearly not for L61.

Some other sediment samples show rather weak but noticeable Verwey transition [17] indicating presence of a magnetite-like phase (Fig. 4). Verwey temperatures range from 121 K (sample SH 8201) corresponding to nearly stoichiometric magnetite [18, 19] to <100 K (sample L73). In sample L05 (Fig. 4e, f), two generations of magnetite appear to be present: one fairly close to stoichiometry, with $T_V = 113$ K, and a strongly cation-substituted phase not having the Verwey transition but manifesting itself in an irreversibility of the SIRM (300 K) cooling-warming cycle starting around 30 K. It must be noted, however, that in all cases magnetite coexists with a much more volumetrically abundant magnetically hard phase(s). Furthermore, magnetite, including cation-substituted varieties, is not detectable by high-temperature thermomagnetic analysis because of severe alteration imposed by heating.

Low-temperature thermomagnetic curves of bricks are smooth and featureless (Fig. 5). This indicates that neither magnetite nor hematite is present in the samples. FC SIRM curves run barely higher than ZFC ones implying that, to no surprise, magnetically hard iron hydroxides are not present either. Together with the high-temperature thermomagnetic curves, this suggests that dominant magnetic phases in these bricks are maghemite, probably cation substituted, and an elusive HCLT phase.

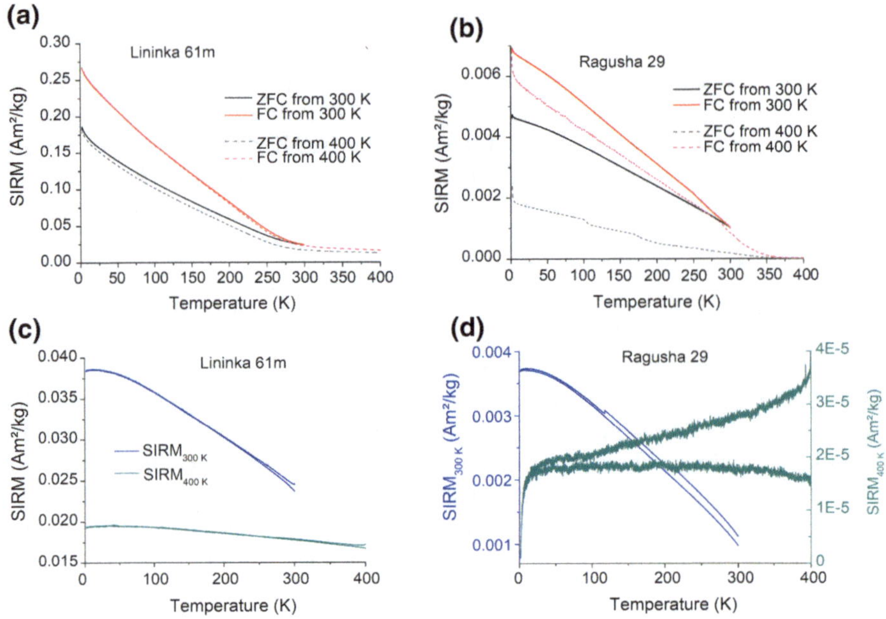

Fig. 3 **a**, **b** Temperature demagnetization curves of SIRM (5 T) acquired at 1.8 K after cooling in zero field (ZFC, black) and in 5 T field (FC, red) from 300/400 K (solid/dashed curves); **c**, **d** zero-field cycles to 1.8 K and back of SIRM (5 T), acquired at 300 K and 400 K, respectively. The difference between the magnetizations is most likely due to the presence of a goethite-like phase with a Néel temperature of ~300 K in sample L61m and ~360 K in R29

3.3 High-Field Hysteresis

Examples of hysteresis loops measured on sediment samples are shown in Fig. 6. As would be expected for hematite and/or goethite-bearing samples, magnetization in the hysteresis cycle remains irreversible up to fields of 3–4 T, and in sample R29 (Fig. 6a) up to the maximum field available to us, 7 T. The total magnetization generally does not reach saturation in a field of 7 T. Accordingly, the paramagnetic susceptibility was calculated by extrapolating the dependence of the differential susceptibility dM/dH on 1/H to a field of 20 T, as outlined in Starunov et al. [20]. Note that even this method is likely to overestimate the paramagnetic susceptibility for samples where goethite is a dominant magnetic phase, since saturation is not reached in this mineral even in much stronger fields, up to 57 T [21].

The shape of hysteresis loops suggests that in the sample R29 (Fig. 6a) the predominant magnetic phase is goethite, confirmed also by the magnetic behavior at cryogenic temperatures. This sample is characterized by a huge coercive force, over 1 T; coercivity of remanence is 1.78 T. It should be further noted that these values underestimate the true ones, since they are obtained from a minor rather than major hysteresis loop.

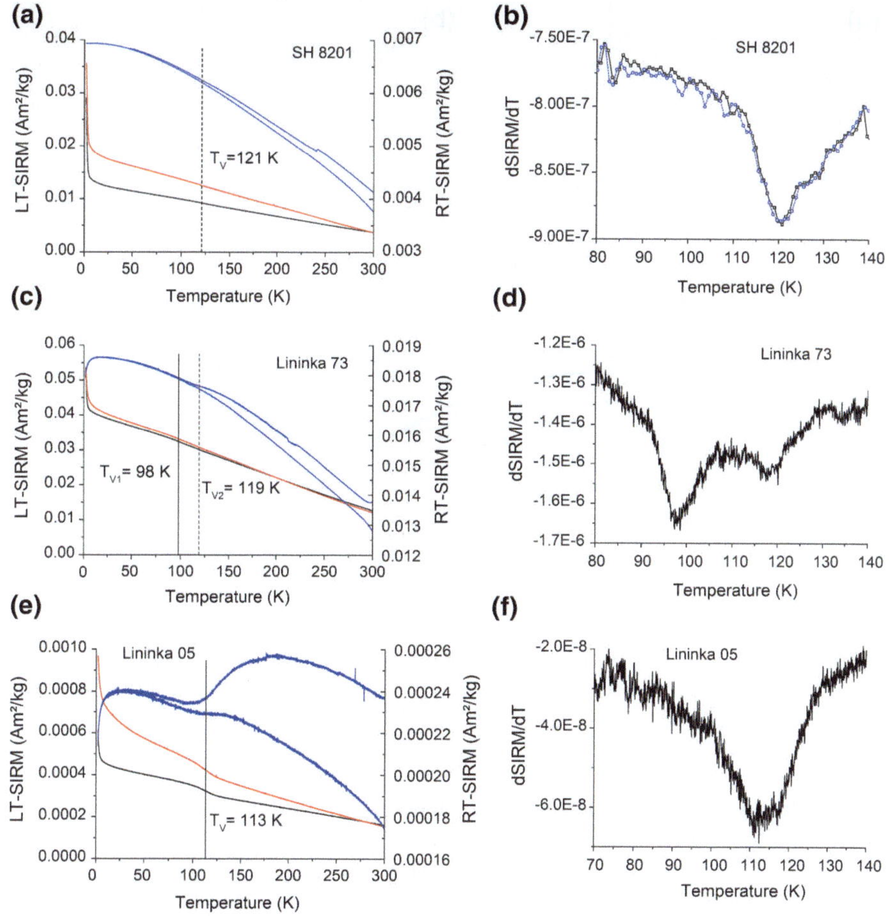

Fig. 4 Temperature demagnetization curves of SIRM (5 T) acquired at 1.8 K after cooling in zero field (ZFC, black) and in 5 T field (FC, red) from 300 K and zero-field cycles to 1.8 K and back (blue) of SIRM (5 T) acquired at 300 K for redbed samples (**a**, **c**, **e**) and temperature derivatives of ZFC SIRM warming curves showing evidence for the Verwey transition (**b**, **d**, **f**). For sample SH8201 ZFC/FC experiment has been repeated twice, to prove the existence of the Verwey transition which is very weak in this sample

In the redbed samples the main magnetic mineral is expectedly hematite. However, lowered Curie temperatures (where these can be determined) and absence of the Morin transition [22] indicate that in all our samples hematite is noticeably cation-substituted, most likely by aluminum. Two types of hysteresis loops of different shape are observed: 'normal' (Fig. 6b, c) and wasp-waisted (Fig. 6d). Loops remain open up to the fields of 3–4 T, dependent on a sample, emphasizing the need to use the highest possible field to trace the major loop. Coercive force ranges from 80 to 250 mT for the 'normal' samples, and from 20 to 110 mT for those showing wasp-

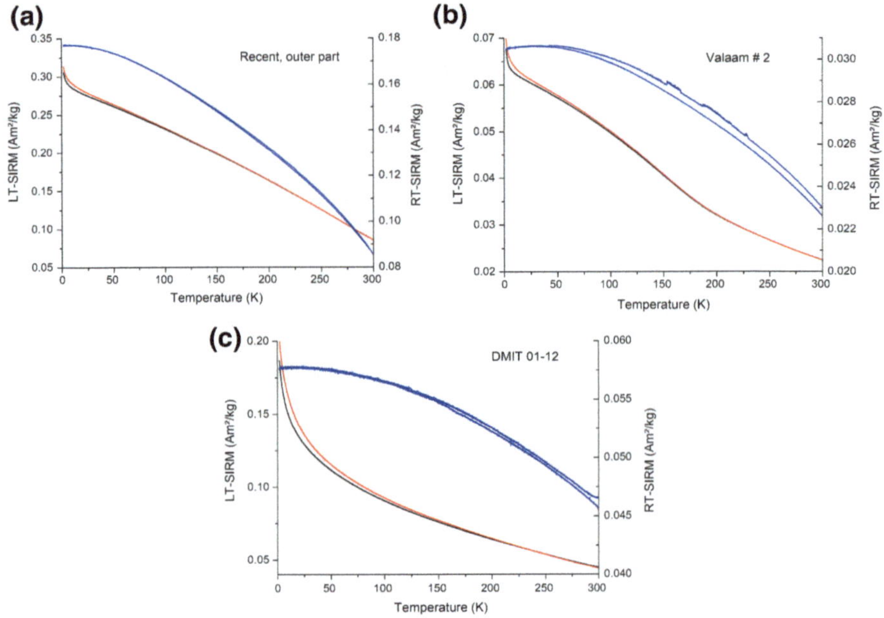

Fig. 5 Temperature demagnetization curves of SIRM (5 T) acquired at 1.8 K after cooling in zero field (ZFC, black) and in 5 T field (FC, red) from 300 K and zero-field cycles to 1.8 K and back (blue) of SIRM (5 T) acquired at 300 K for brick samples

waisted loops, while coercivity of remanence is similar in both cases and ranges rather widely, from about 100 to over 600 mT.

In the case of bricks, hysteresis loops (Fig. 7) all have a pronouncedly wasp-waisted shape, indicating the coexistence of a magnetically soft and hard phase, respectively. As for the redbed samples, loops are open up to about 3 T field. Coercive force, however, varies widely, from 7 mT for the sample DMIT 01–12 from Yaroslavl' region up to about 210 mT in the modern brick sample. The magnetically soft phase may be naturally associated with an oxidized magnetite/maghemite identified by the thermomagnetic analysis. In archaeological ceramics, the hard phase saturation only in 3–4 T field is commonly attributed to hematite. In our samples, however, saturation magnetization reaches as high as 0.3 Am^2/kg. Were it due to hematite, this mineral would constitute up to 20% of the sample by weight. This is clearly incompatible with either microscopic observations or results of thermomagnetic analysis. Furthermore, it is worth noting that soft and hard phases appear to present in different proportion, amount of the hard phase decreasing in more ancient bricks. Our collection is obviously too small to draw definite conclusions, however, one can speculate that this pattern might be due to differences in the brick production technology (for example, different burning temperatures) or else an instability of the magnetically hard phase against time.

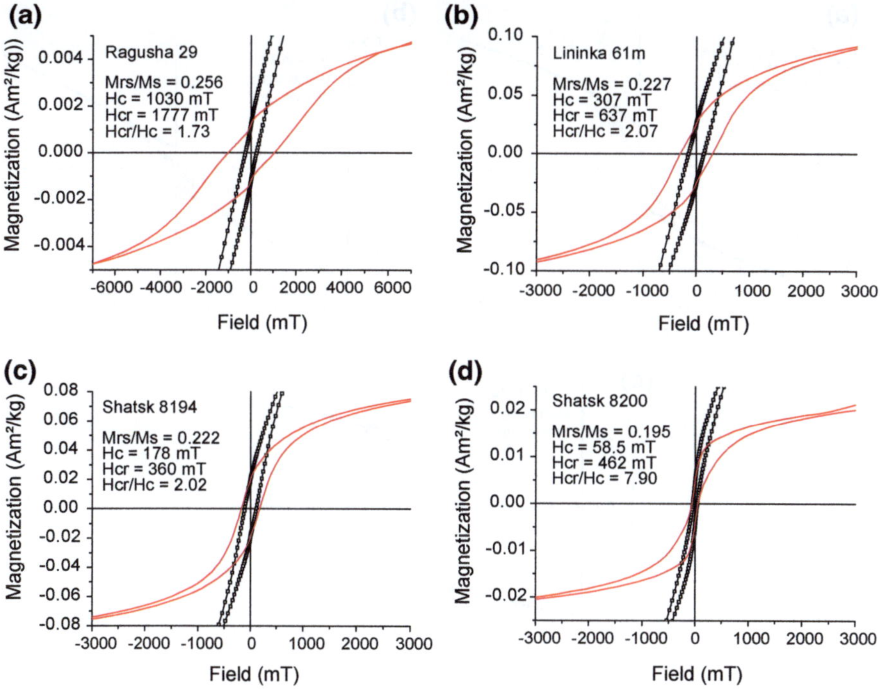

Fig. 6 Examples of hysteresis loops of sedimentary rocks measured in a maximum field of 7 T. Measured curves are shown in black, and corrected for the paramagnetic contribution in red. For clarity, only the central part of the loops in a ±3 T field is shown, except for the sample R29 (**a**), where goethite is the main magnetic phase. All other samples are dominated by hematite

In Fig. 8, we compare hysteresis loops measured in 7 and 1.8 T maximum fields, respectively. In magnetically hard samples saturation is not entirely reached in the latter field, and the respective M_s and M_{rs} values underestimate the true ones, though not as much as might be expected from the appearance of the loops themselves. However, the coercive force is affected to a considerably greater degree. High-field susceptibility also cannot be correctly evaluated from the undersaturated loops.

On the Day-Dunlop diagram (Fig. 9), the representative points corresponding to samples with 'normal' loops fall into the pseudo-single domain range. Moreover, they lie practically on the trend, derived for the case of magnetite, describing the pseudo-single domain behavior as a superposition of single- and multidomain contributions [9]. The goethite-dominated sample R29 also fall on this trend though it may simply be fortuitous, bearing in mind that even the 7 T loop for this sample should be considered as minor. Nevertheless, a close correspondence of the hysteresis parameters obtained from major loops measured on high-coercivity samples to the SD-MD trend as introduced in [9] hints that this trend is underlain by a universal, yet undiscovered, physical mechanism.

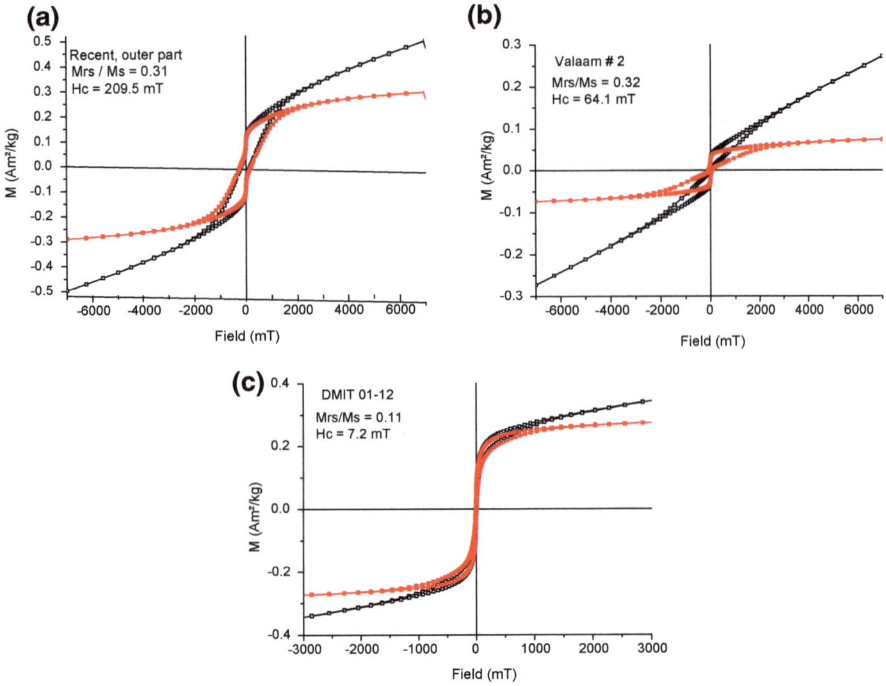

Fig. 7 Hysteresis loops of bricks measured in a maximum field of 7 T. Notation as in the previous figure. For the sample DMIT 01-12 (**c**), only the central part of the loop in a ±3 T field is shown

Representative points of the samples with wasp-waisted loops expectedly lie rather far to the right of the SD-MD trend. M_{rs}/M_s ratios for these samples are like the "normal" ones, while H_{cr}/H_c ratios are much higher. This is readily explained by the presence of two phases with contrasting coercivities. Coercive force of these samples appears largely controlled by the soft phase, while coercivity of remanence is governed by the hard phase and is therefore much higher. In redbeds, bearing in mind that their coercive force is quite high, both soft and hard phases are likely hematite populations of different grain size, e.g. detrital hematite versus pigment. A contribution of magnetite to the soft phase in some of the samples (e.g. L05) also cannot be ruled out. In bricks, the two phases are apparently oxidized magnetite/maghemite and a HCLT phase [4, 5], respectively. However, the identity of the latter still remains enigmatic. Epsilon iron oxide [6, 7] is a possible candidate according to its magnetic properties; still, it must be noted that ε-Fe_2O_3 has not been positively identified in our samples either microscopically or by X-ray diffraction.

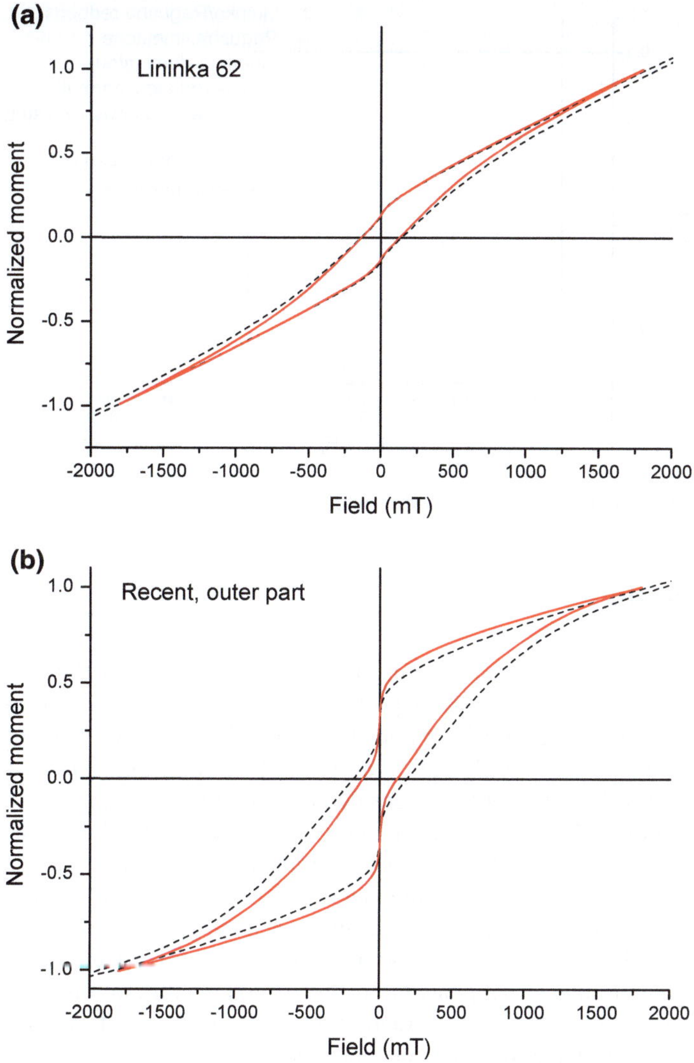

Fig. 8 A comparison of hysteresis loops measured in 7 T (black dashed curves) and in 1.8 T (red solid curves), respectively. **a**—hematite-bearing clay; **b**—brick probably containing HCLT (ε-Fe_2O_3?) phase. Magnetic moment is normalized to its value in the 1.8 T field

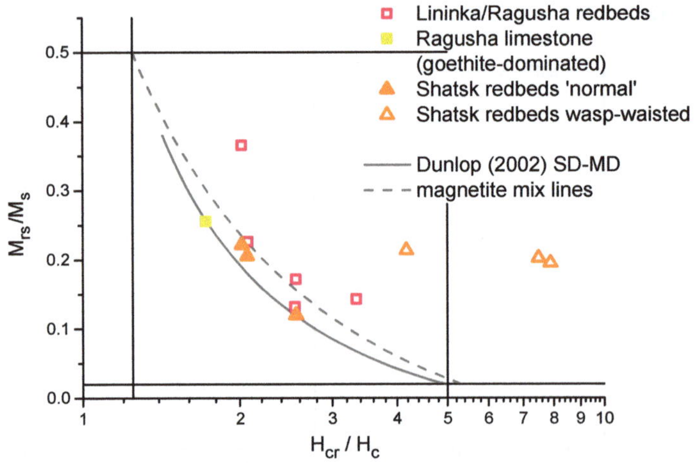

Fig. 9 Day-Dunlop diagram [8, 9] for sedimentary rock samples

4 Conclusion

Hysteresis loops alone appear to be insufficient to distinguish goethite and hematite in a rock, although it should be expected that the higher the content of goethite in the sample, the magnetically harder it will be. The coercive force of goethite-containing samples can reach and even exceed 1 T. An HCLT phase [4, 5] could be distinguished from hematite by considerably larger saturation magnetization and by singularities in the thermomagnetic curves around 200 °C, though the latter are not always sufficiently pronounced. Low-temperature magnetic measurements are often very helpful in establishing the magnetic mineralogy of rocks, although they require specialized equipment and are time-consuming. Obtaining new reference data on magnetic hysteresis and thermomagnetic curves at cryogenic temperatures on hematite, goethite, and HCLT phase (ε-Fe_2O_3?) samples with controlled composition, morphology, and grain size would greatly enhance the understanding of the magnetic behavior of rocks and archaeological ceramics.

Acknowledgements Natalya Salnaya (Institute of Physics of the Earth, Russian Academy of Sciences) greatly assisted in sampling at Lininka and Ragusha in 2017, and donated the samples of bricks from Yaroslavl. Bricks from Valaam Island were donated by Vladimir Karpinsky (Earth Physics Department, St. Petersburg State University). Measurements were carried out at the resource centers of the Scientific Park of St. Petersburg State University: Centre for Geo-Environmental Research and Modelling (GEOMODEL), Centre for Diagnostics of Functional Materials for Medicine, Pharmacology and Nanoelectronics, Centre for Innovative Technologies of Composite Nanomaterials, Centre for Microscopy and Microanalysis, and Centre for X-ray Diffraction Studies. The study was partially supported by the Russian Foundation for Basic Research via grants 16-05-00603a and 18-05-00626a.
This paper benefited from the reviews by Aleksey Smirnov and Mike Jackson.

References

1. Khramov, A.N.: Paleomagnetology. Springer, Heidelberg (1987)
2. Kodama, K.P.: Paleomagnetism of Sediments and Sedimentary Rocks: Process and Interpretation. Wiley, Chichester, West Sussex, Hoboken, NJ (2012)
3. McIntosh, G., Kovacheva, M., Catanzariti, G., Osete, M.L., Casas, L.: Widespread occurrence of a novel high coercivity, thermally stable, low unblocking temperature magnetic phase in heated archeological material. Geophys. Res. Lett. **34**, L21302 (2007). https://doi.org/10.1029/2007GL031168
4. McIntosh, G., Kovacheva, M., Catanzariti, G., Donadini, F., Osete Lopez, M.L.: High coercivity remanence in baked clay materials used in archeomagnetism. Geochem. Geophys. Geosyst. **12**, Q02003 (2011). https://doi.org/10.1029/2010GC003310
5. López-Sánchez, J., McIntosh, G., Osete, M.L., del Campo, A., Villalaín, J.J., Pérez, L., Kovacheva, M., Rodríguez de la Fuente, O.: Epsilon iron oxide: origin of the high coercivity stable low Curie temperature magnetic phase found in heated archeological materials. Geochem. Geophys. Geosyst. **18**(7), 2646–2656 (2017)
6. Jin, J., Hashimoto, K., Ohkoshi, S.-I.: Formation of spherical and rod-shaped ε-Fe_2O_3 nanocrystals with a large coercive field. J. Mater. Chem. **15**, 1067–1071 (2005)
7. Gich, M., et al.: High- and low-temperature crystal and magnetic structures of ε-Fe_2O_3 and their correlation to its magnetic properties. Chem. Mater. **18**(16), 3889–3897 (2006)
8. Day, R., Fuller, M., Schmidt, V.A.: Hysteresis properties of titanomagnetites: grain-size and compositional dependence. Phys. Earth Planet. Inter. **13**, 260–267 (1977)
9. Dunlop, D.J.: Theory and application of the Day plot (M_{rs}/M_s versus H_{cr}/H_c) 1. Theoretical curves and tests using titanomagnetite data. J. Geophys. Res. **107** (2002). https://doi.org/10.1029/2001jb000486
10. Dunlop, D.J.: Theory and application of the Day plot (M_{rs}/M_s versus H_{cr}/H_c) 2. Application to data for rocks, sediments, and soils. J. Geophys. Res. **107** (2002). https://doi.org/10.1029/2001jb000487
11. Roberts, A.P., Tauxe, L., Heslop, D., Zhao, X., Jiang Z.: A critical appraisal of the "Day" diagram. J. Geophys. Res. Solid Earth **123**(4), 2618–2644 (2018)
12. Iosifidi, A.G., Mikhailova, V.A., Popov, V.V., Sergienko, E.S., Danilova, A.V., Otmas, N.M., Zhuravlev, A.V.: Carboniferous of the Russian platform: paleomagnetic data. In: Nurgaliev, D., Shcherbakov, V., Kosterov, A., Spassov, S. (eds.) Recent Advances in Rock Magnetism, Environmental Magnetism and Paleomagnetism, pp. 37–54. Springer International Publishing, Cham (2019)
13. Iosifidi, A.G., Sergienko, E.S., Sal'naya, N.V., Otmas, N.M., Mikhailova, V.A., Danilova, A.V.: Paleomagnetic studies of late Visean deposits from Moscow syneclise (Leningrad region, rivers Lininka and Ragusha) [Paleomagnitnye issledovaniya pozdnevizeyskikh otlozheniy Moskovskoy sineklizy (Leningradskaya obl., r. Lininka, r. Ragusha)]. In: Proceedings of the 12th School-Conference "Problems of Geocosmos", St. Petersburg, 8–12 Oct 2018, pp. 101–104 (2018) (in Russian)
14. Pettijohn, F.J.: Sedimentary Rocks, 3rd edn, xii, 628 p. Harper & Row, New York (1975)
15. Özdemir, Ö., Dunlop, D.J.: Thermoremanence and Néel temperature of goethite. Geophys. Res. Lett. **23**, 921–924 (1996)
16. Guyodo, Y., Mostrom, A., Lee Penn, R., Banerjee, S.K.: From nanodots to nanorods: oriented aggregation and magnetic evolution of nanocrystalline goethite. Geophys. Res. Lett. **30**, 1512 (2003). https://doi.org/10.1029/2003GL017021
17. Verwey, E.J.W.: Electronic conduction of magnetite (Fe_3O_4) and its transition point at low temperatures. Nature **144**, 327–328 (1939)
18. Aragón, R., Buttrey, D.J., Shepherd, J.P., Honig, J.M.: Influence of nonstoichiometry on the Verwey transition. Phys. Rev. B **31**, 430–436 (1985)
19. Özdemir, Ö., Dunlop, D.J., Moskowitz, B.M.: The effect of oxidation on the Verwey transition in magnetite. Geophys. Res. Lett. **20**, 1671–1674 (1993)

20. Starunov, V.A., Kosterov, A., Sergienko, E.S., Yanson, S.Y., Markov, G.P., Kharitonskii, P.V., Sakhatskii, A.S., Lezova, I.E., Shevchenko, E.V.: Magnetic properties of tektite-like impact glasses from Zhamanshin astrobleme, Kazakhstan. In: Nurgaliev, D., Shcherbakov, V., Kosterov, A., Spassov, S. (eds.) Recent Advances in Rock Magnetism, Environmental Magnetism and Paleomagnetism, pp. 445–465. Springer International Publishing, Cham (2019)
21. Rochette, P., Mathé, P.-E., Esteban, L., Rakoto, H., Bouchez, J.-L., Liu, Q., Torrent, J.: Non-saturation of the defect moment of goethite and fine-grained hematite up to 57 Teslas. Geophys. Res. Lett. **32**, L22309 (2005). https://doi.org/10.1029/2005GL024196
22. Morin, F.J.: Magnetic susceptibility of αFe_2O_3 and αFe_2O_3 with added titanium. Phys. Rev. **78**, 819–820 (1950)

Preliminary Data on the Magnetic Properties of Fallow Soils (Zelenodolsky District, Republic of Tatarstan)

L. A. Fattakhova, L. R. Kosareva, V. V. Antonenko, A. V. Fattakhov and R. D. Akhmerov

Abstract Over the last few years, there has been an expansion of fallow lands in Russia due to issues in agricultural economics. Soils become fallow when left uncultivated for a time after successive crops. The recovery process depends on the duration of continuous arable farming in the area, soil type, area size and preferred land use methodology. Therefore, the importance of a comprehensive study of the processes occurring in fallow soils is reflected in their enormous expansion. In the course of soil formation, the transformation and migration of iron compounds change the magnetic susceptibility of soil. This suggests that magnetic susceptibility reflects the process of soil formation and may be considered a diagnostic indicator. This article presents the research results on the magnetic properties of fallow soils in Zelenodolsky district (Kazan). Samples of fallow sod-podzolic soils were used in this work. The magnetic properties of fallow soils were studied, and changes in the magnetic characteristics of fallow sod-podzolic soils were analyzed. The soil is characterized by the eluvial/illuvial type of profile patterns of magnetic susceptibility distribution. Evaluation of the contribution of the dia- /paramagnetic, superparamagnetic and ferromagnetic components from the coercive spectra shows that the increase in the magnetic susceptibility of sod-podzolic soils is due to the contribution of the ferromagnetic component.

L. A. Fattakhova (✉) · L. R. Kosareva · V. V. Antonenko · A. V. Fattakhov · R. D. Akhmerov
Department of Geophysics and Geoinformation Technologies, Kazan Federal University, Kazan, Russia
e-mail: l.a.fattakhova@yandex.ru

L. R. Kosareva
e-mail: lina.kosareva@mail.ru

V. V. Antonenko
e-mail: stay_uzeless@mail.ru

A. V. Fattakhov
e-mail: avfattakhov@gmail.com

R. D. Akhmerov
e-mail: rinaz.akhmerov@mail.ru

T. B. Yanovskaya et al. (eds.), *Problems of Geocosmos—2018*, Springer Proceedings in Earth and Environmental Sciences, https://doi.org/10.1007/978-3-030-21788-4_11

Keywords Fallow soil · Sod-podzolic soil · Magnetic susceptibility · Coercive spectrometry

1 Introduction

Today, arable lands are being taken out of production in most countries in accordance with common land management practices. This causes changes in soil formation processes and, subsequently, in the evolution of arable soils [1].

Plowing removes the vegetative cover and changes the composition and properties of soil. Soil becomes fallow as soon as it is left uncultivated for more than a year. The future plant succession and soil composition depend on the soil type, the duration of continuous arable farming in the area and the area size.

The importance of a comprehensive study of the processes occurring in fallow soils is reflected in their enormous expansion. They are the foundation of the agricultural land reserves.

In the past two decades, the study of environmental magnetism, including the magnetic characteristics of soils, had gained much ground worldwide [2–7]. Magnetic susceptibility patterns are unique for each soil type and depend on the magnetic susceptibility of parent rocks, weathering, and specific properties of the soil itself: organic content, mineral composition, grain size distribution, water and redox regime, composition of newly formed ferrous minerals, biogenic activity, etc. [2, 8]. Magnetic characteristics, such as magnetic susceptibility, are determined primarily by the content of iron compounds, their phase composition and dispersion. It was established that magnetic properties of zonal soils can regularly and continually change with depth, forming a magnetic profile that reflects type and intensity of the soil-forming process [2]. Also, it seems that the assumptions about soil being able to self-recover to its original state under naturally growing plant successions are unfounded. Therefore, it might be interesting to analyze the magnetic profiles of long-fallow (>10 years) sod-podzolic soils.

The purpose of this work is to study the magnetic properties of fallow soils, and to analyze received magnetic characteristics of fallow sod-podzolic soils.

2 Materials and Methods

For this study, long-fallow (11 and 12 years) loamy sod-podzolic soils of Zelenodolsky Municipal District (Republic of Tatarstan, Volga-Vyatsky region) were selected. The fallow vegetation is represented by fir and pine trees, grass and mixed herbs. The area belongs to the boreal (sub-taiga) type. Weathered dolomitized limestones and sandstones, clays, and marls (Upper Kazan stage of the Upper Permian) form flat and slightly wavy watersheds. The area under study is covered by unconsolidated Quaternary deposits represented by light and medium loam, sandy loam, and sand.

Fig. 1 Sampling procedure

Sampling sites formed a hexagonal grid. The distance between the central sampling point and each of the peripheral ones was 50 m (Fig. 1). The samples were taken every 10 cm to a depth of 70 cm (including sod). At the central point, sampling was conducted to a depth of 120 cm.

Preliminary sample preparation was performed in accordance with the guidelines provided by the International Organization for Standardization [9]. AGICO susceptibility bridge MFK1-FA was used to measure the magnetic susceptibility (χ). Prior to this, all samples were ground in an agate mortar.

Coercive spectra can be used to determine the contribution of the dia/paramagnetic (χ_p) and ferromagnetic (χ_f) components to the magnetic susceptibility. Coercive spectra of isothermal magnetization in the magnetic fields of up to 0.5 T were obtained using a coercive spectrometer ("J_meter") [10, 11], which allows separate recording of remanent and induced magnetization at room temperature. Samples were magnetized from their natural state. The following parameters were derived from the magnetization curves: saturation remanent magnetization (Jrs), saturation magnetization corrected for the paramagnetic component (Js), bulk coercive force corrected for the paramagnetic component (Bc), coercivity of remanence (Bcr).

3 Results and Discussions

In the course of soil formation, transformation and migration of iron compounds changes the magnetic susceptibility of soil. This suggests that magnetic susceptibility reflects the process of soil formation and may be considered a diagnostic indicator [2, 12–15]. Using the measured values of magnetic susceptibility, the curves of changes with depth were constructed. Figure 2 shows variations of different components of

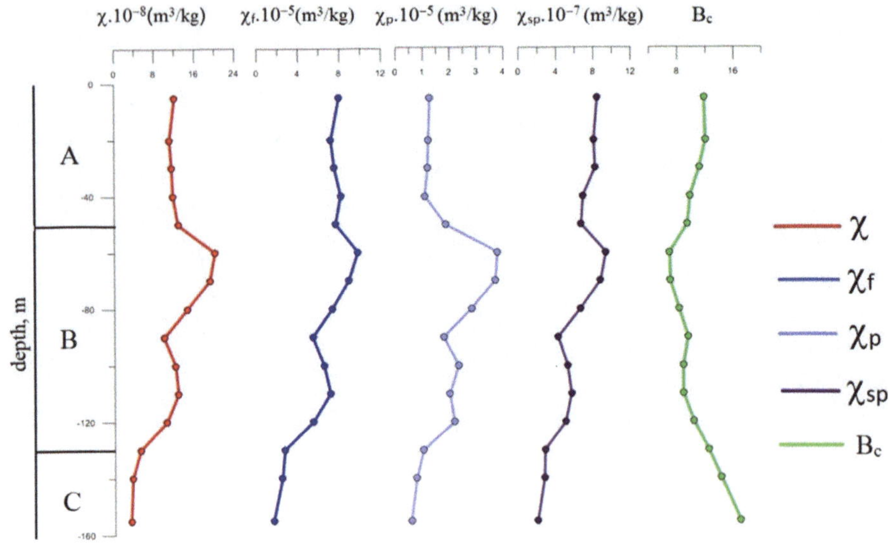

Fig. 2 Magnetic properties as a function of depth (central soil profile)

magnetic susceptibility χ and of bulk coercive force B_c with depth. In our case, the magnetic susceptibility distribution follows the patterns characteristic of the eluvial/illuvial type. This means that the upper part of the soil profile contains relatively small amount of magnetic materials, which increase in the middle part. However, the magnetic mineral content decreases again towards the underlaying soil-forming rock and eventually becomes equal to that of the humus layer.

Coercive spectra can be used to determine the contribution of the dia/paramagnetic (χ_p) and ferromagnetic (χ_f) components to the magnetic susceptibility [16]. The shape of the curves for the paramagnetic, ferromagnetic and superparamagnetic components (Fig. 2) and a direct correlation between the total magnetic susceptibility and its ferromagnetic component (Fig. 3) clearly show that an increase in magnetic susceptibility in the illuvial layer is due to the ferromagnetic component.

Since one profile covers depths to 160 cm, and another six profiles cover depths to 80 cm, it might be interesting to compare the magnetic susceptibility between them. Figure 4 shows that the points 1, 2 and 6 reflect the upper part (0–80 cm) of the central profile, while the points 3, 4 and 5 reflect its lower part (80–160 cm). This indicates the irregular shaped relief. There is a slope gently falling off to the south-east, which means that organic and mineral elements are transported down the slope. This will primarily affect the growing vegetation in the study area. In general, all the magnetic profiles are of the eluvial/illuvial type. B-horizon is rich not only in clay, but also in magnetite, maghemite, hematite and goethite.

Magnetic rigidity and domain state of the magnetic grains were determined using the magnetic hysteresis parameters (*Bc, Bcr, Js, Jrs*), which depend on the composition, magnetic mineral content, shape and size of magnetic grains. *Bcr*/*Bc*

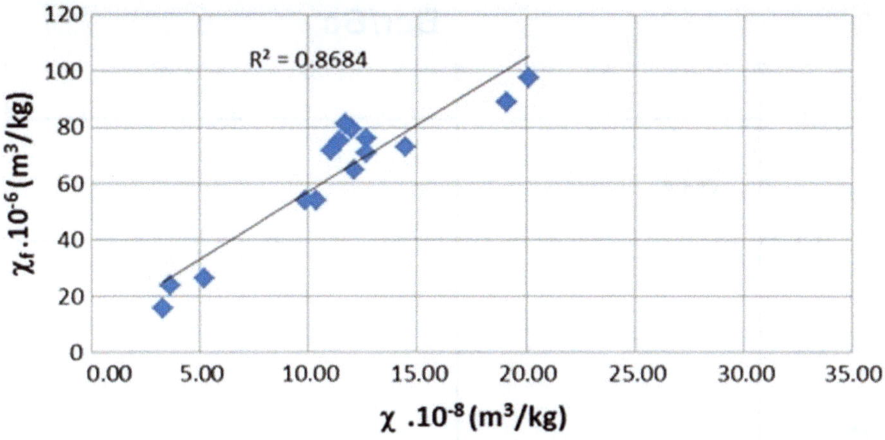

Fig. 3 Relationship between the magnetic susceptibility and its ferromagnetic component (central soil profile)

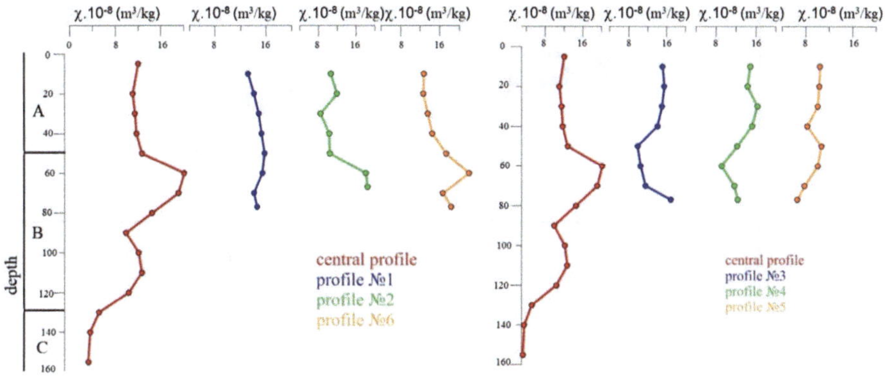

Fig. 4 Magnetic susceptibility as a function of depth (various profiles)

and *Jrs/Js* ratios reflect prevailing grain size and ratios between magnetic fractions with different domain structures. The Day plot [17] is usually used to visualize these characteristics. It shows (Fig. 5) that the magnetic grains in the soil under study are pseudo-single domain particles or a mixture of single domain and multi domain grains.

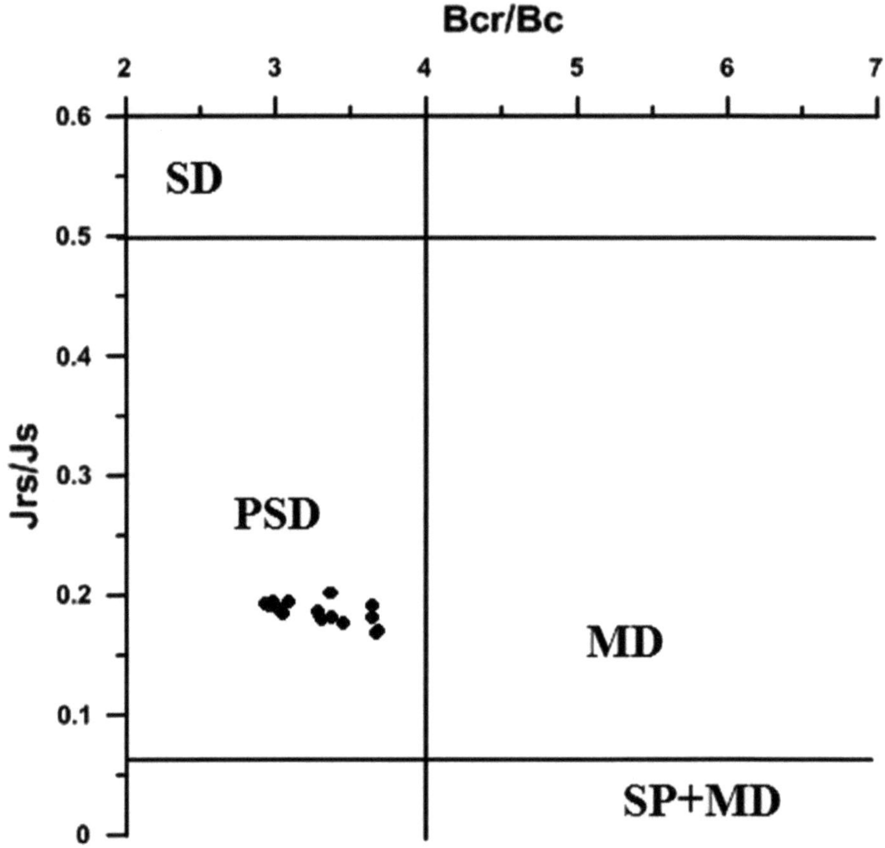

Fig. 5 Day plot

4 Conclusions

Fallow sod-podzolic soils of the Volga-Vyatsky region were studied using a complex of modern magnetic analysis methods, including magnetometry and coercive spectrometry. Magnetic susceptibility, being the ultimate characteristic of magnetic properties, depends primarily on the ferromagnetic component. The shape of the magnetic susceptibility distribution suggests that fine-grained superparamagnetic mineral particles were formed due to the structural and material transformations of iron-containing minerals introduced into the soil from the parent rock. The increase of χ_f in the profiles coincides with the increase of χ_{sp}. So, the χ_f values are enhanced by the presence of superparamagnetic particles that follows the "pedogenic" mechanism of rock-magnetic changes in soils [18] which suggest χ_{sp} as an indicator of pedogenic degree. A gentle slope in the study area indicates different conditions for the formation of sustainable plant successions. It is intended in the future to study soil samples and separate <2.5 μm specimens using the differential thermomagnetic analysis (DTMA) to gain more information on the magnetic and mineralogical features of the studied soils.

Acknowledgements The work is performed according to the Russian Government Program of Competitive Growth of Kazan Federal University.
We would like to thank reviewers Dr. A. Yu. Kazansky and Dr. T. Magiera, and editor Dr. A. A. Kosterov for their valuable advice.

References

1. Aleksandrovsky, A.L.: Evolution of Soils and Geographical Environment, 223 pp. Nauka Publishers, Moscow (2005) (in Russian)
2. Babanin, V.F., Trukhin V.I., Karpachevskiy L.O., Ivanov F.V., Morozov V.V.: Soil Magnetism, 202 pp. Yaroslavl: YaGTU, Moscow (1995) (in Russian)
3. Liu, Q., Robersts, A.P., Larrasoaña, J.C., Banerjee, S.K., Guyodo, Y., Tauxe, L.: Oldfield F. Environmental magnetism: principles and applications. Rev. Geophys. **50**(4), RG4002 (2012). https://doi.org/10.1029/2012RG000393
4. Blundell, A., Dearing, J.A., Boyle, J.F., Hannam, J.A.: Controlling factors for the spatial variability of soil magnetic susceptibility across England and Wales. Earth-Sci. Rev. **95**, 158–188 (2009)
5. de Jong, E., Pennock, D.J., Nestor, P.A.: Magnetic susceptibility of soils in different slope positions in Saskatchewan, Canada. Catena **40**, 291–305 (2000)
6. Maher, B.A., Alekseev, A., Alekseeva, T.: Magnetic mineralogy of soils across the Russian steppe: climatic dependence of pedogenic magnetite formation. Palaeogeogr. Palaeoclimatol. Palaeoecol. **201**, 321–341 (2003)
7. Jordanova N (2017) Soil Magnetism: Applications in Pedology, Environmental Science and Agriculture, 466 pp. Academic Press (2017)
8. Ivanov, A.V.: Magnetic and valence state of iron in the solid phase of the soil: Avtoref. Dis. Dr. biol. Sciences, 25 pp. Moscow (2003) (in Russian)
9. ISO 11464:1994: Soil Quality—Pretreatment of Samples for Physico-Chemical Analysis, p. 11
10. Yasonov, P.G., Nurgaliev, D.K., Burov, B.V., Heller, F.: A modernized coercivity spectrometer. Geol. Carpath. **49**, 224–226 (1998)
11. Nurgaliev, D.K., Yasonov, P.G.: Coercivity spectrometer. Patent of the Russian Federation for a utility model. No 81805. Bul. FIPS No 9 (2009) (in Russian)
12. Lukshin, A.A., Kovrigo, V.P., Rumyantseva, T.I.: About magnetic properties of soils. In: Issues of genesis and rac. use of soil and fertilizer: Cr. to communicate and reference on report 4th Interstitial conference of Soil and agrochemists of the Middle Volga and South Urals, p. 92. Kazan (1968) (in Russian)
13. Boyle, J.F., Dearing, J.A., Blundell, A.: Testing competing hypotheses for soil magnetic susceptibility using a new chemical kinetic model. Geology **38**, 1059–1062 (2010)
14. Alekseev, A.O., Ryskov, Ya.G.: Magnetic susceptibility of soils as an indicator of the directivity and speed of development of steppe landscapes in the Holocene, pp. 16–20. Nature and Anthropogenic Evolution of Soils. Pushchino (1988) (in Russian)
15. Dearing, J.A., Maher, B.A., Oldfield, F.: Geomorphological linkages between soils and sediments: the role of magnetic measurements. In: Richards, K.S., Arnett, R.R., Ellis, S. (eds.) Geomorphology and Soils, pp. 245–266. Allen and Unwin, London (1985)

16. Kosareva, L.R., Nourgaliev, D.K., Kuzina, D.M., Spassov, S., Fattakhov, A.V.: Ferromagnetic, dia-/paramagnetic and superparamagnetic components of Aral Sea sediments: significance for paleoenvironmental reconstruction. ARPN J. Earth Sci. **4**, 1–6 (2015)
17. Day, R., Fuller, M., Schmidt, V.A.: Hysteresis properties of titanomagnetites: grain size and composition dependence. Phys. Earth Planet. Inter. **13**, 260–267 (1977)
18. Evans, M.E., Heller, F.: Environmental magnetism: principles and applications of enviromagnetics, 311 pp. Academic Press, San Diego (2003)

Influence of the Choice of a Geomagnetic Polarity Time Scale on Results of the Geochronological and Geohistorical Analysis of the Marine Magnetic Anomalies

Sergei A. Merkuryev, Sergei A. Ivanov and Charles DeMets

Abstract Based on the Vine-Matthews hypothesis, dozens of magnetic reversal time scales have been constructed over the past 50 years, covering the Cenozoic and part of the Mesozoic eras. Using these time scales, marine magnetic anomalies have been identified, the grids of ocean floor ages have been constructed, and plate rotation parameters have been estimated from reconstructions of the edges of magnetic isochrons. Here, we study how the choice of a geomagnetic time scale influences geochronological and geohistorical interpretations of magnetic anomalies using well-mapped marine magnetic anomalies from the Indian Ocean. For the analysis, 15 of the most widely used Cenozoic time scales were chosen. For chrons C1–C5 and C23–C26, we quantify the degree of influence of the reversal time scales on the geohistorical and kinematic analysis.

Keywords Magnetic field · Paleomagnetism · Earth's magnetic field reversals · Geomagnetic time scale · Marine magnetic anomalies

1 Introduction

Study of the history of changes in the geomagnetic field is one of the fundamental problems of geophysics and is a central problem in paleomagnetism. One of the

S. A. Merkuryev (✉) · S. A. Ivanov
Pushkov Institute of Terrestrial Magnetism of the Russian
Academy of Sciences, St. Petersburg Filial. 1 Mendeleevskaya liniya, St. Petersburg 199034, Russia
e-mail: sam_hg@hotmail.com

S. A. Ivanov
e-mail: sergei.a.ivanov@mail.ru

S. A. Merkuryev
Institute of Earth Sciences, Saint Petersburg State University, Universitetskaya nab., 7-9, St. Petersburg 199034, Russia

C. DeMets
Department of Geoscience, University of Wisconsin-Madison, Madison, WI 53706, USA
e-mail: dcdemets@wisc.edu

T. B. Yanovskaya et al. (eds.), *Problems of Geocosmos—2018*, Springer Proceedings in Earth and Environmental Sciences, https://doi.org/10.1007/978-3-030-21788-4_12

main characteristics of the Earth's magnetic field are polarity reversals, which have occurred many times in the past. Development and improvement of geomagnetic polarity time scales (GPTS), which are based on the record of field reversals that are preserved on land and in the ocean basins, is one of the important problems of geophysics. Refinements in GPTS have been nearly continuous as new data have accumulated, and improved methods of analysis have been developed.

Progress in solving a large number of geological and geophysical problems such identifying marine magnetic anomalies, determining ocean floor ages, determining the kinematic parameters of tectonic plate motion, and calculating sea floor spreading rates [1] depends critically on improvements in our knowledge of the history of

Table 1 Reversal scales considered in the present study

Time Scale Author, Date, and References	Abbreviation	Number of reversals/the shortest polarity interval, Ma	Number of calibrating points
Heirtzler et al., 1968 [5]	HDHPL68	170/0.04	2
Tarling and Mitchell, 1976 [6]	TM76	206/0.01	2
LaBrecque et al., 1977 [7]	LKC77	192/0.02	2
Mankinen and Dalrymple, 1979 [8]	MD79	196/0.02	2
Ness et al., 1980 [9]	NLC80	198/0.02	4
Lowrie and Alvarez, 1981 [10]	LA81	176/0.03	11
Harland et al., 1982 [11]	GTS82	196/0.02	6
Berggren et al., 1985 [12]	BKFV85	139/0.02	6
Kent and Gradstein, 1986 [13]	KG86	142/0.02	6
Harland et al., 1989 [14]	GTS89	196/0.02	6
Cande and Kent, 1982 [15]	CK92	186/0.01	9
Cande and Kent, 1995 [16]	CK95	350/0.007	9
Berggren et al., 1995 [17]	IMBTS95	188/0.008	9
Huestis and Acton, 1997 [18]	HA97	130/0.008	9
Gradstein et al., 2012 [19]	GTS2012	278/0.009	9

geomagnetic field. Studies of all these problems require the use of geomagnetic polarity time scales and thus depend on the GPTS that is used for a given study. Qualitative comparisons of geomagnetic polarity time scales have been carried out by many previous authors [2–4], but no previous study has fully treated the question of how the choice of a GPTS affects the estimates of seafloor spreading rates and seafloor ages when solving inverse problems. In this paper, we compare how interpretations of young and ancient sequences of magnetic anomalies in the northwestern Indian Ocean, both formed during periods of steady spreading rates, vary depending on the magnetic reversal time scale that is adopted.

Fifteen of the best known Cenozoic time scales are used for our analysis, ranging from the earliest time scale of Heirtzler et al. [5] to the most recent geologic time scale of Gradstein et al. [19]. For convenience, we have used generally accepted abbreviation of the time scales [4, 19] and literature references according to the following Table 1.

These time scales, which are depicted in Fig. 1, differ in many characteristics, including the spreading model that was used to derive them, the nomenclature of the magnetic anomalies that were included or excluded from each time scale (see, for example [7]), the tie-point dating method used for their calibration curve, the radiometric decay constant used for radiometric ages, and extrapolation method used to find ages for magnetic reversals between the tie points.

2 History

The earliest reversal time scales were based on absolute ages for reversals determined from radiometric dating of rocks using the potassium-argon method, of which the well-known Cox's time scale [20] for the past 5 million years is an example. Based on absolute age estimates for some reversals, magnetic anomaly time scales for the Cenozoic and Cretaceous periods were constructed by identifying reversals in lineated marine magnetic anomalies flanking the mid-ocean ridges. The first such time scale, which was published in 1968 [5], was derived from seafloor spreading magnetic anomalies in the South Atlantic. The authors identified magnetic anomalies C1 near the axis of seafloor spreading, to C32, with an approximate age of 80 million years. In fact, when constructing anomaly timescales, the sequence of reversals is based on a sequence of magnetic anomalies, and their age is determined by biostratigraphic data from the most ancient sediments covering the basalt basement and (or) with estimates of the absolute age of basalts [21].

Refinements to geomagnetic polarity scales since 1968 and the emergence of new scales were based on several methodological and other advances. First, the resolution of the time scales increased as short-duration (<20 k.y.) geomagnetic events (cryptochrons) were identified [22]. Second, the relative spacing of seafloor spreading magnetic anomalies was revised, leading to improvements in the structure of the time scale. Third, improved methods for calibrating time scales (the number of points and the smoothing method) were adopted. During the past two decades, the

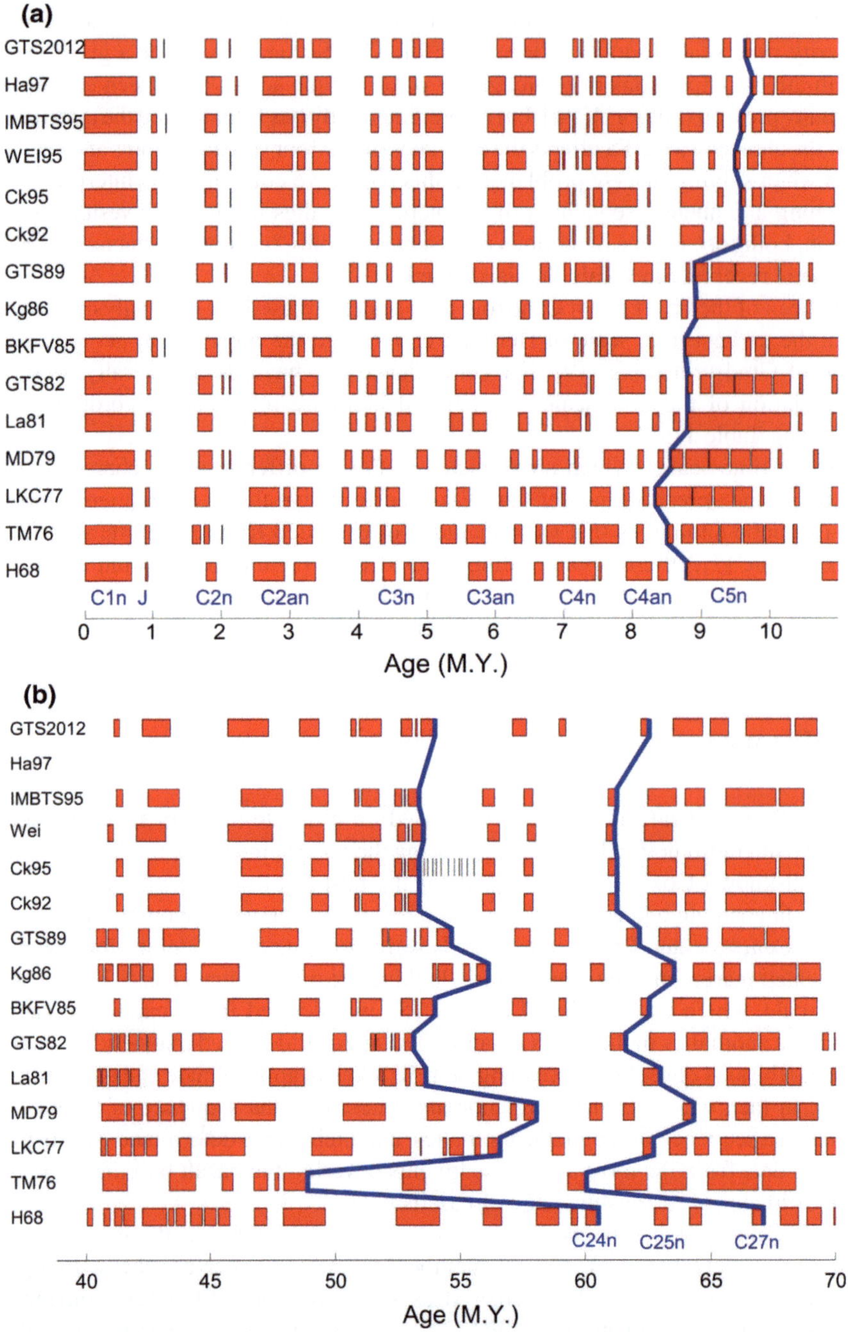

Fig. 1 Correlated paleomagnetic intervals from the studied Cenozoic time scales for 0–10 m.a. (top) and 40–70 m.a. (bottom). Red indicates intervals of normal polarity; white indicates intervals of reversed polarity. Blue tie lines are drawn between some selected anomaly boundaries

combination of radioisotopic dating and astrochronological tuning have improved the accuracy of estimated reversal ages and reduced their uncertainties, and the use of statistical methods for constructing optimal time scales have also significantly reduced uncertainties in the ages of individual chrons and their relationship with each other [23–25].

3 The Method and the Data

According to the hypothesis of Vine and Matthews, the reversal-spreading mechanism of formation of the magnetoactive layer of oceanic lithosphere provides a record of the geomagnetic field reversal history. This record can be deciphered by examining marine magnetic anomalies measured in the ocean basins. From systematic studies of marine magnetic anomalies in the northwestern part of the Indian Ocean, we have identified two sequences of marine magnetic anomalies (A1–A6 and A23–A26) that record two distinct stages of spreading. For both sequences the anomalies can be identified with high confidence using geomagnetic time scales. Based on these identifications, the age of the oceanic lithosphere for the two sequences is 0–20 Ma for A1–A6 and 53–63 Ma for A23–A26 [26, 27].

The conjugate magnetic lineations and fracture zones on each side of the ridge are used to calculate poles of rotation of the Indian and Somali plates. For the sequence A1–A6 of young anomalies (0–20 million years), a detailed kinematic model was proposed and seafloor spreading rates were calculated at one million year intervals [26]. This study shows that spreading rates decreased steadily from 20 to 10 million years and have remained steady for the last 10 million years [26]. Analysis of anomalies C24–C26 from the Arabian and Somali basins showed that seafloor spreading rates were rapid and remained steady from 63 to 53 Ma.

For this analysis, we use the same two sequences of anomalies, when spreading rates and the pole of rotation were both stable. A magnetic anomaly grid of the northwestern Indian Ocean is given on Fig. 2.

Geochronological interpretations of marine magnetic anomalies depend on matching observed and synthetic magnetic profiles, the latter calculated using a spreading block model that explores a range of possible spreading rates and different subsets of the magnetic reversal sequence. The choice of the best spreading rate and reversal sequence is based on the goodness of fit of the model to the observed profile, which is usually evaluated by eye. To carry out a quantitative comparison of all fifteen reversal time scales and assess their influence on the geochronological interpretation of a given observed profile, we calculated modeled magnetic profiles for each GPTS while varying the range of possible spreading rates and the assumed sequence of magnetic reversals in each modeled profile. A correlation coefficient between the observed magnetic profile and each possible modeled profile is calculated to quantify the goodness-of-fit between the two.

For each of the 15 reversal time scales listed in the Table, we repeated the above procedure to identify the optimal parameters from the wide range of values that

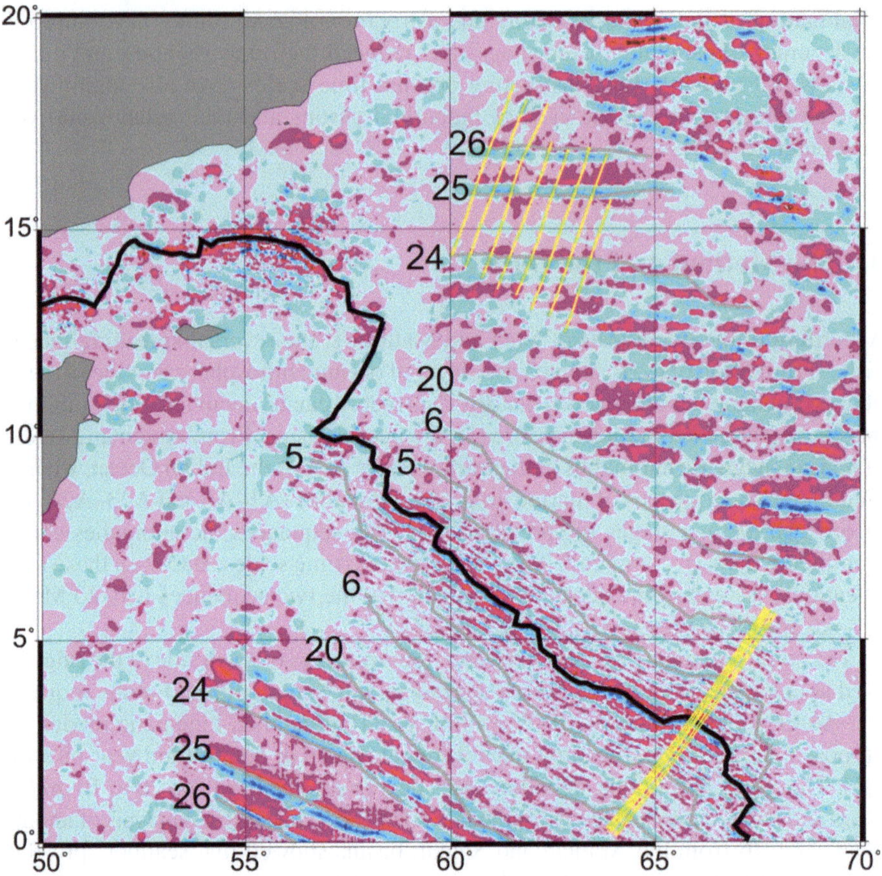

Fig. 2 Magnetic anomaly grid of the northwestern Indian Ocean. Map of linear magnetic anomalies according to [26, 27] and magnetic anomaly profiles, crossing anomalies A1–A6 and A24–A26. Depicted are the plate boundary (black bold line), magnetic lineations (light gray lines) and chron identifiers (5, 6 24–26), and magnetic profiles used for the analysis described below (yellow)

were explored for our synthetic modeling. Spreading rates were varied from 8 to 20 mm/year and possible reversal sequences between ages of 0–80 million year were explored. Readers may also refer to previous studies [27, 29] in which spreading rates and anomaly ages were estimated for these data.

Figure 3 illustrates the process of selecting the spreading rate for A1–A6 anomalies and the subset of magnetic reversals for which the correlation coefficient between the observed profile and the synthetic one is maximized. The same approach was applied to find best-fitting parameters for the A23–A26 anomaly sequence for each of the 15 GPTS included in our earlier table. Figures 4 and 5 compare the best-fitting synthetic magnetic anomaly profiles for all 15 GPTS included in our study to the observed profile for A1–A5 (Fig. 4) and A23–A26 (Fig. 5).

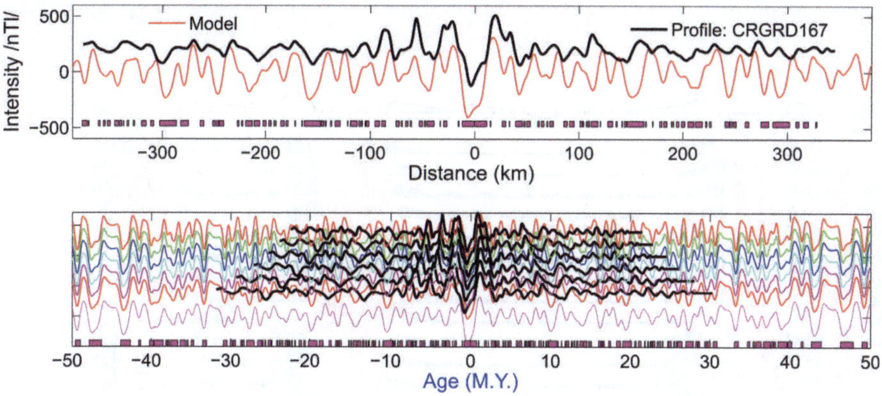

Fig. 3 The fitting procedure for a reduced-to-pole representative profile on the Carlsberg ridge for A1–A6. Theoretical magnetic anomalies calculated in the range of ages (0–50 million years) and half-spreading rate of 8–20 mm/year with a step of 1 mm/year

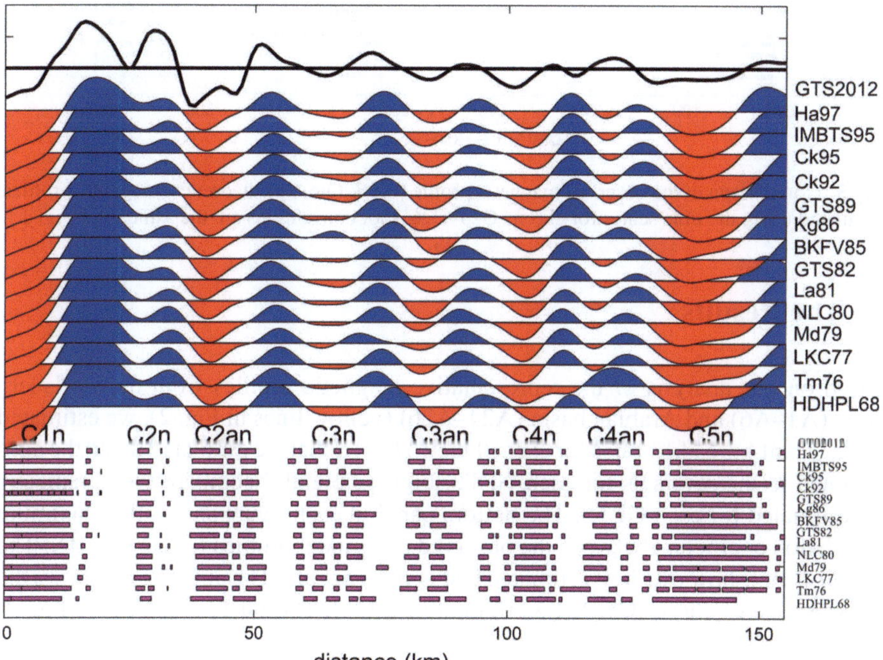

Fig. 4 Comparison of the observed representative magnetic profile (black line) and theoretical magnetic profiles, calculated using the time scales (below) for chrons C1–C5. For each reversal time scale the best spreading rate was found by fitting of the observed profile to the synthetic profiles. Blocks of normal magnetic polarity are shown in purple, reverse polarity—in white. The assigned thickness of the magnetic source layer is 0.5 km. Magnetization of the blocks is ±4 A/m with the centermost block at the ridge crest being ± 8 A/m. The names of the corresponding time scales are on the right-hand side of the figure. The chron names are found above the reversal time scales

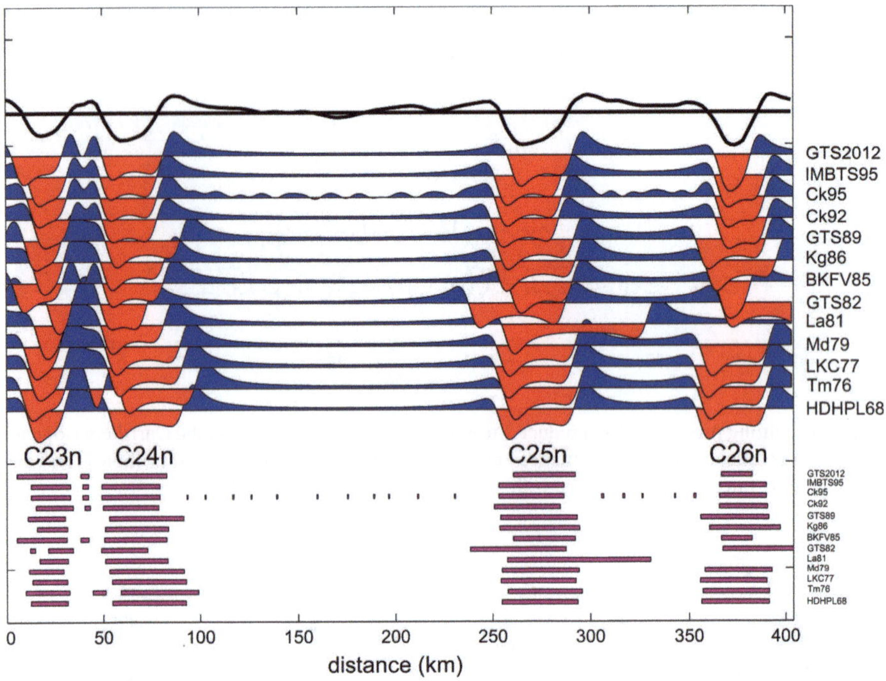

Fig. 5 Comparison of the observed magnetic profile (black line) and theoretical magnetic profiles, calculated using the time scales for chrons C23–C26. Plot conventions are the same as in Fig. 4

4 The Results

Based on our analysis of the representative magnetic profiles from the Carlsberg ridge (A1–A6) and Arabian basin (A24–A26) (yellow lines in Fig. 2), we estimated average half-spreading rates for each of the 15 GPTS that we evaluated using the ages assigned to the reversals for each GPTS. Figures 6 and 7 compare the parameters we estimated for each GPTS. The bottom panel in each figure shows the maximum correlation coefficient between the observed and model profiles for each GPTS. The middle panel shows the best estimated half spreading rate. The upper panel shows the assigned age of chrons C5n1 (Fig. 6) or chrons C23n1, C24n1, and, C26n1 (Fig. 7) for each GPTS.

Finally, we analyzed the influence of two particular reversal time scales on the history of the derived interval spreading rates (Fig. 8). For one time scale, the reversal ages were estimated using astronomical age calibrations [19] (upper). For the other, reversal ages were derived from the width of magnetic anomalies (bottom) [14]. To find the India-Somalia interval spreading rates, we used kinematic parameters from [26]. The variation in interval spreading rates over the last 10 million years is smaller for the astronomically calibrated time scale. Since the simpler spreading-rate history

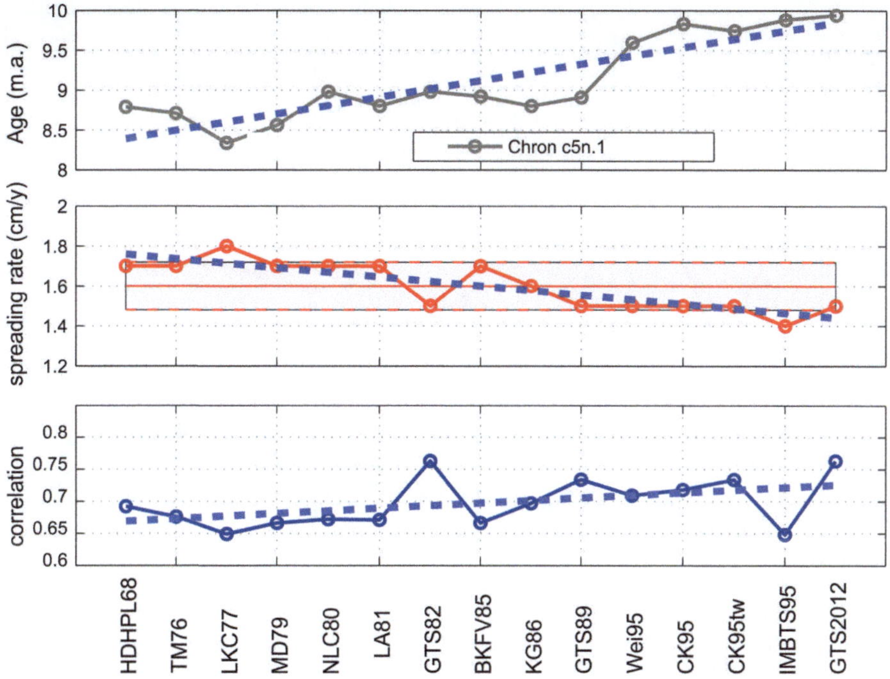

Fig. 6 The age of chron C5n.1 (top), average half-spreading rate for A1–A5 (middle) and correlation coefficient (bottom) between the observed and model profiles for each GPTS

is preferable, this suggests that reversal ages that rely on astronomical tuning are more accurate.

A similar, earlier analysis [28] used magnetic data from Pacific basin spreading centers, where interval spreading rates were estimated from high-resolution plate kinematic models using a reversal time scale based mostly on radiometric age dating and an alternative time scale with astronomical age calibration, in which the estimated reversal age uncertainties are likely to be smaller than 0.02 million years. It has been shown that the spreading rate can remain constant for several million years even for fast moving plates and the magnetic anomaly interpolation method used in [15] can cause variations in spreading rates.

In the region considered in this study, the spreading rate for anomalies A23–A27 was steady, but fast, whereas the spreading rate for A1–A5 is steady, but slow. The transition from fast to slow spreading rates explains the decrease in the spreading rate from A6 (20 m.y.) to A5 (10 m.y.), as seen in Fig. 8.

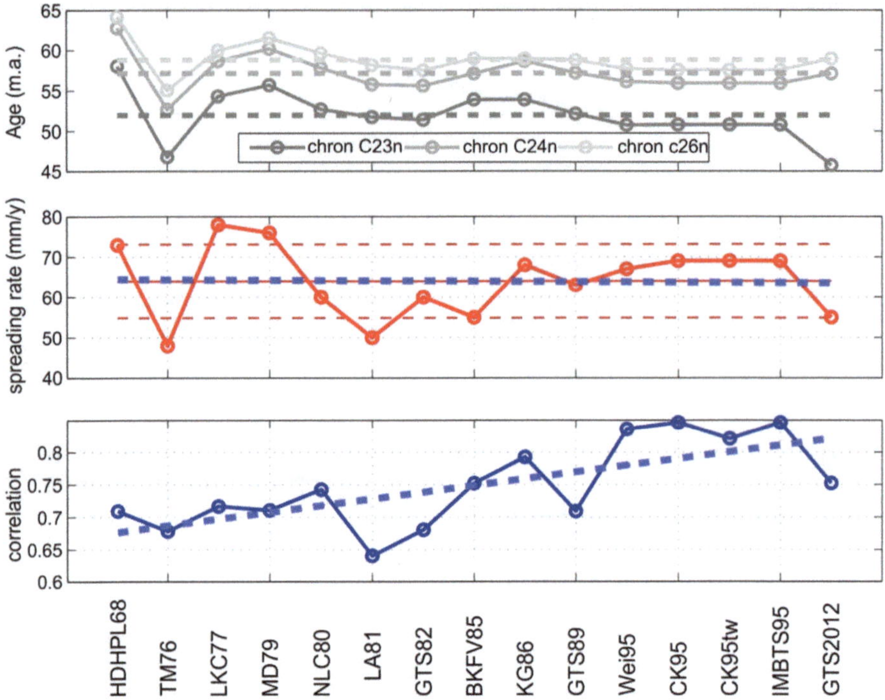

Fig. 7 The ages of chrons C23n1, C24n1, and, C26n1 (top), half-spreading rate (middle) and correlation coefficient (bottom) between the observed and model profiles for each GPTS

5 Conclusion

Ages estimated for marine magnetic reversals in different, published geomagnetic reversal time scales may differ by 15%. For younger anomalies (chrons C1n–C5n), more recently published reversal time scales suggest ages for C5 n.l that are close to 10 m.y. whereas older time scales suggest ages as young as 8.5 m.y. These imply significantly different seafloor spreading rates depending on which is more accurate. Differences in the ages estimated for magnetic reversals older than ~50 m.y. are even greater than for younger reversals.

Based on goodness-of-fit comparisons between observed magnetic profiles and synthetic magnetic profiles for 15 different reversal time scales, we show that the goodness-of-fit as measured by a correlation coefficient varies significantly depending on which reversal time scale is used to create the synthetic magnetic profile. In general, synthetic magnetic profiles that are created using more recently published reversal time scales give higher correlation coefficients between the observed and modeled profiles than for older reversal time scales, particularly for reversals younger than 20 m.y. For reversals older than 50 m.y., the highest correlation coefficients are not always determined from synthetic magnetic profiles based on the more recently

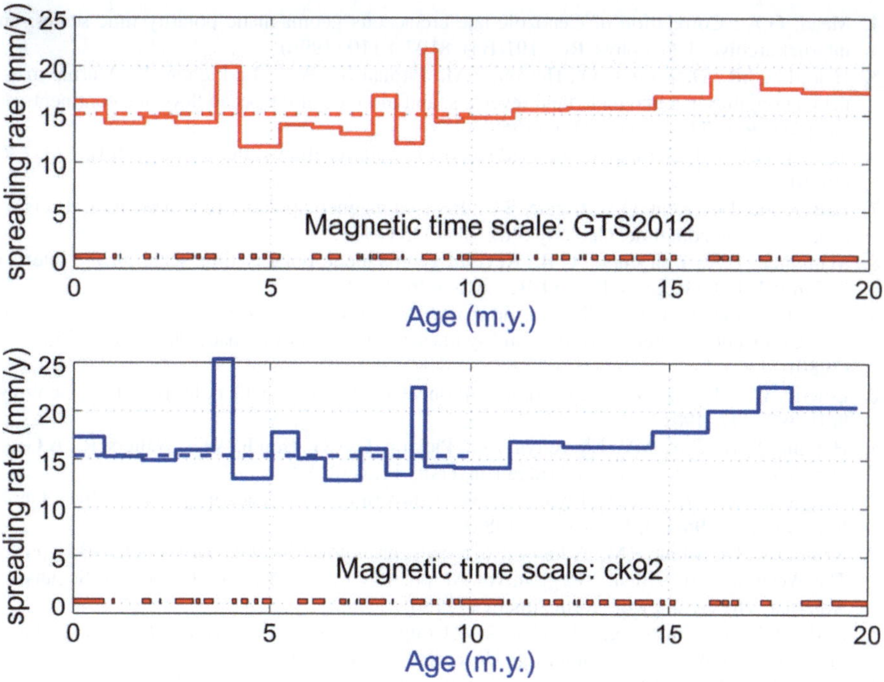

Fig. 8 India-Somalia interval spreading rates calculated by the kinematic parameters [26] for reversal time scales with astronomical age calibrations (upper) and based on the width of magnetic anomalies (lower) [30]. Red blocks indicate the interval of normal polarity, white ones indicate the interval of reverse polarity

published reversal time scales, suggesting that much work remains to improve the accuracy of ages estimated for reversals older than 50 m.y. Finally, the spreading rate history between the India and Somalia plates for the past 10 m.y. is simpler when it is derived using astronomically-tuned reversal ages. Astronomical tuning of reversal ages thus appears to improve the accuracy of reversal age estimates [19].

Acknowledgements In preparing the drawings, the GMT software package [31] was used. We are grateful to Prof. A. L. Piskarev for critically reading the manuscript.

References

1. Karasik, A.M.: Magnetic anomalies of the ocean and the ocean floor spreading hypothesis. Geotectonika 2, 3–18 (1971) (in Russian)
2. Heirtzler, J.R.: The paleomagnetic field as inferred from marine magnetic studies. J. Geomag. Geoelectr. **22**(1–2), 197–211 (1970)
3. Shreyder, A.A.: Chronology of linear marine magnetic anomalies. Izvestiya Phys. Solid Earth **29**(4), 305–319 (1993)

4. Mead, G.A.: Correlation of Cenozoic-late cretaceous geomagnetic polarity time scales: an internet archive. J. Geophys. Res. **101**(B4), 8107–8110 (1996)
5. Heirtzler, J.R., Dickson, G.O., Herron, E.M., Pitman III, W.C., Le Pichon, X.: Marine magnetic anomalies, geomagnetic field reversals, and motions of the ocean floor and continents. J. Geophys. Res. **73**(6), 2119–2136 (1968)
6. Tarling, D.H., Mitchell, J.G.: Revised Cenozoic polarity time scale. Geology **4**(3), 133–136 (1976)
7. LaBrecque, J.L., Kent, D.V., Cande, S.C.: Revised magnetic polarity time scale for Late cretaceous and cenozoic time. Geology **5**(6), 330–335 (1977)
8. Mankinen, E.A., Dalrymple, G.B.: Revised geomagnetic polarity time scale for the interval 0–5 m.y. B.P. J. Geophys. Res. **84**(B2), 615–626 (1979)
9. Ness, G., Levi, S., Couch, R.: Marine magnetic anomaly time scales for the Cenozoic and Late Cretaceous: a Precis, critique, and synthesis. Rev. Geophys. Space Phys. **18**(4), 753–770 (1980)
10. Lowrie, W., Alvarez, W.: One hundred million years of geomagnetic polarity history. Geology **9**(9), 392–397 (1981)
11. Harland, W.B., Cox, A.V., Llewellyn, P.G., Pickton, C.A.G., Smith, A.G., Walters, R.: A Geologic Time Scale, 131 pp. Cambridge University Press (1982)
12. Berggren, W.A., Kent, D.V., Flynn, J.J., Van Couvering, J.A.: Cenozoic geochronology. Geol. Soc. Am. Bull. **96**(11), 1499–1510 (1985)
13. Kent, D.V., Gradstein, F.M.: A Jurassic to recent chronology. In: Vogt, P., Tucholke, B.E. (eds.) The Western North Atlantic Region, vol. M, pp. 45–50, Boulder, CO, Geological Society of America, The Geology of North America (1986)
14. Harland, W.B., Armstrong, R.L., Cox, A.V., Craig, L.E., Smith, A.G., Smith, D.G.: A Geologic Time Scale 1989, 263 pp. Cambridge University Press (1990)
15. Cande, S.C., Kent, D.V.: A new geomagnetic polarity time scale for the Late Cretaceous and Cenozoic. J. Geophys. Res. **97**(B10), 13917–13951 (1992)
16. Cande, S.C., Kent, D.V.: Revised calibration of the geomagnetic polarity time scale for the Late Cretaceous and Cenozoic. J. Geophys. Res. **100**(B4), 6093–6095 (1995)
17. Berggren, W.A., Kent, D.V., Swisher III, C.C., Aubry, M.-P.: A revised cenozoic gochronology and chronostratigraphy. In: Berggren, W.A., Kent, D.V., Aubry, M.-P., Hardenbol, J. (eds.) Geochronology time scales and global stratigraphic correlation, pp. 129–212. SEPM, Tulsa, Oklahoma (1995)
18. Huestis, S.P., Acton, G.D.: On the construction of geomagnetic timescales from non-prejudicial treatment of magnetic anomaly data from multiple ridges. Geophys. J. Int. **129**, 176–182 (1997)
19. Gradstein, F.M., Ogg, J.G., Schmitz, M., Ogg, G.: The Geologic Time Scale 2012, 1144 pp. Boston, Elsevier (2012)
20. Cox, A.: Geomagnetic reversals. Science **163**, 237–245 (1969)
21. Gordin, V.M.: Selected Works and Memories, 138 pp. IFZ RAN (2007) (in Russian)
22. Cande, S.C., LaBrecque, J.L.: Behavior of the Earth's paleomagnetic field from small scale marine magnetic anomalies. Nature **247**, 26–28 (1974)
23. Agrinier, P., Gallet, Y., Lewin, E.: On the age calibration of the geomagnetic polarity time scale. Geophys. J. Int. **137**, 81–90 (1999)
24. Malinverno, A., Hildebrandt, J., Tominaga, M., Channell, J.E.T.: M-sequence geomagnetic polarity time scale (MHTC12) that steadies global spreading rates and incorporates astrochronology constraints. J. Geophys. Res. **117**(B6), B06104 (2012). https://doi.org/10.1029/2012jb009260
25. De Vleeschouwer, D., Parnell, A.C.: Reducing time-scale uncertainty for the Devonian by integrating astrochronology and Bayesian statistics. Geology **42**, 491–494 (2014)
26. Merkouriev, S., DeMets, C.: Constraints on Indian plate motion since 20 Ma from dense Russian magnetic data: implications for Indian plate dynamics. Geochem. Geophys. Geosyst. **7**, Q02002 (2006). https://doi.org/10.1029/2005gc001079
27. Seton, M., et al.: Community infrastructure and repository for marine magnetic identifications. Geochem. Geophys. Geosyst. **15**, 1629–1641 (2014)

28. Wilson, D.S.: Confirmation of the astronomical calibration of the magnetic polarity timescale from sea-floor spreading rates. Nature **364**, 788–790 (1993)
29. Ivanov, S.A., Merkuriev, S.A.: Preliminary results of the geohistorical and paleomagnetic analysis of marine magnetic anomalies in the Northwestern Indian Ocean. In: Nurgaliev, D., Shcherbakov, V., Kosterov A., Spassov, S., (eds.) Recent Advances in Rock Magnetism, Environmental Magnetism and Paleomagnetism, pp. 479–490. Springer, Cham (2019)
30. Lourens, L., Hilgen, F.J., Laskar, J., Shackleton, N.J., Wilson, D.: The neogene period. In: Gradstein, F., Ogg, J., Smith, A. (eds.) A Geologic Time Scale 2004, pp. 409–440. Cambridge University Press, New York (2004)
31. Wessel, P., Smith, W.H.F.: Free software helps map and display data. Eos Trans. AGU **72**, 441–446 (1991)

Preliminary Estimation of the Non-dipole Part of the Geomagnetic Field in the Quaternary Period Based on the Investigation of Marine Magnetic Anomalies on the Carlsberg Ridge

Sergei A. Merkuriev, Irina M. Demina and Sergei A. Ivanov

Abstract Marine magnetic anomalies are continuous records of the history of geomagnetic field reversals preserved in oceanic crust over the past ~160 Myr. In addition, the shape of the magnetic anomalies depends on the effective inclinations of the external magnetic field vector I_0' and the remanent magnetization vector I_r' can be numerically expressed by the skewness parameter $\theta = I_r' + I_0' - \pi$. By determining the values of theta from magnetic profiles the inclination of the remanent magnetization can be found which leads to estimates of the contribution of the averaged non-dipole field. We investigated a series of closely spaced magnetic profiles intersecting the axis of the Carlsberg Ridge near the equator, and intersecting the axial anomaly region (Brunhes normal polarity chron C1n) and adjacent anomalies (subchron C2, Matuyama chron). The observed inclination anomaly near the equator can be viewed as an evidence of presence of the quadrupole component in the averaged non-dipole field.

Keywords Geomagnetic field · Non-dipole part · Paleomagnetism · Marine magnetic anomalies

1 Introduction

One of the main principles of paleomagnetism, on which paleotectonic interpretation of paleomagnetic data is based, is the hypothesis of a geocentric axial dipole (GAD). The GAD hypothesis states that the time-averaged geomagnetic field is well

S. A. Merkuriev (✉) · I. M. Demina · S. A. Ivanov
Pushkov Institute of Terrestrial Magnetism of the Russian Academy of Sciences, St Petersburg Filial, 1 Mendeleevskaya liniya, St. Petersburg 199034, Russia
e-mail: sam_hg@hotmail.com

I. M. Demina
e-mail: irina_demina@mail.ru

S. A. Ivanov
e-mail: sergei.a.ivanov@mail.ru

S. A. Merkuriev
Institute of Earth Sciences, St. Petersburg State University, St. Petersburg, Russia

T. B. Yanovskaya et al. (eds.), *Problems of Geocosmos—2018*, Springer Proceedings in Earth and Environmental Sciences, https://doi.org/10.1007/978-3-030-21788-4_13

approximated by the field of a central axial dipole which is located in the center of the Earth and aligned with its rotation axis. More detailed studies have shown that the time-averaged field is in fact asymmetric and long-term deviations in the field geometry from the field of a geocentric axial dipole are observed [19]. This non-dipole field is a few percent of the geocentric axial dipole, which implies that non-dipole field contributions cannot be completely ignored in tectonic studies [9]. A classical method to study these effects is exploring the inclination anomaly. This value can be derived, in particular, from deep sea sediments or from marine magnetic anomalies [3]. The inclination anomaly is defined as the difference between the average paleomagnetic inclination reflected by the remanent magnetization and the local GAD inclination

$$\Delta I = I_r - I_{GAD} \tag{1}$$

Because marine magnetic anomalies record the history of the geomagnetic field (GMF), we can hope to extract information about the non-dipole field from these anomalies. Indeed, the shape of the magnetic anomalies depends on the effective inclinations of the ambient magnetic field and the remanent magnetization vectors, and can be written by the skewness (asymmetry) parameter θ, which is expressed through the paleo $\left(I_r^{'}\right)$ and ambient $\left(I_0^{'}\right)$ effective inclination as follows [20–22]:

$$\theta = I_r^{'} + I_0^{'} - \pi . \tag{2}$$

The effective inclination is defined as the angle between the Z-axis and the projection of the magnetic field on the plane normal to the strike [5].

Note that the geomagnetic intensity variations during the Bruhnes epoch recorded in the central anomaly are a symmetrical function with respect to the anomaly axis and do not disturb the symmetry of this anomaly.

If θ is known, the remanent inclination is uniquely determined and the contribution of the averaged non-dipole field can be estimated.

We have studied 28 magnetic profiles crossing the axis of the Carlsberg Ridge near the equator and covering the axial anomaly (Chron C1n, the Brunhes normal polarity epoch) and the adjacent anomalies 2 (Olduvai C2n subchron during the Matuyama epoch) [12]. A study of the anomalies along the axis of the Ridge showed that its shape varies significantly from profile to profile resulting in a high variability of the corresponding inclination anomaly.

The dispersion of the parameter θ for profiles located within the same segment, where the change in inclination has to be small, is apparently related to the effect of the bottom relief and reflects the influence of tectonic longitudinal and irregular transversal deformations with respect to the ridge axis. In order to obtain more reliable θ values we have used two approaches: (1) stacking using a newly developed algorithm [8] and (2) estimation of the values θ for each profile of this segment and averaging.

2 Possible Approaches to the Inclination Anomaly Analysis

A convenient method for studying non-dipole effects is to study the inclination anomaly [3], defined as the difference between the observed inclination of the samples and the inclination of the geocentric dipole, Eq. (1). Magnetic scalar potential is a harmonic function and, therefore, when $a \leq r$ is written in the well-known form [14]

$$\Psi(r,\theta,\varphi,t) = \sum_{l=1}^{\infty}\sum_{m=0}^{l}\left(\frac{a}{r}\right)^{l+1} \cdot \left(g_l^m(t)\cos m\varphi + b_l^m(t)\sin m\varphi\right)P_l^m(\cos\theta). \quad (3)$$

In the case of an axially symmetric paleomagnetic field, we have $m = 0$, moreover, the field on the surface can be quite well described by the sum of three terms

$$\Psi(\theta) = g_1^0 P_1(\cos\theta) + g_2^0 P_2(\cos\theta) + g_3^0 P_3(\cos\theta), \quad (4)$$

where the coefficient g_2^0 corresponds to an axial quadrupole and g_3^0 corresponds to an axial octupole. By explicitly writing down the Legendre polynomials, we have

$$\Psi(\theta) = g_1^0(\cos\theta) + g_2^0\left(3\cos^2\theta - 1)/2\right) + g_3^0\left(5\cos^3\theta - 3\cos\theta)/2\right), \quad (5)$$

The vertical component of the field is $B_z = -\frac{\partial\Psi}{\partial r}$, and the horizontal one is $B_h = -\frac{\partial\Psi}{\partial\theta}$, what gives the inclination [2, 9]

$$\tan(I) = \frac{\left[2\cos\theta + G2\left(\frac{9}{2}\cos^2\theta - \frac{3}{2}\right) + G3\left(10\cos^3\theta - 6\cos\theta\right)\right]}{\left[\sin\theta + G2(3\sin\theta\cos\theta) + \mathrm{G3}\left(\frac{15}{2}\sin\theta\cos^2\theta - \frac{3}{2}\sin\theta\right)\right]}, \quad (6)$$

For the latest IGRF model [10] we obtain the following numerical values of the coefficients appearing in (6):

$$G2 = g_2^0/g_1^0 = 244.5.1/29{,}442 \approx 0.08$$
$$G3 = g_3^0/g_1^0 = 1350.7/29{,}442 \approx 0.05. \quad (7)$$

The inclination anomaly function of the axial quadrupole field is symmetric with respect to the equator, while the contribution of the axial octupole field is antisymmetric, as shown in Fig. 1.

To estimate the magnitude of the inclination anomaly, there are several methods, among which are the following:

1. Spherical harmonic analysis of paleomagnetic data sets [2, 13]. The main disadvantage of this method is that to construct an expansion into harmonic functions one needs samples from many sites reasonably well distributed over the globe.

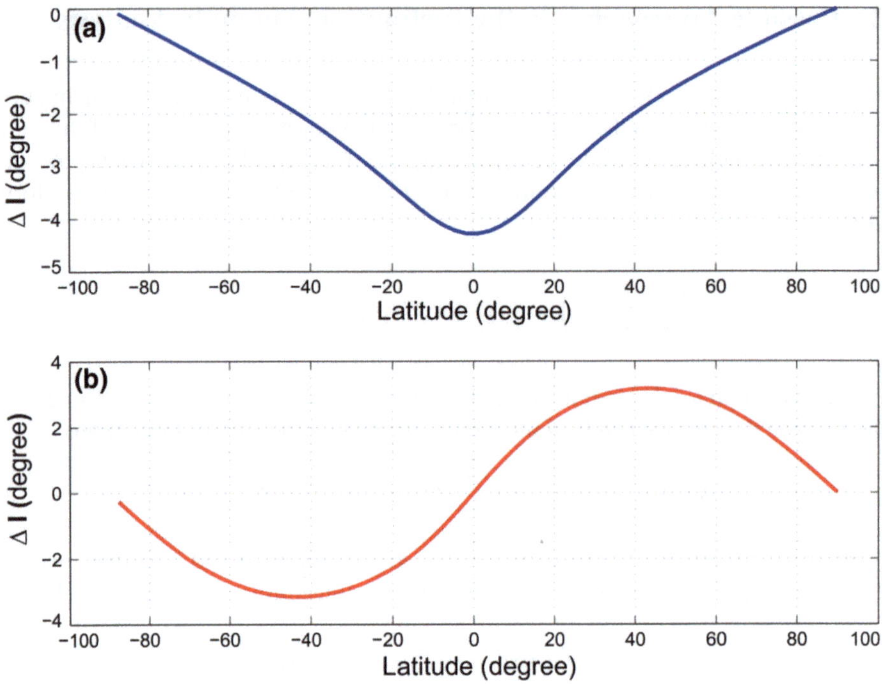

Fig. 1 Inclination anomaly as a function of latitude. **a** axial quadrupole NDF having a g_2^0/g_1^0 ratio of 0.08, **b** axial octupole NDF having a g_3^0/g_1^0 ratio of 0.05

2. The study of non-dipole effects recorded in deep-sea sediments [15, 18]. The advantage of this method is the direct measurement of the inclination of the palaeomagnetic field. On the other hand the technique of this method is more complicated and expensive than the previous one and a few data sets are available.
3. Assessment of the long-term non-dipole field using marine magnetic anomalies (skewness method) [1, 17].

In this study we have applied the last approach because we have collected the large set of marine magnetic profiles crossing the Carlsberg Ridge near the equator. Previously, these data have been used only for the geohistorical analysis.

3 Data

The magnetic field data used in this study were collected by satellite-navigated Soviet ships during a series of systematic regional surveys in the 1980s. The total intensity of the magnetic field was recorded with surface-towed standard proton precession magnetometers. The magnetic field anomaly was calculated by removing from each

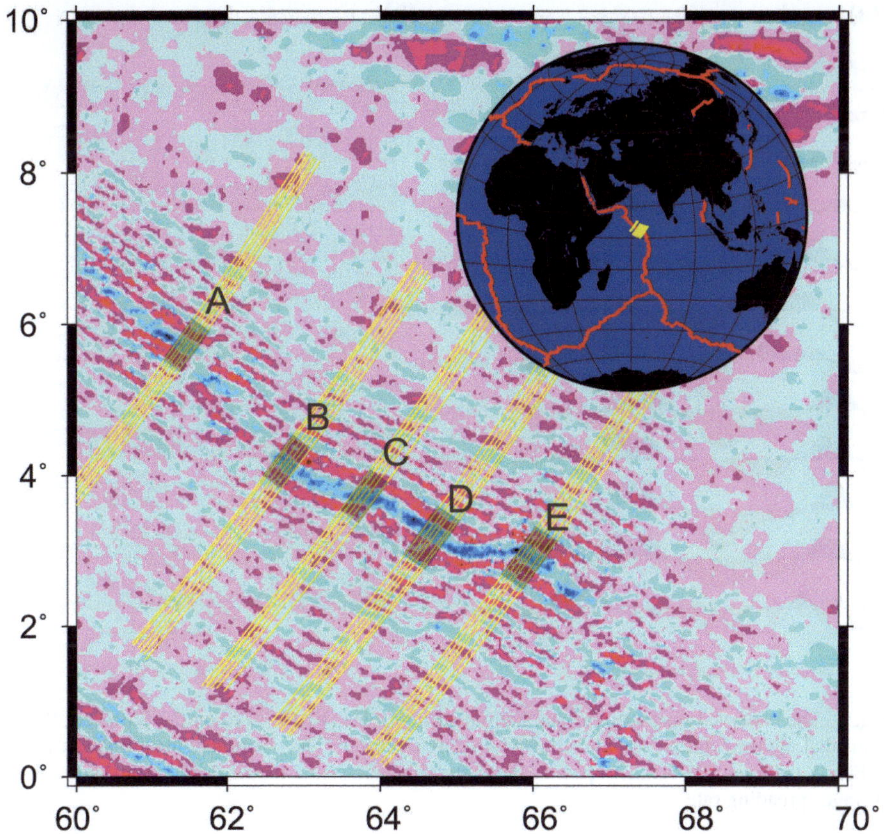

Fig. 2 Magnetic anomaly grid of the northwestern Indian Ocean in the region of the Carlsberg Ridge. Light yellow color lines indicate the position of the long profiles. The part of the profiles, including the axial anomaly and the adjacent A2 anomalies, are highlighted in grey boxes. A–E indicate midpoints of segments defined by magnetic and bathymetric results

measurement the regional magnetic field predicted by the best available International Geomagnetic Reference Field [6]. A magnetic anomaly grid has been compiled [11] and 320 ridge-normal profiles were extracted and have been studied by Merkouriev and DeMets [12] for kinematic analysis.

In Fig. 2 a map illustrating the anomalous magnetic field and the position of the profiles intersecting the axis of the ridge is given. This map demonstrates linear pattern of magnetic anomalies over the axial zone of the ridge which are symmetric with respect to the spreading axis. For a preliminary study and algorithm testing, 28 magnetic profiles have been selected.

We have studied the profiles from all spreading segments shown in Fig. 2 and below we describe our approach for the segment "E". Figure 3 shows the magnetic and bathymetric profiles observed on the Carlsberg Ridge, aligned along the ridge

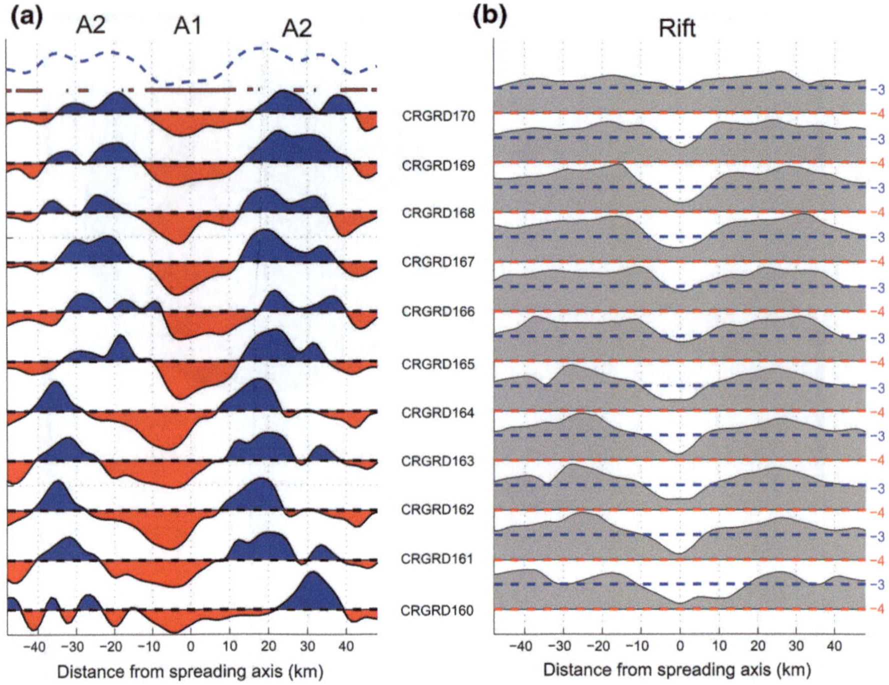

Fig. 3 The observed magnetic (**a**) and bathymetric profiles (**b**) 160–170 on the Carlsberg Ridge, aligned along the ridge axis. Anomalies A1–A2 are indicated. The position of the profiles is shown in Fig. 2. (segment E). At the top of **a**, the dotted line shows the model profile calculated for the average spreading rate

axis and attached to it. In the upper part of Fig. 3a above the magnetic profiles, a model profile is shown, calculated at the constant spreading rate. This spreading rate is average of ones obtained by the method of correlation coefficient maximizing. As one can see, on profiles with a length of about 100 km, it is possible to trace the axial anomaly located in the center, and anomalies 2 to the right and to the left from the center. The observed profiles lying in the middle of the segment have a characteristic shape and correspond to the model anomaly. From the other side the shape of the anomalies near to the faults differs from the model one. So, the correlation coefficient between the model and observed profiles decreases from 0.9 in the middle to 0.6 at the segment boundary.

This is due to the fact that the axis of the ridge is divided by numerous transform and non-transform offsets, even for an individual axial segment. The localized disturbances in the magnetic field attributable to the axial offsets causes the seafloor spreading magnetic anomalies to lose their characteristic shape and thus differ from the model profiles, which are calculated based on a simple 2D-structure. In addition, lateral displacements of the entire anomaly sequence along transform faults are observed.

The bathymetric profiles crossing the axis of the Carlsberg Ridge are shown in Fig. 3b. The relief of the axial part of the ridge has a typical slow-spreading ridge structure with a deep and wide rift valley and rift mountains that flank the axial valley. The bathymetric profiles demonstrate that the ocean floor in general is symmetrical with respect to the axis of the ridge, which on average is located in the center of the rift valley.

Magnetic and bathymetric profiles in most cases can be used for paleomagnetic interpretation except near fracture zones and transform faults, where the magnetic anomaly loses linearity and is significantly distorted. A number of further reasons can change the shape of the magnetic anomalies observed on profiles crossing the ridge axis. The shapes of model magnetic anomalies, including axial, are mainly determined by (1) the spreading rate, (2) the ridge strike, (3) the paleolatitude of the ridge position and (4) its strike during the formation of the observed anomaly. During the last 1.2 Ma (age of anomalies) no significant changes in the above parameters have likely occurred, and they can be set to their modern values. The variation in the spreading rate along the segment is less than 1% and is not so significant as to lead to a change in the shape of the anomalies.

For geochronological and paleomagnetic interpretation of marine magnetic anomalies, it becomes necessary to select profiles to be interpreted, as well as the application of procedures that reduce the noise superimposed on a useful paleomagnetic signal.

4 Method

The geochronological and paleomagnetic interpretation of marine magnetic anomalies is partly based on an interactive method: from several synthetic magnetic profiles calculated using a theoretical spreading model, we select one that best matches the observed profile. The model profiles are calculated for the given spreading rates, the trend of anomalies and the GMF parameters (inclination and declination) at the observation site. In the case of paleomagnetic analysis of anomalies, the desired parameter is also a paleomagnetic inclination, information about which can be obtained by analyzing the shape of magnetic anomalies.

The shape of marine magnetic anomalies, or its asymmetry (skewness), is determined by the parameter θ, which is expressed through the paleo and ambient effective inclination by Eq. (2).

The ambient effective inclination is defined as

$$I_0' = \tan^{-1}(\tan I_0 / \sin(A - D)) \tag{8}$$

where I_0 and D are respectively the inclination and declination of the GMF at the observation point which can be estimated from the International Geomagnetic Reference Field [10]. A, the azimuth of the strike of the linear magnetic anomalies, is measured from the direction to the north and can be determined by the magnetic map or calculated as an angle of 90° to the direction of the spreading, which can be estimated from [4]. The effective remanent inclination is not known, but we can estimate θ (see below) and find it from Eq. (2) and ultimately estimate the magnitude of the inclination anomaly

$$\Delta I = I_r - I_{GAD} = \tan^{-1}\left(\tan I_r' \cdot \sin(A_r - D_r)\right) - \tan^{-1}(2\tan\varphi). \quad (9)$$

Here I_{GAD} is the inclination of the geocentric axial dipole. To obtain the inclination anomaly we assume that the ancient azimuth and remanent declination coincide with the modern ones.

Initially, several spreading segments were identified on the Carlsberg Ridge, where a high correlation between the magnetic profiles was observed. Paleomagnetic analysis of the profiles belonging to each of these segments was carried out according to the following scheme. The first step was to align the profiles to obtain an averaged profile (stack) with the best signal-to-noise ratio [7]. Then both the observed profiles and the obtained stack were analyzed, but only in those parts that cover the anomalies under investigation (A1 and A2). Changing the position of the profiles and of the stack relative to the model profile at different spreading rates, these were tied to that part of the geomagnetic time scale where the correlation coefficient between the stack and the model profile was maximal for both the spreading rate and the time scale portion [4].

Figure 4 shows an example of the selection of a model profile for one of the profiles observed in the spreading segment "E" during which the time scale section and the spreading rate and asymmetry parameter that best fit the observed profile were determined. The search for the desired parameters was carried out in a wide range of values: a portion of the time scale for the search was chosen from 0 to 7 Ma, the half spreading rate varied from 9 to 21 mm/year with a step of 1 mm/year, and the skewness parameter from 140° to 220° with a step of 5°.

According to preliminary theoretical estimates [4] the full spreading rate, in the considered section of the Carlsberg Ridge, is 25–30 mm/year, so that the axial anomaly width should be 20–24 km. The observed width of the axial anomaly is slightly higher and varies from 25 to 35 km. This discrepancy probably results from a shift of the reversal boundary caused by a non-verticality of the boundaries of the magnetized blocks, which we estimate on average to be 3.1 km [12]. Although the age of the axial anomaly does not cause doubts, a more extended part of the time scale was chosen for the search. For all values of the spreading rate the axial anomaly was clearly identified.

In the following we describe the details of the phase shifting procedure, shown in the Fig. 5. On Fig. 5a we plot the model profiles (grey) at the pole and phase shifted observed profiles (red). We optimize the phase shift parameter by three methods: (1) by maximizing the correlation coefficient, (2) by minimizing the standard deviation, and (3) by a "pseudogravity" method [16].

The curves in the top panel of Fig. 5b demonstrate that different methods yield different values. In the end, we chose the average of these values, which was usually

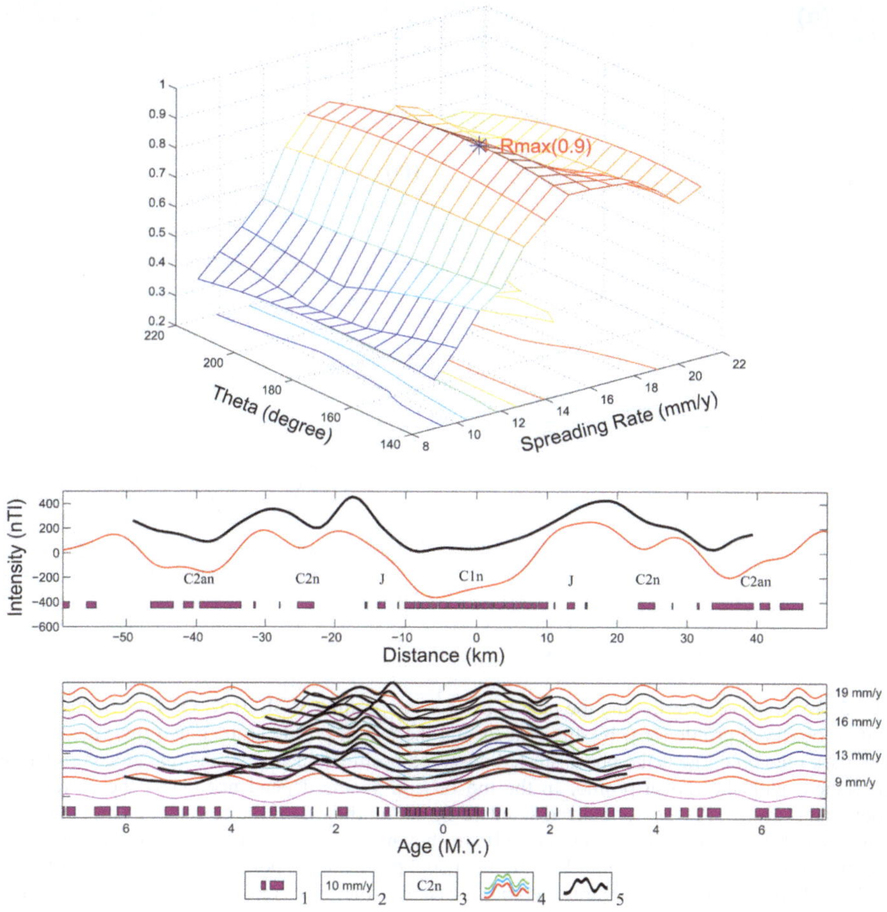

Fig. 4 The procedure for selecting the average half-spreading rate for a representative profile CRGRD167 on the Carlsberg Ridge for anomalies 1–2. The lower panel shows examples of theoretical magnetic anomalies were calculated for 7 Ma to the present assuming half-spreading rates of 8–21 mm/year in increments of 1 mm/year. For each spreading rate, a model profile was calculated for the best fitted θ (bottom). The middle panel displays the modeled profile for the best-fitting spreading rate and θ. The upper panel displays a surface plot of the correlation coefficient of the model and observed profile as a function of the spreading rate and θ. 1—time scale where magenta indicates an interval of normal polarity; white indicates an interval of reversed polarity, 2—spreading rate, 3—chron number, 4—model profiles calculated for the given spreading rate, 5—best fitting model profile for the given spreading rate

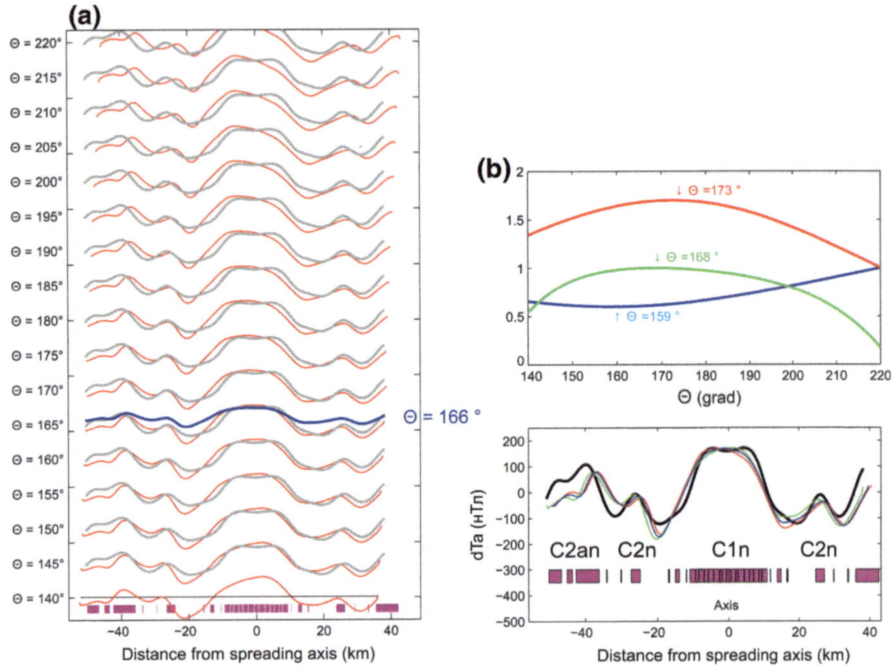

Fig. 5 Example of determining the phase shift for the profile CRGRD167. On Fig. 5a the phase-shifted observed profile (red) by different θs and the model profile at the pole (grey) are shown. The dark blue bold line is the phase-shifted observed profile, which gives the maximal correlation coefficient with the model profile. In the top of Fig. 5b the dependence (1) of the correlation coefficient (green), (2) the normalized standard deviation (blue), (3) the correlation coefficient for functions of the pseudogravity method (red) are shown for the phase-shifted observed profile and the model profile. In the bottom of Fig. 5b the model profile at the pole (black) and the best fitting deskewed observed profiles for three methods are shown (the colors red, green, and blue correspond to ones at the top). Also, the blocks of direct polarity with the names of the chrons are shown

close to the value determined visually. In the bottom panel of Fig. 5b the model profile at the pole and the best fitting deskewed observed profiles are shown for three methods.

To date, there exists no universal automatic method for determining the phase shift parameter. This is due to difficulties in developing an objective criterion for the symmetry of anomalies and the development of a corresponding algorithm [8]. The development of such a criterion and the corresponding algorithm is a separate task that is beyond the scope of this study.

For each segment of the Carlsberg Ridge the procedure for determining the phase shift was applied to all profiles belonging to this segment. Two estimates of inclination anomaly were obtained for each segment: (1) the average inclination anomaly of the profiles belonging to a given segment, and (2) the inclination anomaly for the stack of these profiles. Figure 6 displays the inclination anomaly of all profiles of Segment

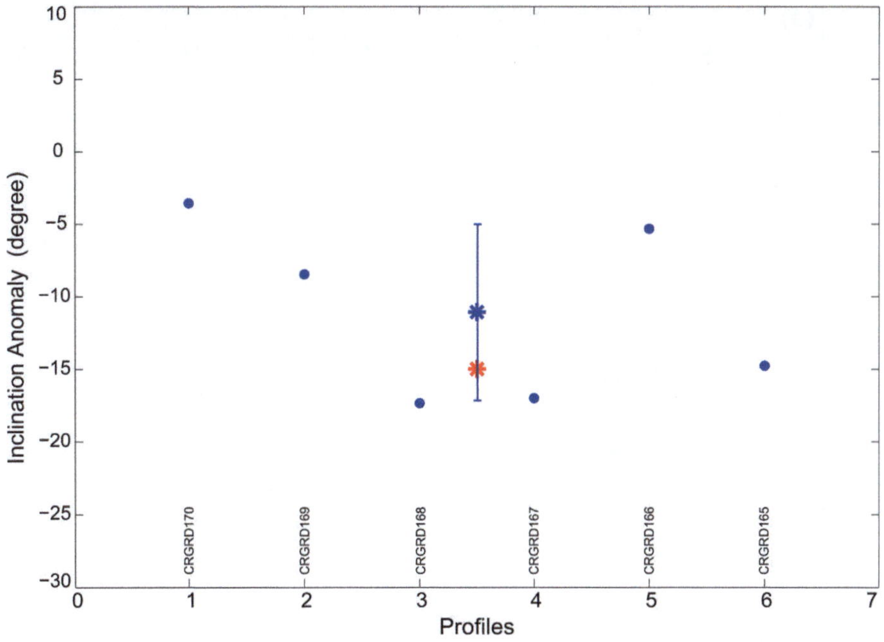

Fig. 6 The inclination anomaly for the segment "E". The blue circles indicate the inclination anomaly of profiles CRGRD165-CRGRD170. The blue asterisk indicates the average inclination anomaly. The red asterisk shows the inclination anomaly for the stack

'E', the average inclination anomaly, and the inclination anomaly for the stack profile of this segment.

5 Results

We have carried out a paleomagnetic interpretation of magnetic profiles intersecting the Carlsberg Ridge near the equator. For profiles from each segment, the magnitude of the phase shift was determined, from which the values of the inclination anomaly were calculated. The results of the preliminary study are presented for all segments shown in Fig. 7. The red asterisk shows the average value of the inclination anomaly of the segments. Also on the bottom, these values and its average are shown. The latitude of the segment was determined as the average of the points of intersections of the observed profiles with the axis of the Carlsberg Ridge. The blue line is the inclination anomaly of the field with the quadrupole contribution (see Eq. 4) giving the inclination anomaly at the equator equal to of 12.5° (compare with Fig. 1). Thus, it can be assumed that the quadrupole component makes a significant contribution to the non-dipole part. The estimate of the ratio $G2 = g_2^0/g_1^0$ is 0.15 ± 0.03 exceeding

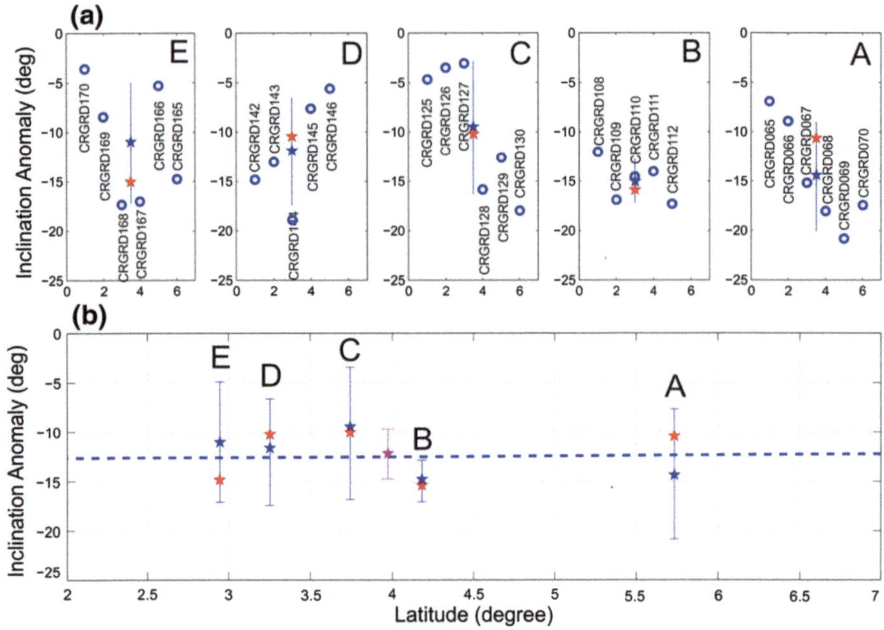

Fig. 7 The inclination anomaly. **a** For each segment (A–E), the inclination anomaly is shown for all profiles of these segments. The blue asterisk shows the average inclination anomaly of the segments. The red asterisk shows the inclination anomaly for the stack. **b** Average inclination anomaly for all 5 segments (blue asterisk) and the stack average (magenta asterisk) are shown with errors. The dashed blue line shows the theoretical inclination anomaly from Eq. 4, which equals -12.5° at the equator.

its present-day value equal to 0.08 [10]. Note that it is impossible to estimate the octupole coefficient g_3^0 from observations at Carlsberg Ridge, because the octupole field contribution to the inclination anomaly is antisymmetric with respect to the equator and therefore should be very small at low latitudes.

Our preliminary estimates of the inclination anomaly thus reveal the presence of a quadrupole component in the time-averaged non-dipole field. Our estimates of the inclination anomaly have the same sign as in [13], but its magnitude is significantly larger than obtained from previous paleomagnetic investigations of sediments in the Indian Ocean [18] and magnetic anomaly profiles from the Galapagos Ridge [17].

The errors of the inclination anomaly, which we obtained from the marine magnetic anomalies are about the same as obtained by a similar method at the Galapagos spreading center [17]. Further research will focus on the development of an effective method for estimating the magnitude of the phase shift to minimize scatter. This will include an attempt to take into account the distorting effect of the sea-floor relief.

Acknowledgements In preparing the plots, the GMT program package [23] was used. We are grateful to Dr. Karl Fabian and Dr. Andrei Kosterov for critically reading the manuscript and constructive reviews.

References

1. Acton, G.D., Petronotis, K.E., Cape, C.D., Ilg, S.R., Gordon, R.G., Bryan, P.C.: A test of the geocentric axial dipole hypothesis from an analysis of the skewness of the central marine magnetic anomaly. Earth Planet Sci. Lett. **144**, 337–346 (1996)
2. Ben'kova, N.P., Khramov, A.N., Cherevko, T.N., Adam, N.V.: Spherical harmonic analysis of the paleomagnetic field. Earth Planet. Sci. Lett. **18**, 141–147 (1973)
3. Cox, A.: The frequency of geomagnetic reversals and the symmetry of the non dipole field. Rev. Geophys. **13**(3), 35–51 (1975)
4. DeMets, C., Gordon, R.G., Argus, D.F.: Geologically current plate motions. Geophys. J. Int. **181**(1), 1–80 (2010)
5. Gay, S.P.: Standard curves for interpretation of magnetic anomalies over long tabular bodies. Geophysics **28**, 161–200 (1963)
6. International Association of Geomagnetism and Aeronomy (IAGA): Division V, Working group 8, international geomagnetic reference field 2000. Geophys. J. Int. **141**, 259–262 (2000)
7. Ivanov, S.A., Merkuryev, S.A.: Some remarks on resolving and interpretation of short wavelength marine magnetic anomalies. In: Proceedings of the 11th International School and Conference "Problems of Geocosmos", SPb.: VVM Publishing, pp. 37–44 (2016)
8. Ivanov, S.A., Merkuriev, S.A.: Preliminary results of the Geohistorical and Paleomagnetic analysis of marine magnetic anomalies in the northwestern Indian Ocean. In: Nurgaliev, D. et al. (eds.) Recent Advances in Rock Magnetism, Environmental Magnetism and Paleomagnetism. International Conference on Geomagnetism, Paleomagnetism and Rock Magnetism (Kazan, Russia), pp. 479–490. Springer International Publishing (2019)
9. Langel, R.A., Estes, R.H.: A geomagnetic field spectrum. Geophys. Res. Lett. **9**, 250–253 (1982)
10. Macmillan, S., Maus, S.: International geomagnetic reference field-the tenth generation. Earth Planets Space **57**, 1135–1140 (2015)
11. Merkouriev, S.A., Sotchevanova, N.A.: Structure and evolution of the Carlsberg Ridge: evidence for nonstationary spreading on old and modern spreading centers. Current Sci. **85**, 334–338 (2003)
12. Merkouriev, S., DeMets, C.: Constraints on Indian plate motion since 20 Ma from dense Russian magnetic data: implications for Indian plate dynamics. Geochem. Geophys. Geosyst. **7**, Q02002 (2006). https://doi.org/10.1029/2005gc001079
13. Merrill, R.T., McElhinny, M.W.: Anomalies in the time averaged paleomagnetic field and their implications for the lower mantle. Rev. Geophys. Space Phys. **15**, 309–323 (1977)
14. Merrill, R.T., McElhinny, M.W., McFadden, P.L.: The Magnetic Field of the Earth, p. 531. Academic Press, New York (1996)
15. Opdyke, N.D.: Paleomagnetism of deep-sea cores. Rev. Geophys. **10**(1), 213–249 (1972)
16. Roest, W.R., Arkani-Hamed, J., Verhoef, J.: The seafloor spreading rate dependence of the anomalous skewness of marine magnetic anomalies. Geophys. J. Int. **109**(3), 653–669 (1992)
17. Schneider, D.A.: An estimate of the long-term non-dipole field from marine magnetic anomalies. Geophys. Res. Lett. **15**, 1105–1108 (1988)
18. Schneider, D.A., Kent, D.V.: Inclination anomalies from Indian Ocean sediments and the possibility of a standing nondipole field. J. Geophys. Res. **93**(B10), 11621–11630 (1988)
19. Schneider, D.A., Kent, D.V.: The time-averaged paleomagnetic field. Rev. Geophys. **28**(1), 71–96 (1990)
20. Schouten, H.: A fundamental analysis of magnetic anomalies over oceanic ridges. Mar. Geophys. Res. **1**(2), 111–144 (1971)
21. Schouten, H., McCamy, K.: Filtering marine magnetic anomalies. J. Geophys. Res. **77**, 7089–7099 (1972)
22. Schouten, H., Cande, S.C.: Palaeomagnetic poles from marine magnetic anomalies. Geophys. J. R. Astron. Soc. **44**, 567–575 (1976)
23. Wessel, P., Smith, W.H.F.: Free software helps map and display data. Eos Trans. AGU **72**, 441–446 (1991)

Simple Planetary Convection and Magnetism Estimations via Scaling and Observations

S. V. Starchenko

Abstract Scaling laws describing planetary convection are considered with the purpose to obtain a simple formulation of equations governing turbulent convection in the critical regime close to the convection onset. This justifies, in a certain way, the available numerical models of the planetary convection and MHD dynamos that are too simple and nevertheless successful. Such dynamos are considered here for known terrestrial planets and moons. The arithmetic mean magnetic field (AMF) in the dynamo region is ~1 mT, and AMF vector follows closely the rotation axis for the Earth, Jupiter and Saturn. The Uranus and Neptune have large inclinations and AMF ~0.1 mT. Similar value is believed to hold for Ganymede. The smallest AMF ~0.01 mT is in Mercury. The geodynamo root-mean-squared field (RMF) to AMF ratio is estimated from 10 to 100 from scaling laws. The same order values may hold for Jupiter and Saturn while other planets might have smaller ratios. RMF/AMF ratio can also be viewed as a simple measure of the internal planetary magnetic field intermittency.

Keywords Planetary magnetism and convection · Turbulent transport · Scaling laws · Observed planetary magnetism · Geodynamo

1 Introduction

Satellite geomagnetic and planetary observations only cover a few decades; observatory measurements span 185 years. Reliable direct historical observations lasted less than 500 years [1–3]. However, the geodynamo naturally operates at larger time scales [4], leaving the archaeomagnetic [5, 6] and paleomagnetic methods [7–9] an only option to study the evolution of the geomagnetic field. Unfortunately, these methods provide us with a rather limited information that is not sufficient to solve

---

S. V. Starchenko (✉)
Pushkov Institute of Terrestrial Magnetism, Ionosphere and Radio Wave Propagation, Russian Academy of Sciences (IZMIRAN), Kaluzhskoe hwy 4, Troitsk, Moscow 108840, Russia
e-mail: sstarchenko@mail.ru

T. B. Yanovskaya et al. (eds.), *Problems of Geocosmos—2018*, Springer Proceedings in Earth and Environmental Sciences, https://doi.org/10.1007/978-3-030-21788-4_14

geodynamo problem. In a larger part, a solution to this problem has unexpectedly come from the numerical geodynamo models that remained rather unrealistic almost up to the end of the 20th century. The situation has changed radically with the publication of [10] where a first-principle numerical model showing a close similarity to the geodynamo has been constructed, despite the key parameters being far from a planetary MHD dynamo. To date, hundreds of similar models have been developed [1, 11–17]. These models are widely used in modeling/forecasting the dynamics and morphology of both the observed and reconstructed planetary magnetic field. The fact that simplistic numerical dynamo simulations produce an Earth-like magnetic field despite not all of their parameters are realistic has already been documented [12, 17]. The researchers found important to have realistic relations between timescales to obtain an Earth-like dynamo behaviour. The first goal of this paper is to partly justify this fact, base on that turbulent planetary dynamo parameters should be near-critical for the convection to start.

In the part 2 the planetary-type convection scaling and critical solutions from [18] are used to make new turbulent estimations. This allows obtaining simple estimates for turbulent [19]. Reynolds-type transport coefficients; those are significantly higher than the commonly used molecular values. This sufficiently removes the deviation of the cited numerical models from the parameters related to viscosity and diffusion when the turbulent convection is not as strongly supercritical as was believed so far.

Based on Sect. 2 results and well-known satellite magnetic observations new estimates (the second goal of the paper) are presented in Sect. 3 for typical and critical convective parameters, for the degree of turbulence and corresponding transport coefficients characteristic of the Earth's core and of cores of the known terrestrial planets. Some [20–22] scenarios of convectively driven dynamos in the Earth and presumably in the Moon, Venus and Mars are discussed. A special type of convection possibly diving Mercury and Ganymede dynamos is considered that could be supplemented or/and dominated by strong tidal effects from the Sun or Jupiter respectively.

The third goal of this work presented in the Sect. 4 is to relate an observed magnetic planetary dipole to the internal arithmetic mean magnetic field (AMF hereafter) vector hidden in the region where MHD dynamo is working. The simple (and exact!) relation (5) between the two easily derived from a textbook formula (e.g., Davidson [23]) has been apparently overlooked so far. The Earth, Jupiter and Saturn have those mean fields of the order of 1 mT, and deviation of their dipoles from the axis of rotation is relatively small. It is thus not surprising that the geodynamo-type models are successful for all these otherwise rather different planets [1, 10, 12–17, 24–27]. In contrast, Uranus and Neptune have AMF ~ 0.1 mT and their dipoles are strongly inclined [24, 28] requiring some modifications in modeling their dynamos [1, 2, 12, 14, 15, 24, 25, 29–31]. A similar AMF value is found for Ganymede (moon of Jupiter), but its observed magnetic field is almost symmetric with respect to the rotation axis and rather asymmetric with respect to the equator [20, 24]. A like configuration is observed for Mercury where AMF ~ 0.01 mT [11, 20, 24].

The root-mean-squared field (RMF hereafter) in the dynamo region is estimated in Sect. 5. The RMF for geodynamo is about a few milliTesla as estimated from the well-known Christensen-Aubert scaling law [1] confirmed and modified in recent studies [2, 12–17, 22, 27]. This yields the geodynamo RMF/AMF ratio of the order of 10. The same ratio may be expected for Jupiter and Saturn, however, a recent scaling law from [2] yields the ratio of the order of 100 for such a system. This is consistent with some older geodynamo/planetary-dynamo models [32] and with independent observational estimates [33]. Uranus, Neptune and Ganymede perhaps have RMF just slightly (up to a few times) exceeding AMF field in accordance with [34] and some other scaling laws [2]. Intermittency of the internal planetary magnetic field is also briefly introduced and defined as measured by the RMF/AMF ratio employing a simple relation between the latter and turbulence characteristics obtained via observations and scaling.

2 Scaling, Criticality and Turbulence for Convection

Assuming planetary parameters as in [35, 36] for reaching the extreme range, I consider a convection in a plane layer based on [18]. The liquid layer with thickness D rotates with angular velocity Ω in the gravitational field with acceleration g anti-parallel to z-axis. The liquid has kinematic viscosity ν, diffusion λ, diffusivity κ, coefficients of thermal and compositional expansion α and β (for the Earth's core, about $\pm 10^{-5}$/K and $+0.6$, respectively [26, 37, 38]). The initial temperature and concentration are respectively $(D-z)T'$ and $(D-z)\Xi'$. As in the Earth, the convection instability is dominated by compositional effects characterized by Prandtl number μ. Effect of temperature either leads to an additional instability (at $\alpha > 0$) or stabilizes convection (at $\alpha < 0$). Using relatively too small [36, 38] transport coefficients and very fast rotation [18] represented by a typical velocity U, new (δ, r, s) and classical (E, Ra) similarity factors can be written as

$$U = \Omega D \delta^2,\ \delta = \frac{\sqrt{g\beta\Xi'}}{\Omega} = E\sqrt{\frac{Ra}{\mu}},\ r = \frac{\alpha T'}{\beta\,\Xi'}$$

$$s = \frac{\nu\,\Omega^2/D^2}{(g\beta\Xi')^{3/2}} = E^{-2}\left(\frac{\mu}{Ra}\right)^{3/2} \tag{1}$$

Here the new factors are r for the thermal/composition effect ratio, s for the effective dissipation/generation ratio and δ for the relative size perpendicular to the rotation axis. The classical Ekman number $E = \nu/(\Omega D^2)$, taken ~10^{-15} for the molecular viscosity value in the Earth's core as determined from the first principles [43]. The classical Rayleigh number $Ra = g\beta\,\Xi' D^4/(\nu\lambda)$, is determined by the compositional convection, and is ~10^{27} for the commonly accepted molecular diffusion of the light admixture in the Earth's core [20, 38]. With such extreme values of these key geodynamo similarity factors, the real geodynamo will cannot be modeled because of computational limitations. Use of new similarity coefficients $\delta \sim 10^{-3}$ (instead of Ekman number $E \sim 10^{-15}$) and $s \sim 10^{-6}$ (instead of $1/Ra \sim 10^{-27}$) from (1) as prefactors for higher order derivatives appears more promising. In this approach, the solution is likely within reach of sufficiently powerful computers even for the real geodynamo parameters. This simplified system could be easily generalized on a planetary spherical layer based on the approach first suggested in [39].

Following new results of this paper are based on already somewhat standard [40, 41] and inspired by [19] estimation of unified turbulent transport coefficient

$$kUD\delta \approx \nu \approx \kappa \approx \lambda. \tag{2}$$

Here k is a numerical coefficient. The new optimization here is that due to (1–2) this k is simply equal to s. Therefore, those numbers are estimated as

$$0 < k = s < s_{cr}. \tag{3}$$

Here, s_{cr} is a critical number. Convection exists at $s < s_{cr}$ only. Critical (index 'cr') values from [18] are related (σ is the thermal Prandtl number) as

$$s_{cr}^2 = \frac{4}{27}\frac{(\mu + r\sigma)^3}{\pi^4} = \frac{\mu^3}{E^4 Ra_{cr}^3}, \tag{4}$$

For pure compositional ($r = 0$) and turbulent ($\mu = 1$) convection $s_{cr} = 0.04$. In a realistic spherical shell geometry, more parameters and complexities are added determined by the convective instability profile and the relative inner/outer shell radii. However, for the typical terrestrial-like turbulent parameters s_{cr} is still about a few hundredths, as follows, for example, from the scaling $Ra_{cr} \sim E^{-4/3}$ [39, 42] for the case where a terrestrial-like convection was investigated in spherical shells. In this case, the supercriticality of turbulent convection is only slightly (by a factor of $\sim s_{cr}/s$) greater than the critical level. This is perhaps one of the important arguments in favor of the successful (22-year after [10]) numerical modeling of the geodynamo-like systems. Another important fact is that the coefficients of turbulent transport (2) are several orders of magnitude higher than the corresponding molecular quantities. For example, s is of the order of 10^{-2}, turbulent viscosity/diffusion is 0.02 m^2/s and turbulent Ekman number is $E = 3 \times 10^{-11}$. The latter can be achieved in the nearest future for the known geodynamo-like systems using a sufficiently powerful supercomputer. At the same time, the corresponding simplified convective system with the realistic turbulent $s \approx 0.01$ and even molecular $s \approx 10^{-6}$ parameters of the geodynamo-like systems and other planets can in principle be numerically solved even now.

3 Critical Convection Estimations for the Terrestrial Planets

The Earth's thermal Prandtl number σ is 10^{-2} and the compositional Prandtl number μ is 10^3 [37, 38, 43]. The modulus of r is less than one because compositional convection dominates over the thermal one in the Earth and similar terrestrial planets. Due to $\mu >> \sigma$ the uncertainty in r barely affects the critical molecular value of (4) which is $s_{cr} = 1200$. This enormous value for the critical dissipation/generation ratio indicates that convection is extremely easily generated in the Earth's core and is highly nonlinear as the realistic molecular $s \sim 10^{-6}$ is by nine (!) orders of magnitude lower. Thus, the small-scale convection can be naturally considered as extremely chaotic or turbulent with huge Reynolds numbers $\mathrm{Re} = UD/\kappa = 10^9$. Corresponding large-scale turbulent values σ and μ are about 1 as follows from (2). If one neglects the thermal contribution assuming $r = 0$, the critical turbulent value from (4) will be $s_{cr} = 0.04$. Similar moderate values are also obtained using all other probable values of r from (1). They are likely to be significantly higher than -1, but rather smaller than 1 in the cores of planets similar to the Earth. However, on approaching $r = -1$, non-standard weakly nonlinear or almost critical modes with nearly laminar flows become possible. This extremely strong stabilization (r is slightly above -1) is possible in the interiors of some terrestrial planets. For Mercury having an unusually configured magnetic field, this is partly supported by numerical [11] and asymptotic [44] modeling. A similar situation is also probable for the Jovian satellite Ganymede. Thus, Mercury and Ganymede dynamos could have a special type of 'unconventional' convection where the compositional instability is almost stabilized by thermal effects. For the present-day Moon, Venus, Mars, and most of the large moons, the r value is probably about -1, and in this case convection cannot be excited at all. In the deep past, a non-Earth-like compositional convection could possibly support rather strong MHD dynamos with small $|r|$ in Moon and Mars, while too slowly rotating Venus could never develop the convection intensity necessary to initiate the dynamo [20, 22].

4 Observed Dynamo AMF (Arithmetic Mean Magnetic Field)

It is natural to define the vector $< \mathbf{B} >$ of AMF in the planetary MHD dynamo region W as a correspondent volume integral divided by the spherical region's radius c. According to the textbook definition [23] of the magnetic dipole vector $\mathbf{M}$ via the internal magnetic field $\mathbf{B}$ AMF is equal to the observed $\mathbf{M}$ divided by the volume of W multiplied by $3/2\mu_0$:

$$< \mathbf{B} > \equiv \frac{\int_W \mathbf{B} d^3 r}{4\pi\ c^3/3} = \frac{2\mu_0 \mathbf{M}}{4\pi\ c^3}\,. \tag{5}$$

Table 1 Arithmetic mean magnetic field in planets and moons

Planet/moon	Dipole moment M, A m^2	Surface a, m	Dynamo c, m	< B >, T
Earth	8.0E+22	6.37E+06	3.50E+06	3.7E−04
Jupiter	1.6E+27	6.99E+07	6.22E+07	1.3E−03
Saturn	4.7E+25	5.82E+07	2.91E+07	3.8E−04
Mercury	5.0E+19	2.44E+06	2.00E+06	1.2E−06
Ganymede	1.9E+20	2.63E+06	7.90E+05	7.6E−05
Uranus	3.8E+24	2.54E+07	2.03E+07	9.1E−05
Neptune	2.0E+24	2.46E+07	1.97E+07	5.2E−05
Venus	≤4.0E+18	6.05E+06	3.03E+06	2.9E−08

M is an observable magnetic dipole moment, a—well-known radius of the planet/moon, c—radius of the dynamo region, < B >—AMF modulus evaluated from Formula (5). Values of M and c are taken from [14, 20, 22, 24, 28]

The available observed planetary dipole moments and the estimated radii [14, 20, 22, 24, 28] of dynamo regions were used in Formula (7) to evaluate < B > for all known planets and moons, listed in the Table 1. From geometric considerations, at the outer dynamo region boundary the AMF modulus is larger than the RMS dipole field by a factor 1.4.

From this Table 1, we see that planets and moons with internal magnetic fields form distinct groups. The first "geodynamo-type" group including the Earth, Jupiter and Saturn is characterized by AMF of the order of 1 mT and a small to nearly absent, in the case of Saturn, angle between the dipole axis and the axis of rotation. It is no surprise then that similar dynamo models are successful for these planets [1, 10, 12–17, 24–27].

The second group including Uranus and Neptune is characterized by AMF ~ 0.1 mT and a dipole strongly inclined to the rotation axis [24, 28] requiring significant, and still not clear yet, modifications in the respective dynamo models. Ruzmaikin and Starchenko [29, 31] proposed a thin-layer kinematic mean-field dynamo in these ice giants. This led to a successful prediction of the Neptune's dipole inclination [30]. More recently Christensen, Schubert, Yadav and others [1, 2, 12, 14, 15, 24, 25] have developed more sophisticated models capturing the essential physics of dynamo action.

AMF is roughly about 0.1 mT or maybe less for Ganymede. Its observed field is almost symmetric with respect to the rotation axis but rather asymmetric with respect to equator, so that the effective dipole appears quite strongly off-centered [24, 28]. Very similar spatial configuration of the magnetic field is observed for Mercury [11, 24, 28], but its AMF ~ 0.01 mT is the smallest of all planets and moons having the internal magnetic field. This poses serious problems for modeling exacerbated by large uncertainties in configurations of dynamo regions for these objects [11, 20, 24].

5 RMS Magnetic Field (RMF) from Scaling Laws

The AMF is typically sufficiently lower than the RMF in the entire range of known planetary-type MHD dynamo models [1–6, 10–18, 21, 22, 24–27]. This could be attributed to the cancelation of differently directed internal and somewhat chaotic vector fields **B** during the integration in (5).

The geodynamo RMF typical value B could be about a few milliTeslas as estimated from the well-known and relatively well-tested Christensen and Aubert [1] scaling law:

$$B = \sqrt{\mu_0 \rho}\ \sqrt[3]{HF} \tag{6}$$

Here μ_0 is the magnetic permeability of vacuum (SI constant), ρ—average density of the liquid dynamo region (10 Mg/m^3 for the Earth), H—half-thickness of the region (1 Mm in the Earth) and $F = P/M$—relative convection power driving the dynamo, expressed via its total power P (0.3–3 TW for the Earth) and mass of the conducting liquid M. For the Earth, Formula (6) gives the RMF/AMF ratio ~ 10. The same value might be expected for Jupiter and Saturn. However, it is not entirely certain that (6) holds for all geodynamo-type systems [2, 20, 24, 26, 36, 37]. An alternative recent scaling law [2] expresses B via the angular rotation rate Ω as

$$B = \sqrt{\mu_0 \rho}\ \sqrt[5]{H^3 \Omega^2 F} \tag{7}$$

The correspondent RMF/AMF ratio is ~100 for Earth- or Jupiter-like systems. This is consistent with some older geodynamo/planetary dynamo models [32] and with independent observational estimations [33].

According to Starchenko [2], the scaling laws (6) and (7) yield the magnetic field that is considerably larger than the fields expected from other laws suitable for non-geodynamo-type planets. Thus, these planets could have a smaller RMF/AMF ratio in comparison with the geodynamo type planets. Uranus, Neptune and Ganymede perhaps have RMF just slightly (up to a few times) exceeding AMF. This is consistent with some scaling laws applicable to these planets from [2, 12, 34].

Besides, the intermittency of the internal planetary magnetic field could be defined and measured by the respective RMF/AMF ratio. At the moment there is seems no better instrument to evaluate a planetary magnetic field intermittency. Thus, the above analyses suggest a strongly intermittent or extremely turbulent magnetic field for the geodynamo-type planets.

6 Discussion and Conclusions

The estimates of "turbulent" quantities like (2) assume isotropy stemming from the stochasticity of turbulence which, however is not a very well defined and understood concept [19, 40, 41, 45, 46]. By contrast, rigorous asymptotic methods applied to idealized small-scale flows yield full anisotropic tensors of turbulent diffusion [46, 47] (as well as of the magnetic and kinematic α-effects), which can sometimes drive large-scale instabilities rather than enhance molecular diffusivity [48, 49] (the so-called phenomenon of negative eddy diffusivity). All these and similar issues were not addressed in this paper because it focused on the simplest possible estimations. However, an idea of optimal scaling, linking exact critical evaluations with approximate turbulent estimations could be implemented in the mentioned rigorous asymptotic methods. Here, this allowed obtaining simple estimates for turbulent transport coefficients; those are significantly higher than the commonly used molecular values. This decreases somewhat a gap between viscosity and diffusion values used in existing numerical models, and the realistic ones. Further simplification is achieved due to the fact that turbulent convection is not as strongly above its critical level as was believed so far. The major remaining problem is estimation of a characteristic turbulent value of s from (1) and (3).

It is demonstrated here that the observed magnetic planetary dipole is related to the internal AMF vector hidden in the region where MHD dynamo is operating. The simple and exact relation (5) has been apparently overlooked so far. Using it, planets/moons with internal magnetic fields are found to form distinct groups that could be investigated further. The RMF in the dynamo region is roughly estimated in this paper as well for the Earth, Uranus, Neptune and Ganymede. Further work is necessary to find better estimations those could be used to evaluate an intermittency of the internal planetary magnetic field. At present, it is only possible to suggest that intermittency is rather large for the geodynamo-type systems, and moderate for the other planetary dynamo-systems. This could be linked to the extremely turbulent state of geodynamo-type systems and not so for the others. This paper is perhaps for the first time presents combined observational and modeling evidence for an extremely turbulent geodynamo. The suggested intermittency measure RMF/AMF cannot not be used when quadrupole dominating over dipole. It is because AMF would be too small in this case.

Let me summarize the key results of this work.

1. Introduced here similarity factors and correspondent scaling laws describing planetary convection are considered with the purpose to obtain a simple formulation of equations governing turbulent convection. The latter is argued to operate in a close to critical regime for the convection onset. This vindicates, in a certain way, the available numerical models of the planetary convection and magnetic dynamos that are too simple and nevertheless successful. Such dynamos are only possible at sufficiently powerful convection, which is here considered for the terrestrial planets and moons of the Solar System.
2. The arithmetic mean magnetic field (AMF) vector in the dynamo region is ~1 mT and only slightly deviates from the rotation axis for the Earth, Jupiter and Saturn. The Uranus and Neptune have AMFs ~ 0.1 mT and large dipole inclinations. Similar AMF value is obtained for Ganymede, while the smallest AMF ~ 0.01 mT is in Mercury.

3. The geodynamo RMS field is ~1 mT as estimated from Christensen and Aubert scaling [1]. That gives RMS to arithmetic mean field ratio ~10, while scaling from [2] gives ~100. Similar values are likely for Jupiter and Saturn, while the ice giants, Ganymede, and Mercury are expected to show a smaller RMF/AMF ratio. Intermittency of the internal planetary magnetic field could also simply be defined and estimated from RMF/AMF ratio.

Acknowledgements I am very grateful to the anonymous referee who not only gave me very valuable remarks to improve this work, but also shared with me valuable references and a few pieces of text I enjoyed to use. This work was basically supported by IZMIRAN budget. Partial supports in planetary magnetism were by 28th program of the Presidium of Russian Academy of Sciences. My special thanks to the editor of this volume Dr. Andrei Kosterov for his efforts to improve my text.

References

1. Christensen, U.R., Aubert, J.: Scaling properties of convection-driven dynamos in rotating spherical shells and application to planetary magnetic fields. Geophys. J. Int. **166**, 97–114 (2006)
2. Starchenko, S.V.: Analytic scaling laws in planetary dynamo models. Geophys. Astroph. Fluid Dyn. **113**(1–2), 71–79 (2019)
3. Jackson, A., Jonkers, A.R.T., Walker, M.R.: Four centuries of geomagnetic secular variation from historical records. Phil. Trans. R. Soc. Lond. **A358**, 957–990 (2000)
4. Ogg, J.G., Ogg, G., Gradstein, F.M.: The Concise Geologic Time Scale, 150 pp. Cambridge University Press (2008)
5. Korte, M., Constable, C.G.: Continuous geomagnetic field models for the past 7 millennia: 2. CALS7 K. Geochem. Geophys. Geosyst. 6, Q02H16 (2005). https://doi.org/10.1029/2004GC000801
6. Korte, M., Constable, C.: Improving geomagnetic field reconstructions for 0–3 ka. Phys. Earth Planet. Int. **188**, 247–259 (2011)
7. Khramov, A.N.: Paleomagnetology. In: Tarling, D.H. (ed.) 308 pp. Springer, Berlin (1987)
8. Macouin, M., Valet, J.-P., Besse, J.: Long-term evolution of the geomagnetic dipole moment. Phys. Earth Planet. Int. **147**, 239–246 (2004)
9. Tarduno, J.A., Cottrell, R.D., Davis, W.J., Nimmo, F., Bono, R.K.: A Hadean to Paleoarchean geodynamo recorded by single zircon crystals. Science **349**, 521–524 (2015)
10. Glatzmaier, G.A., Roberts, P.H.: A three-dimensional convective dynamo solution with rotating and finitely conducting inner core and mantle. Phys. Earth Planet. Int. **91**, 63–75 (1995)
11. Christensen, U.R.: A deep dynamo generating Mercury's magnetic field. Nature **444**, 1056–1058 (2006)
12. Christensen, U.R.: Dynamo scaling laws and applications to the planets. Space Sci. Rev. **152**, 565–590 (2010)
13. Christensen, U., Aubert, J., Hulot, G.: Conditions for Earth-like geodynamo models. Earth Planet. Sci. Lett. **296**, 487–496 (2010)
14. Olson, P., Christensen, U.R.: Dipole moment scaling for convection-driven planetary dynamos. Earth Planet. Sci. Lett. **250**, 561–571 (2006)
15. Yadav, R.K., Gastine, T., Christensen, U.R.: Scaling laws in spherical shell dynamos with free-slip boundaries. Icarus **225**, 185–193 (2013)
16. Matsui, H., et al.: (37 authors): performance benchmarks for a next generation numerical dynamo model. Geochem. Geophys. Geosyst. **17**, 1586–1607 (2016)

17. Aubert, J., Gastine, T., Fournier, A.: Spherical convective dynamos in the rapidly rotating asymptotic regime. J. Fluid Mech. **813**, 558–593 (2017)
18. Starchenko, S.V.: Scaling and excitation of combined convection in a rapidly rotating plane layer. J. Exp. Theor. Phys. **124**, 352–357 (2017)
19. Reynolds, O.: On the dynamical theory of incompressible viscous fluids and determination of the criterion. Phil. Trans. Roy. Soc. London **186**, 123–161 (1895)
20. Breuer, D., Rueckriemen, T., Spohn, T.: Iron snow, crystal floats, and inner-core growth: modes of core solidification and implications for dynamos in terrestrial planets and moons. Prog. Earth Planet Sci. **2**(39) (2015). https://doi.org/10.1186/s40645-015-0069-y
21. Starchenko, S.V.: Hypothetical parameters of planetary dynamos deduced from direct observations, scaling laws, paleomagnetic and isotope reconstructions. In: Semenov V.S. et al. (eds.) Proceedings of the 11th International School and Conference "Problems of Geocosmos" (Oct. 03–07, 2016, St. Petersburg, Russia), pp. 64–72 (2016) (in Russian with English abstract)
22. Starchenko, S.V.: Energy geodynamo parameters compatible with analytical, numerical, paleomagnetic models and observations. Izv., Phys. Solid Earth **53**, 908–921 (2017)
23. Davidson, P.A.: An Introduction to Magnetohydrodynamics. Cambridge Texts in Applied Mathematics (2006)
24. Schubert, G., Soderlund, K.M.: Planetary magnetic fields: observations and models. Phys. Earth Planet. Inter. **187**, 92–108 (2011)
25. Starchenko, S.V.: Analytic base of geodynamo-like scaling laws in the planets, geomagnetic periodicities and inversions. Geomag. Aeron. **54**, 694–701 (2014)
26. Starchenko, S.V., Jones, C.A.: Typical velocities and magnetic field strengths in planetary interior. Icarus **157**, 426–435 (2002)
27. Starchenko, S.V., Pushkarev, Yu.D.: Magnetohydrodynamic scaling of geodynamo and a planetary protocore concept. Magnetohydrodynamics **49**, 35–42 (2013)
28. Ness, N.F.: Space exploration of planetary magnetism. Space Sci. Rev. **152**, 5–22 (2010)
29. Ruzmaikin, A.A., Starchenko, S.V.: On the origin of Uranus and Neptune magnetic fields. Icarus **93**, 82–87 (1991)
30. Ruzmaikin, A.A., Starchenko, S.V.: Generation of a large-scale magnetic field in Uranus and Neptune by a turbulent dynamo. Cosmic Res. **27**, 242–247 (1989)
31. Starchenko, S.V.: Generation of nonaxisymmetric magnetic structures in galaxies, the Sun, and planets. Astron. Rep. **39**, 227–231 (1995)
32. Braginsky, S.I., Roberts, P.H.: A model-Z geodynamo. Geophys. Astrophys. Fluid Dyn. **38**, 327–349 (1987)
33. Gillet, N., Jault, D., Finlay, C.C.: Planetary gyre, time-dependent eddies, torsional waves, and equatorial jets at the earth's core surface. J. Geophys. Res. Solid Earth **120**, 3991–4013 (2015)
34. Starchenko, S.V.: Energetic scaling of the geodynamo. Geomag. Aeron. **53**, 243–247 (2013)
35. Cheng, J.S., Stellmach, S., Ribeiro, A., Grannan, A., King, E.M., Aurnou, J.M.: Laboratory-numerical models of rapidly rotating convection in planetary cores. Geophys. J. Int. **201**, 1–17 (2015)
36. Cheng, J.S., Aurnou, J.M.: Tests of diffusion-free scaling behaviors in numerical dynamo datasets. Earth Planet Sci. Lett. **436**, 121–129 (2016)
37. King, E.M., Buffett, B.A.: Flow speeds and length scales in geodynamo models: the role of viscosity. Earth Planet. Sci. Lett. **371**, 156–162 (2013)
38. Terasaki, H., Fischer, R.A.: Physics and Chemistry of the Lower Mantle and Core, 312 p. Wiley (2016)
39. Starchenko, S.V., Kotelnikova, M.S.: Critical stability of almost adiabatic convection in fast rotating and wide spherical shell. J. Exp. Theor. Phys. **116**, 338–345 (2013)
40. Smagorinsky, J.: General circulation experiments with the primitive equations I. The basic experiments. Mon. Weather Rev. **91**(3), 99–164 (1963)
41. Frick, P.G.: Turbulence: Approaches and Models, 2nd edn. RDC "Regular and Chaotic Dynamics", 332 p. Moscow, Izhevsk (2010) (in Russian)
42. Starchenko, S.V., Kotelnikova, M.S., Maslov, I.V.: Marginal stability of almost adiabatic planetary convection. Geophys. Astrophys. Fluid Dyn. **100**, 397–428 (2006)

43. Pozzo, M., Davies, C., Gubbins, D., Alfe, D.: Transport properties for liquid silicon-oxygen-iron mixtures at Earth's core conditions. Phys. Rev. B. **87**, 014110 (2013)
44. Bassom, A.P., Soward, A.M., Starchenko, S.V.: The onset of strongly localized thermal convection in rotating spherical shells. J. Fluid Mech. **689**, 376–416 (2011)
45. Moffatt, H.K.: Magnetic Field Generation in Electrically Conducting Fluids, 343 p. Cambridge University Press, Cambridge (1978)
46. Dubrulle, B., Frisch, U.: Eddy viscosity of parity-invariant flow. Phys. Rev. A **43**, 5355–5364 (1991)
47. Zheligovsky, V.A.: Large-scale perturbations of magnetohydrodynamic regimes: linear and weakly nonlinear stability theory. Lecture Notes in Physics, vol. 829, 330 p. Springer, Heidelberg (2011)
48. Lanotte, A., Noullez, A., Vergassola, M., Wirth, A.: Large-scale dynamo by negative magnetic eddy diffusivities. Geophys. Astrophys. Fluid Dyn. **91**, 131–146 (1999)
49. Rasskazov, A., Chertovskih, R., Zheligovsky, V.: Magnetic field generation by pointwise zero-helicity three-dimensional steady flow of an incompressible electrically conducting fluid. Phys. Rev. E. **97**, 043201 (2018)

Part III
Seismology

Atlas of Aftershock Sequences of Strong Earthquakes

A. V. Guglielmi, O. D. Zotov and A. D. Zavyalov

Abstract This year marks the 150th birthday anniversary of the outstanding seismologist Fusakichi Omori. Our paper is devoted to this significant event. We found that the well-known Omori law may be represented in the form of a differential equation describing the evolution of aftershocks. This allows us to formulate the inverse problem of physics of the earthquake source which is "cooling down" after the main shock. The paper gives the examples of solving the inverse problem and illustrates a possibility to create the atlas of aftershocks after a series of strong earthquakes. The atlas contains a description of the parameters, the original sequence of aftershocks, and the so-called deactivation function for each event. The analysis of the atlas showed a rich variety of the evolution forms of the earthquake source after the main shock.

Keywords Seismology · Omori law · Aftershocks equation · Deactivation function · Earthquake source · Inverse problem · Phase portrait

1 Introduction

Our work is devoted to the memory of the outstanding Japanese scientist Fusakichi Omori (1968–1923). In 1896, he discovered a hyperbolic dependence of the frequency of aftershocks on time:

$$n(t) = k/(c + t). \tag{1}$$

A. V. Guglielmi · A. D. Zavyalov
Institute of Physics of the Earth, RAS, Moscow, Russia
e-mail: guglielmi@mail.ru

A. D. Zavyalov
e-mail: zavyalov@ifz.ru

O. D. Zotov (✉)
Borok Geophysical Observatory of Institute of Physics of the Earth, RAS, Borok, Yaroslavl, Russia
e-mail: ozotov@inbox.ru

T. B. Yanovskaya et al. (eds.), *Problems of Geocosmos—2018*, Springer Proceedings in Earth and Environmental Sciences, https://doi.org/10.1007/978-3-030-21788-4_15

And this was the first empirical law of the physics of earthquakes [1]. Here the parameter k is the most important integral characteristic of the earthquake source that "cools" after the main shock. The parameter c is of no interest to us in the context of this work. In essence, (1) is the one-parameter formula.

So, the Omori law is represented by a rather rigid one-parameter formula. Many authors have worked to make the Omori law more flexible. The idea of R. Hirano and T. Utsu to replace the denominator in Formula (1) with the expression $(c+t)^p$ was widely spread in the literature [2, 3]. Unlike the Omori formula, the Hirano-Utsu formula is two-parameter.

We have chosen a different path, and our idea is this. On the one hand, we do not want to introduce additional phenomenological parameters that do not have a geodynamic meaning. That is, we want to leave the Omori formula with one parameter. On the other hand, we would like to take into account the nonstationarity of rocks in the earthquake source after the main shock. In other words, we must bear in mind that the value of k depends on the time, thereby reflecting the nonstationary relaxation of the of the earthquake source to a new metastable state.

However, it is impossible to replace k with $k(t)$ in Formula (1). We will do otherwise. First, we make the following substitution $k = 1/\sigma$. This is just a redefinition. We will call σ the deactivation coefficient. The value of σ tells us how fast the earthquake source loses its ability to excite the aftershocks. Now we use the evolution equation of aftershocks

$$dn/dt + \sigma n^2 = 0, \tag{2}$$

which is completely equivalent to the Omori law (1), if sigma does not change over time. The advantage of the law in the form of a differential equation (2) is that now nothing prevents us from taking into account the nonstationarity of the geological environment after the main shock. We solve the equation and obtain a generalization of the law of aftershocks evolution:

$$n(t) = n_0 \left[1 + n_0 \int_0^t \sigma(t') dt' \right]^{-1} \tag{3}$$

Our version (3) preserves the hyperbolic structure of the Omori law, but it takes into account that time in the earthquake source, figuratively speaking, flows unevenly [4]. If $\sigma = \text{const}$, then in this and only in this case the generalized Formula (3) coincides with the classical Omori Formula (1) up to notation.

2 Inverse Problem of the Aftershocks Physics

We show that the generalized law of evolution (3) is useful for processing and analyzing of the aftershock sequences. Let us rewrite (3) in the form

$$\int_0^t \sigma(t')dt' = \frac{1}{n(t)} - \frac{1}{n_0} \equiv g(t). \tag{4}$$

Here $g(t)$ is an auxiliary function. It is known from experiment. This mathematical expression resembles the Volterra integral equation of the first type (with trivial kernel). This circumstance prompted us the idea to set and solve the inverse problem, i.e. to find the unknown deactivation factor σ by using the known frequency of aftershocks n.

Like almost every inverse problem, our inverse problem is incorrectly posed. Regularization consists in smoothing the auxiliary function: $g(t) \to \langle g(t) \rangle$. Solution of the problem is of the form

$$\sigma(t) = d\langle g(t) \rangle / dt. \tag{5}$$

3 Atlas of Aftershocks

According to the results of solving the inverse problem, we plan to release an "Atlas of Aftershocks" (ATAS). Currently, about a dozen solutions have been accumulated and in this section we want to show the first results. Let us say that we conceived ATAS as a collective project.

Figure 1 gives an idea about the content of the atlas [5]. Here are three events that occurred in California according to the catalogs http://www.data.scec.org and http://www.ncedc.org. Aftershock frequencies $n(t)$ are shown in the upper rows. Here are the dates and magnitudes of the main shocks. The middle panels show auxiliary functions $g(t)$ before and after regularization. The bottom panels give an idea of the variety of forms of the deactivation function $\sigma(t)$. We observe growing, decreasing and oscillating fragments of $\sigma(t)$. Let us pay special attention to the so-called Omori epoch, i.e. for long time intervals, when the value of σ is constant and, thus, the Omori law (1) is strictly satisfied. The duration of Omori epoch varies from case to case in a wide range. According to the presence or absence of the Omori epoch and other characteristic properties of the function $\sigma(t)$ one can try to make a classification of earthquake sources in the course of further work.

One more interesting event is shown in Fig. 2. In Southern California at a close (~1 km) distance from each other, three rather strong earthquakes occurred: the first with M = 5.4 at 22 h 39 m 59 s 1995.08.17, the second with M = 5.8 after 34 days

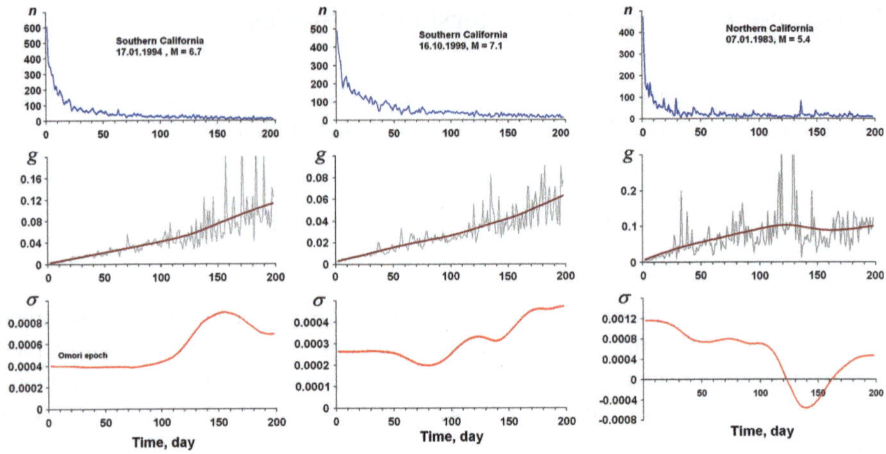

Fig. 1 The first three sheets of an aftershocks atlas (please, see the text)

at a distance of about 1 km, the third with M = 5.8 after 120 days at a distance of about 1 km from the second.

In the framework of the ATAS project, we want to consider the earthquake source as a dynamic system. It is known that the construction of phase portraits is an effective means in the study of dynamical systems. There is an idea to represent the evolution of aftershocks in the form of a trajectory in the phase planes. Examples of phase portraits of the earthquake source are presented in Fig. 3.

Here we used the data on earthquakes presented in Fig. 2. The top panel in Fig. 3 gives an idea of the portrait in the phase plane (g, σ). The point on the phase trajectory shows the initial state of the dynamical system, and the arrow indicates the direction of the system's movement along the trajectory. In the bottom panel we see the phase portrait in the plane $(\sigma, \dot{\sigma})$, where $\dot{\sigma} = d\sigma/dt$. We hope that the phase portraits of such sort will make it possible to produce a specific classification of the earthquake sources.

4 Discussion

It is clear that observations are the main source of our knowledge, and we also understand that the selection of empirical formulas plays a key role in systematizing our experience. Sometimes the empirical formula turns out to be fundamental.

Omori proposed the one-parameter formula to describe the flow of aftershocks. However, another empirical formula, namely, the two-parameter Hirano-Utsu formula, has become widely spread in seismology. The simple and elegant Omori Formula (1) was forgotten, its hidden possibilities were not realized and were not used.

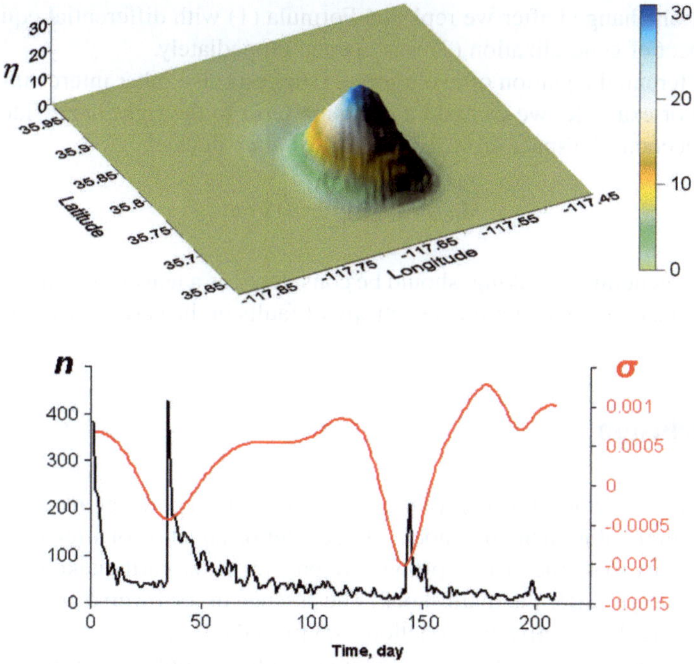

Fig. 2 Earthquakes in Southern California. Top panel shows the distribution of the aftershocks number on the earth's surface. The bottom panel shows three strong earthquakes and their aftershock sequences in the epicentral zone with a radius of 0.2° from August 17, 1995 till March 14, 1996 (black line) and the corresponding deactivation function σ(t) (red line)

Fig. 3 Phase portraits of the earthquake source in the planes (g, σ) and $(\sigma, \dot{\sigma})$ (top and bottom panels correspondingly)

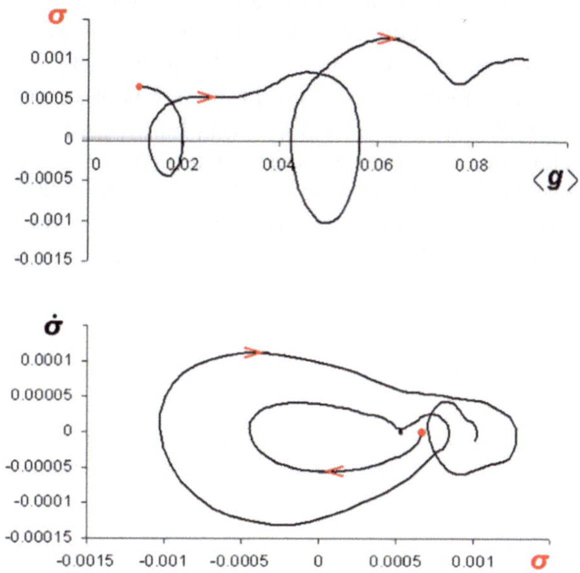

The situation changed after we replaced Formula (1) with differential equation (2). The prospect of generalization (3) was opened immediately.

The differential equation of evolution (2) suggests also other interesting generalizations. For example, we can add a diffusion term to the right-hand side of (2) to take into account the space-time distribution of aftershocks:

$$\partial n/\partial t + \sigma n^2 = \hat{D}\nabla^2 n. \tag{6}$$

Here $\hat{D}$, generally speaking, should be considered as a tensor $\hat{D} = \mathrm{diag}\,(D_{\parallel}, D_{\perp})$ in order to take into account the anisotropy of faults in the earth's crust, $D_{\parallel} \gg D_{\perp}$.

5 Conclusion

The well-known Omori law (1) we presented in the form of differential equation (2) describing the evolution of aftershocks. The evolution equation of aftershocks gave us the opportunity to set the inverse problem of physics of the earthquake source, which is "cooling down" after the main shock. The essence of the inverse problem consists in determining the deactivation function $\sigma(t)$ from the frequency of aftershocks $n(t)$ known from observation. We presented a concrete example of solving the inverse problem. There is no doubt that the use of our aftershock processing technique opens up new experimental possibilities.

We are planning to create an atlas of aftershocks after the strong main shocks. The atlas will contain a description of the basic parameters, the initial sequence of aftershocks, the function of deactivating the source $\sigma(t)$ and the phase portrait for each event. A preliminary analysis revealed the exceptional wealth of the evolutionary forms of earthquake source after the main shock.

The authors thank the staff of Northern California Earthquake Data Center (NCEDC) and Southern California Earthquake Center (SCEC) for providing the catalogues of earthquakes. This research was partially supported by the Russian Foundation for Basic Research (Grant No. 18-05-00096), the Program 28 of the Presidium of RAS, and Russian Governmental assignment Programs 0144-2014-0097, 0144-2014-00103. There are no potential conflicts of interest to disclose.

References

1. Omori, F.: On the aftershocks of earthquake. J. Coll. Sci. Imp. Univ. Tokyo **7**, 111–200 (1894)
2. Hirano, R.: Investigation of aftershocks of the great Kanto earthquake at Kumagaya. Kishoshushi, Ser. 2 **2**, 77–83 (1924)
3. Utsu, T.: A statistical study on the occurrence of aftershocks. Geophys. Mag. **30**, 521–605 (1961)
4. Guglielmi, A.V.: Omori law: a note on the history of geophysics. Phys. Usp. **60**, 319–324 (2017)
5. Guglielmi, A.V., Zavyalov, A.D.: Omori law: to the 150th anniversary of Fusakichi Omori. J. Volcanol. Seismol. **12**(5), 353–358 (2018)

On the Spatial-Temporal Structure of Aftershock Sequences

O. D. Zotov, A. D. Zavyalov and B. I. Klain

Abstract The work presents results of statistical analysis of the spatial-temporal distribution of strong aftershocks. Data from the world catalog of earthquakes USGS/NEIC from 1973 to 2014 (https://earthquake.usgs.gov) and from the regional earthquakes catalog of Northern California from 1968 to 2007 (http://www.ncedc.org) were used. Short time intervals were considered—no more than 10 h before and after main shock and distance from the epicenter of the main shock—up to 5°. The distribution of the aftershocks number in the time-distance coordinates from the main shocks was investigated. The main shocks with magnitude Mms $\geq$ 7.5 and aftershocks with 6 $\leq$ Maft < Mms for the world catalog NEIC USGS and the main shocks Mms $\geq$ 5 and aftershocks with 4 $\leq$ Maft < Mms for Northern California catalog were considered. It is shown that the spatial-temporal distribution of strong aftershocks forms a similarity of a wave structure.

Keywords Main shock · Aftershock · Spatial-temporal structure

1 Introduction

The work [1] shows that that the maximum of the dependence of the aftershocks (aft) number from the distance to the epicenter of the main shock (ms) is observed at a certain distance (from 10 to 40 km). This distance is directly proportional to the magnitude of the main shock.

O. D. Zotov (✉) · B. I. Klain
Borok Geophysical Observatory, Schmidt Institute of Physics of the Earth, Russian Academy of Sciences, Borok, Yaroslavl Region, Russia
e-mail: ozotov@inbox.ru

B. I. Klain
e-mail: klb314@mail.ru

A. D. Zavyalov
Schmidt Institute of Physics of the Earth, Russian Academy of Sciences, Moscow, Russia
e-mail: zavyalov@ifz.ru

T. B. Yanovskaya et al. (eds.), *Problems of Geocosmos—2018*, Springer Proceedings in Earth and Environmental Sciences, https://doi.org/10.1007/978-3-030-21788-4_16

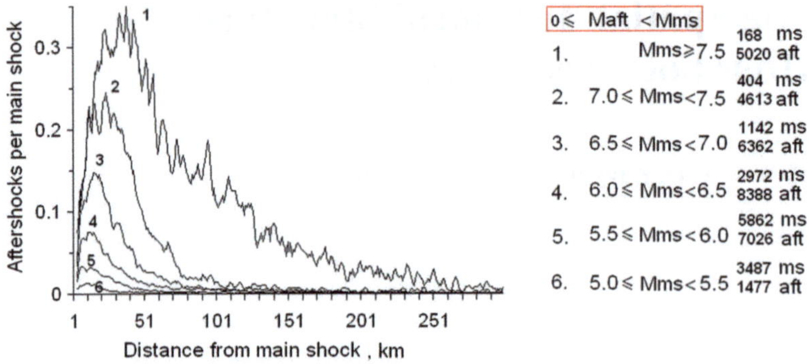

Fig. 1 Dependences of the average number of aftershocks on the distance to the main shock for different ranges of the magnitudes of the main shocks (see information on the right in the plot). The curves top to bottom correspond to the range of magnitudes of the main shocks Mms top to bottom [1]. Global catalog USGS/NEIC from 1973 to 2014

Figure 1 shows the dependences of the average number of aftershocks on the distance from the main shock for different ranges of the magnitudes of the main shocks Mms. The method of superposed epochs was used. In Fig. 1 on the right gives information on the number of main shocks and aftershocks for each range of magnitudes of main shocks. The analysis included aftershocks that occurred within 10 h after the main shocks. The average number of aftershocks was calculated as the number of aftershocks per 1 main shock.

We see that the maximum of the dependence of the average number of aftershocks from the distance is observed at a distance of 10–40 km and this distance is connected to the magnitude of the main shock. The larger the magnitude of the main shocks, the greater this distance.

Also in [1] it is noted that the maximum of the dependence of the aftershocks number from the distance to the epicenter of the main shock is independent of time.

This spatial feature of the aftershock sequences demonstrates Fig. 2, which shows the result of the accumulation of aftershock sequences (Maft < 7.0) after main shocks with Mms $\geq$ 7.0 according to the global catalog USGS/NEIC from 1973 till 2014.

The search for other features of the temporal and spatial dynamics of the aftershock flow is important from the point of view of earthquakes physics. This work continues to research in this direction.

2 Data and Method

Data from the world catalog of earthquakes USGS/NEIC from 1973 to 2014 (https://earthquake.usgs.gov) and from the regional earthquakes catalogs of Northern California from 1968 to 2007 (http://www.ncedc.org) and Southern California from 1983 to 2008 (http://www.scec.org) were used. Short time intervals were considered, no

Fig. 2 USGS/NEIC 1973–2014. Dependence of the aftershocks number from the distance to the epicenter of the main shock in interval of 10 h after the main shocks. Red arrow indicates the position of the maximum

more than 10 h before and after main shock and distance from the epicenter of the main shock—up to 5°. The method of superposition of epochs in the temporal and spatial coordinates was used. The main shocks performed the function of reference mark for the synchronization of sequences of repeated shocks. To each main shock was attributed the conventional time equal to zero, from which the time of aftershock observation was counted. Epicenters of each main shock coincided with the origin of coordinates. The distances from the main shock to the associated aftershocks were calculated for aftershocks synchronization in space.

3 Analysis Results

Figure 3 presents the result of the analysis of strong aftershocks dynamics from the catalog USGS/NEIC 1973–2014. 208 aftershocks with $6 \leq Maft < Mms$ after 190 main shocks with $Mms \geq 7.5$ in 10 h before and after main shock and distance from the epicenter of the main shock 3° were selected.

On Fig. 3 we do not see any structure in temporal and in spatial sequences. But if the same main shocks and aftershocks are located on a time-distance plane, then we see a completely different picture (see Fig. 4). Namely, we see the spatial-temporal structure that is formed by strong aftershocks.

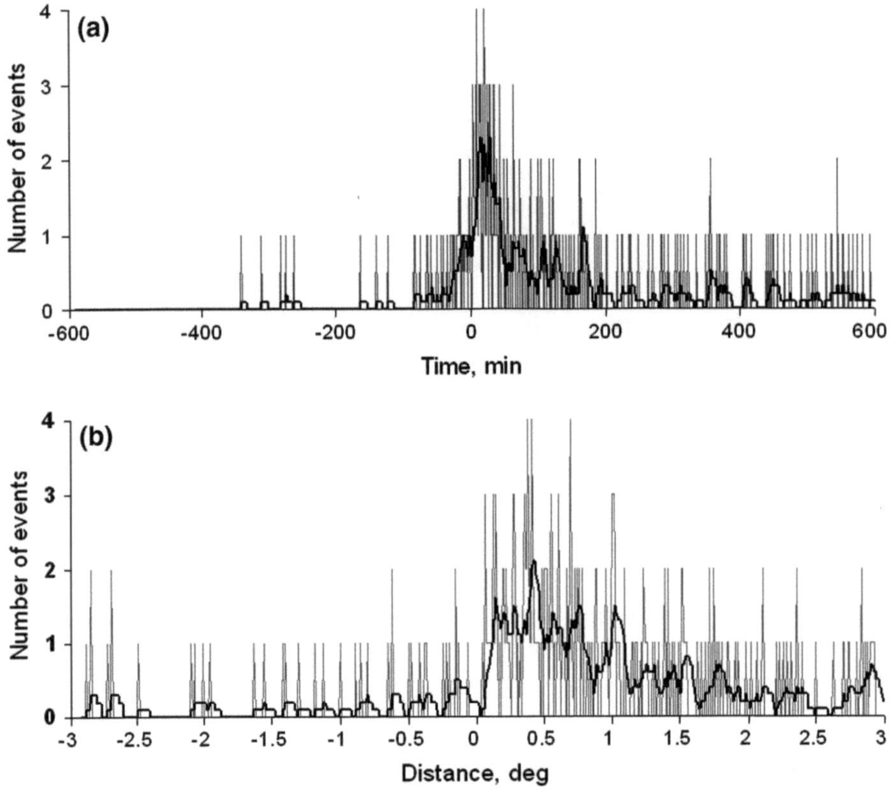

Fig. 3 Temporal (**a**) and spatial (**b**) sequences of strong aftershocks. The gray curve is the time sequence of aftershocks number in minute intervals (**a**) and spatial sequence of aftershocks number in 1 km intervals (**b**). The black curves (**a, b**) were obtained by smoothing the gray curves with a sliding window of 10 points with 1 point shift

Figure 5 presents the result of the analysis of strong aftershocks dynamics from the Northern California catalog from 1968 to 2007. In this case 265 aftershocks with $4.5 \leq$ Maft < Mms after 156 main shock with Mms ≥ 5.5 in 10 h before and after main shock and distance from the epicenter of the main shock 0.5° were selected.

As in Fig. 3 on Fig. 5 we do not see in temporal and in spatial sequences any structure. But if the same main shocks and aftershocks are located on a time-distance plane, then we see a completely different picture (see Fig. 6). Namely, we see the spatial-temporal structure that is formed by strong aftershocks.

Additionally, we give an example of an individual main shock and its aftershocks. We see the spatial-temporal structure that is formed by aftershocks (Fig. 7).

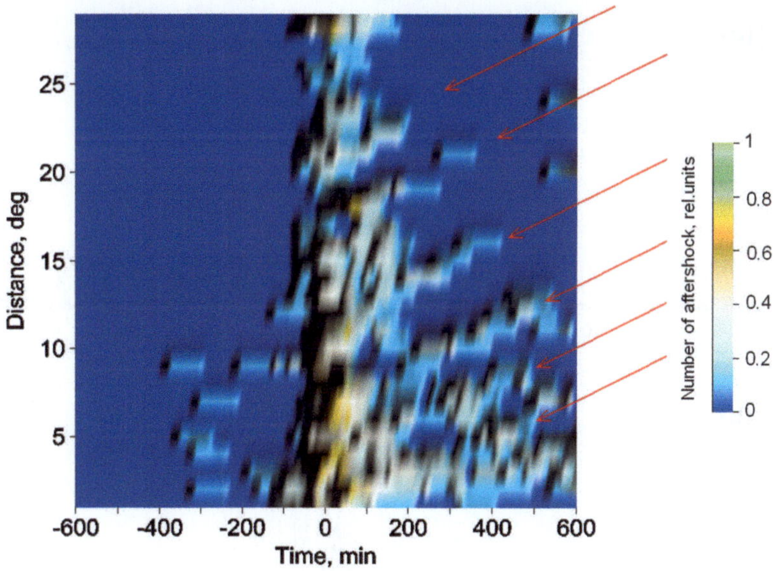

Fig. 4 USGS/NEIC. The distribution of the 208 aftershocks with 6 ≤ Maft < Mms after 190 main shocks with Mms ≥ 7.5 in the time-distance coordinates. Distance is given in degrees of the large circle arc. Red arrows indicate the observed structure

4 Possible Interpretation

Apparently, we were able to find that the spatial-temporal distribution of strong aftershocks forms some kind of wave pattern. In this regard, we would like to remind that the temporal evolution of the aftershocks is described by an ordinary differential equation

$$dn/dt + \sigma n^2 = 0,$$

where n is the frequency of the aftershocks, and σ is the so-called deactivation coefficient of the earthquakes source [2].

A natural generalization of this equation that takes into account the space-time evolution of aftershocks can be represented in the following form [3]:

$$\partial n/\partial t = -\sigma n^2 + D\nabla^2 n.$$

This nonlinear partial differential equation is known in the mathematical literature as the Kolmogorov-Piskunov-Petrovsky equation (KPP). Here D is a phenomenological parameter having the physical meaning of the diffusion coefficient. From our point of view, the KPP equation is of interest in the context of the spatial-temporal distribution of aftershocks. It is especially useful for interpreting the wavelike dis-

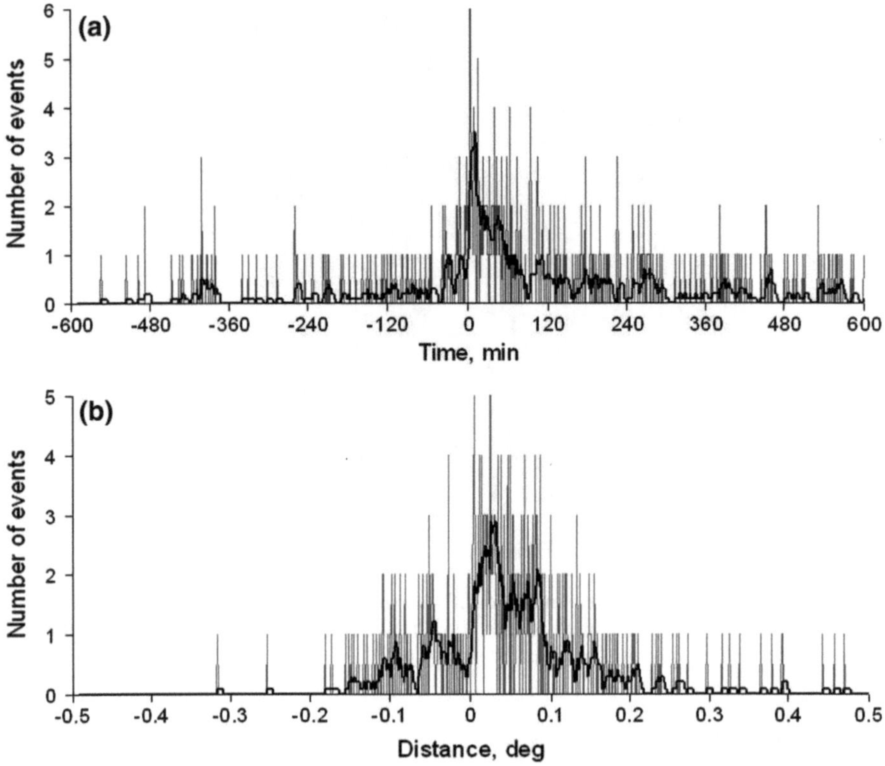

Fig. 5 Temporal (**a**) and spatial (**b**) sequences of strong aftershocks for Northern California. The gray curve is the time sequence of aftershocks number in minute intervals (**a**) and spatial sequence of aftershocks number in 0.1 km intervals (**b**). The black curves (**a, b**) were obtained by smoothing the gray curves with a sliding window of 10 points with 1 point shift

tribution of aftershocks, which we discovered. The solution of the KPP equation, generally speaking, depends on the dimension of the space in which the boundary value problem is considered. Since the experimental space-time distribution of aftershocks is reduced to one-dimensional in space (see Fig. 4), a simple KPP model can be used, which has self-similar solutions in the form of nonlinear wave propagation [4].

The authors thank the staff U.S. Geological Survey and of Northern California Earthquake Data Center for providing the catalogues of earthquakes. We are grateful to Prof. A. Guglielmi for his interest in this work and for valuable comments. The work was partially supported by the Governmental task # 0144-2014-00116 and RFBR project # 18-05-00096. There are no potential conflicts of interest to disclose.

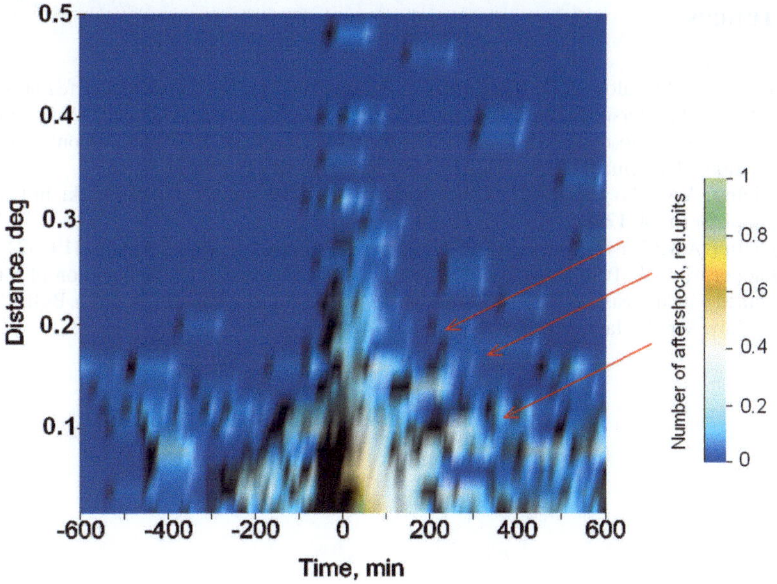

Fig. 6 Northern California. The distribution of the 265 aftershocks with $4.5 \le$ Maft < Mms after 156 main shocks with Mms ≥ 5.5 in the time-distance coordinates. Distance is given in degrees of the large circle arc. Red arrows indicate the observed structure

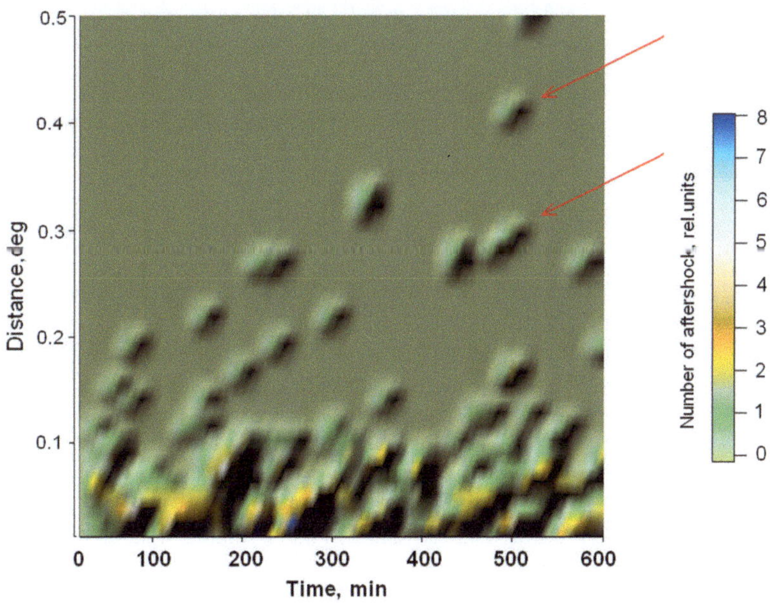

Fig. 7 Spatial-temporal structure of aftershocks with Maft < 6.6 after the main shocks with Mms = 6.6, 1987.11.24, Latitude = 33.02 Longitude = −115.85 in Southern California Red arrows indicate the observed structure

References

1. Zotov, O.D., Zavyalov, A.D., Guglielmi, A.V., Lavrov, I.P.: Spatial-temporal features of the evolution of the aftershock sequences. In: Adushkin, V.V., Kocharyan, G.G. (eds.) Trigger Effects in Geosystems (Moscow, 6–9 June 2017), Proceedings of the IV Russian Conference with International Participation, pp. 121–127. IDG RAS, M. GEOS (2017)
2. Guglielmi, A.V., Zavyalov, A.D.: Omori law: to the 150th anniversary of Fusakichi Omori. J. Volcanol. Seismol. **12**(5), 353–358 (2018)
3. Guglielmi, A.V.: Omori law: a note on the history of geophysics. Phys. Usp. **60**, 319–324 (2017)
4. Kolmogorov, A.N., Petrovsky, I.G., Piskunov, N.S.: Investigation of the equation of diffusion combined with increasing of the substance and its application to a biology problem. Bull. Moscow State Univ. Ser. A Math. Mech. **1**, 1–25 (1937)

Possible Short-Term Ionospheric Precursors of Strong Crustal Earthquakes

L. P. Korsunova, A. D. Legenka and V. V. Hegai

Abstract We have studied changes in the ionosphere prior to crustal earthquakes with magnitudes of $M \geq 6.5$ based on the data from the ground-based stations of vertical ionospheric sounding in Japan for the period 1968–2004 (32 cases). The data are analyzed based on hourly measurements of the virtual height and frequency parameters of the sporadic E layer and critical frequency of the regular $F2$ layer for the period of three days prior to the earthquakes provided that strong geomagnetic disturbances are absent. In the studied intervals of time before all earthquakes, anomalous changes were discovered both in the frequency parameters of the Es and $F2$ ionospheric layers and in the virtual height of the sporadic E layer. The changes were observed on the same day at stations spaced apart by several hundred kilometers. It is found that the lead-time of these ionospheric anomalies preceding the seismic impact is sensitive to the magnitude of the subsequent earthquakes and it tends to increase with the rise of the earthquake magnitude. It is concluded that such ionospheric disturbances can be short-term ionospheric precursors of earthquakes.

Keywords Crustal earthquakes · Ionospheric disturbances · Ionospheric precursors of earthquakes

1 Introduction

Over the last decade, studies of the earthquake preparation processes and their influence on the Earth's ionosphere have significantly progressed due to in situ [11, 15] and GPS satellite measurements [12, 16, 17]. Satellite observations have made it possible to acquire information on the state of the ionosphere in poorly accessible,

L. P. Korsunova (✉) · A. D. Legenka · V. V. Hegai
IZMIRAN, Kaluzhskoe Hwy 4, 108840 Moscow, Troitsk, Russian Federation
e-mail: lpkors@rambler.ru

A. D. Legenka
e-mail: leg@izmiran.ru

V. V. Hegai
e-mail: hegai@izmiran.ru

T. B. Yanovskaya et al. (eds.), *Problems of Geocosmos—2018*, Springer Proceedings in Earth and Environmental Sciences, https://doi.org/10.1007/978-3-030-21788-4_17

seismically active regions, while the dense network of GPS stations has provided almost uninterrupted monitoring of changes in this state over large areas. An important achievement in the study of seismo-ionospheric disturbances based on GPS network data is the refinement of spatial scales of the disturbances for strong earthquakes: several thousand kilometers along the parallel and approximately one thousand and half kilometers along the meridian [18, 19]. However, the main information on seismogenic effects in the ionosphere was obtained from ground based stations of the vertical ionospheric sounding (VIS) [2, 7, 10, 13, 14]. The long-term observations of such effects have made it possible to reveal a number of features typical for seismo-ionospheric disturbances in the period of preparation of earthquakes and after.

The studies of seismogenic disturbances in the ionosphere were conducted in two directions: detailed studies of individual earthquakes and their influence on the ionosphere in different regions of the Earth [2–4] and statistical analysis of the effects of several earthquakes in a specific seismically active zone [7, 8, 10]. While every earthquake is unique, it turned out to be possible to identify the most typical properties of ionospheric disturbances preceding the main impact with some lead-time [4, 8–10].

In the works [7, 13], it was shown that the lead-time (ΔT) of the detected ionospheric disturbances preceding earthquakes (IDPE) is from one day to several weeks, depending on the magnitude of the earthquake and the distance from the observation point to the epicenter. These disturbances can be ionospheric earthquake precursors (IEPs) of various immediacy (in particular, medium-term, MIEPs, or short-term, SIEPs). SIEPs are of the largest practical significance, as they occur from one to three days before the earthquake and can be used for further prediction of major seismic events as part of the general picture of pre-earthquake anomalies in other geophysical fields. At the same time, SIEPs are the least studied, since the "immediacy" of a given disturbance is rather difficult to determine in epignosis. Studies of SIEP were significantly moved forward by the paper of Hobara and Parrot [4], who thoroughly considered the disturbances in the $foF2$ critical frequencies during long intervals of time before and after a strong earthquake with $M = 8.3$, which were detected by multiple vertical-sounding stations at a distance from 400 to 3000 km from the epicenter. Hobara and Parrot [4] showed that possible SIEPs of this earthquake occur on the same day at all stations within the preparation zone ($r \leq 1500$ km, where r is the radius of the earthquake preparation zone on the Earth's surface). Therefore, with comparison of the times of occurrence of seismo-ionospheric disturbances at two or more stations spaced apart from each other within one region, it is possible to identify a SIEPs by the proximity of these times.

Thus, the objective of the present work is to identify possible SIEPs preceding earthquakes with $M \geq 6.5$ with a lead-time from several hours to 3 days based on the data of VIS ground base stations at different positions in the region of Japan.

2 Procedure of Data Analysis

This study used the data from hourly ionospheric measurements at VIS ground based stations Kokobunji ($\varphi = 35.7°$N, $\lambda = 139.5°$E), Akita ($\varphi = 39.7°$N, $\lambda = 140.1°$E), and Wakkanai ($\varphi = 45.4°$N, $\lambda = 141.7°$E) in the time interval of four days, including the day of the earthquake. A requirement for the ionospheric data was the absence of significant geomagnetic planetary disturbances, i.e., the three-hour *Kp* indexes of geomagnetic activity should not exceed the level of 3_+ in the specified time intervals before each earthquake. Only strong crustal earthquakes with $M \geq 6.5$ and a hypocenter depth of $h \leq 60$ km for the period 1968–2004 were considered. Thirty-two earthquakes were selected in this way. We analyzed the daily variations in the $F2$-layer critical frequency ($foF2$) and Es-layer blanketing frequency (f_bEs), which characterize the electron density in the maxima of the corresponding layers; the daily variations in the virtual height of the sporadic E-layer ($h'Es$) were analyzed as well. Further, we calculated the deviations of the above-mentioned ionospheric parameters from monthly median values for each hour of the day at all stations in the preparation zone of a given earthquake; the radius of the zone was determined as $r_D = 10^{0.43M}$ km, in accordance with [1]. The relative variations $\Delta f_bEs/f_bEs_{\mathrm{med}}$ and $\Delta foF2/foF2_{\mathrm{med}}$ were obtained for the frequency parameters. The interquartile range (IQR) was chosen as a measure of scatter due to random deviations. The IQR is the difference between the upper and lower quartiles of a given parameter calculated on a three-day interval for any of the abovementioned parameters ($Y(t_i)$, where t_i is the point of the reference time). In this case, the "noise" bands $K\pm = [\Delta Y(t_i)]_{\mathrm{MED}} \pm 0.75\ IQR$ or $K\pm = [\Delta Y(t_i)/Y_{\mathrm{med}}(t_i)]_{\mathrm{MED}} \pm 0.75\ IQR$ restrict possible variations of the parameters, which are explained by random deviations with a certain degree of probability. As follows from the encyclopedia [5], in the case of the normal distribution of the "error" of $\Delta Y(t_i)$ or $\Delta Y(t_i)/Y_{\mathrm{med}}(t_i)$, the value 0.75 IQR approximately corresponds to one standard deviation. The $\Delta Y(t_i)$ and $\Delta Y(t_i)/Y_{\mathrm{med}}(t_i)$ values that exceed the limits of the corresponding "noise" bands were attributed to anomalous $Y(t_i)$ values if the duration of such a disturbance (τ) was 1 h $\leq \tau \leq$ 3 h (see also [7, 10]).

The pre-earthquake ionospheric disturbances discovered in the investigation below occur on the same day above different stations, although the distances between the stations and the epicenters of the respective earthquakes differ by several hundred kilometers. This fact disagrees with the patterns of MIEP occurrence (see [6, 7]) but is in close agreement with the features of IEP occurrence before a strong earthquake with $M = 8.3$ in the same region [4], which were obtained from the data on the variations in the critical frequency of the $F2$ ionospheric layer.

The features of the occurrence of possible SIEPs before crustal earthquakes are illustrated by Fig. 1, which shows the time variations in the ionospheric parameters at two stations, Kokobunji and Akita, separated from each other by a great-circle distance $R \approx 480$ km. The variations in the planetary Kp-index are shown at the top panel; the time of the earthquake is marked with a dashed line and an arrow; and

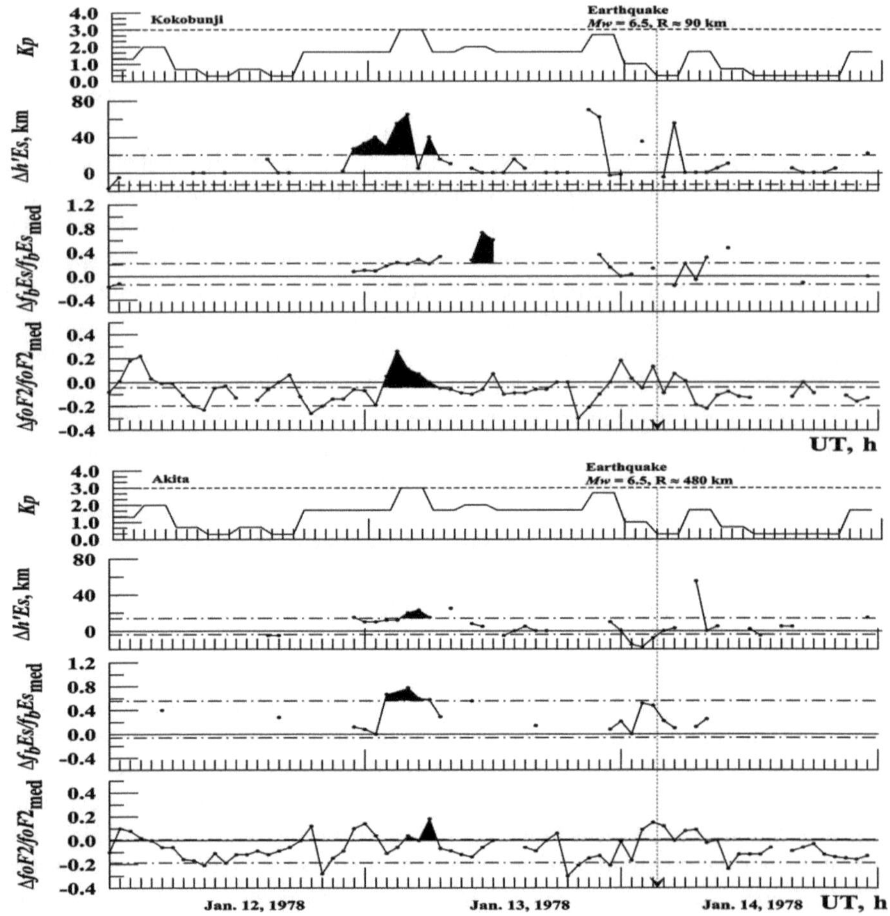

Fig. 1 The time variations in the ionospheric parameters at two stations, Kokobunji and Akita, separated from each other by a great-circle distance $R \approx 480$ km. The black "surges" in all parameters mark possible SIEPs for the earthquake of January 14, 1978 with $M = 6.5$

the dash-dotted lines show the $K\pm$ levels. The black "surges" in all parameters mark possible SIEPs for the earthquake of January 14, 1978 with $M = 6.5$.

A similar analysis was performed for the ionospheric measurement data of VIS ground based stations at different positions before all other crustal earthquakes with $M = 6.5$–7.4 (32 events, as mentioned above). This made it possible to identify pre-earthquake disturbances in the parameters of the *Es* and *F*2 layers observed during the same day, which allowed us to identify them as possible SIEPs.

3 Results and Discussions

Analysis of the identified ionospheric disturbances preceding earthquakes of different magnitudes at VIS ground based stations separated from the epicenters by a distance from 90 to 1000 km showed the following.

1. Possible seismo-ionospheric anomalies in the *Es* layer are characterized by an increase in the virtual height of the layer and in the electron density as compared to the background; the critical frequency of the $F2$ ionospheric layer (and corresponding electron density at the layer maximum) can either increase (positive anomaly) or decrease (negative anomaly); however, negative anomalies are predominant (24 cases out of 32).
2. The lead-times (ΔT) of possible ionospheric anomalies before earthquakes with $M = 6.5$–7.4 vary from 0.5 to 2.5 days, increasing with the rise of the earthquake magnitude.

In order to obtain the quantitative characteristics of the supposed dependence of the lead-times (ΔT) on the magnitude for the identified pre-earthquake ionospheric disturbances (possible SIEPs), we plotted the logarithms of these times (lg[ΔT]) as functions of magnitudes (M) for all 32 earthquakes using the data of Kokobunji station; the times ΔT were determined from the extrema in the parameters $\Delta h'Es$ ($\Delta T_{h'Es}$) and $\Delta foF2/foF2_{\text{med}}$ (ΔT_{foF2}). The linear regressions obtained by the method of least squares (with the corresponding correlation coefficients ρ) have the following form:

$$\lg[\Delta T_{h/Es}] = 0.35M - 2.46, \quad \rho = 0.616 \tag{1}$$

$$\lg\left[\Delta T_{foF2}\right] = 0.47M - 3.28, \quad \rho = 0.728 \tag{2}$$

Additionally, the same procedure was performed for 20 earthquakes from the full set, for which all values of the 3 h planetary *Kp* indexes during three days prior to the corresponding earthquake did not exceed the level of 3_0 (level Q, or "quiet", i.e. no significant geomagnetic disturbances). For geomagnetically quiet conditions (level Q) the following linear regressions were obtained:

$$\lg[\Delta T_{h'Es}] = 0.3M - 2.12, \quad \rho = 0.622 \tag{3}$$

$$\lg\left[\Delta T_{foF2}\right] = 0.47M - 3.24, \quad \rho = 0.803 \tag{4}$$

Figure 2 illustrates the above relationship (4).

That being said, the correlation coefficients for the Q level are higher when similar parameters are compared. It has also turned out that the correlation coefficients for the ΔT_{foF2} dependences are significantly higher than those for the corresponding $\Delta T_{h'Es}$ dependences. This implies that, if the epignosis results are used to predict earthquakes with account of the ionospheric changes, the variations in the $F2$-layer

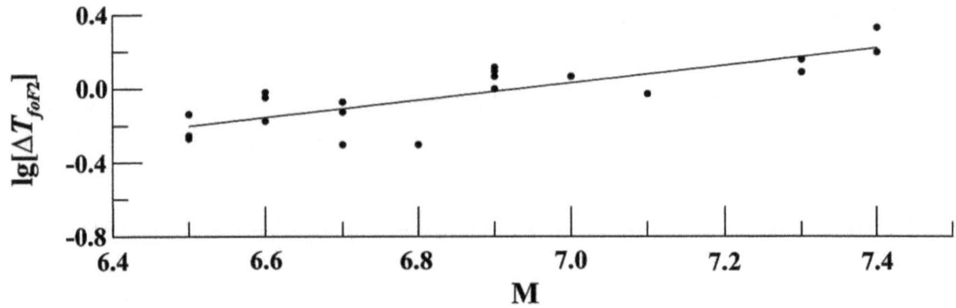

Fig. 2 The "cloud" of $\lg[\Delta T_{foF2}]$ values (points) and the dependence (4) is given by a solid line

critical frequency provide a higher accuracy of the determination of the possible time of the underground shock based on a SIEPs identified by the above-mentioned technique.

It is of interest to compare our results of the determination of the lead-time of a possible SIEPs (ΔT) (dependence (4) with the highest correlation coefficient ρ = 0.803) with those obtained earlier [20] from the electricity measurements in the near-surface atmospheric layer. For the upper limit of the magnitude range, $M = 7.4$, in accordance with Zubkov [20], we obtain $\Delta T(M = 7.4)_{\text{Zubkov}} = 1.84$ day for short-term earthquake precursors, while we have $\Delta T_{\text{foF2}}(M = 7.4) = 1.73$ day in our case. These values are in rather close agreement, which is also evidence that our identification of possible SIEPs from the measurement data of the *Es* and *F*2 layer parameters was correct.

4 Conclusions

Analysis of the changes in the ionospheric parameters of the *Es* and *F*2 layers for three days prior to strong crustal earthquakes with $M = 6.5$–7.4 based on the observations of VIS ground based stations at different positions in the region of Japan allowed us to identify the following features of possible SIEPs:

1. In the case of possible SIEPs, the frequency parameters of the *Es* and *F*2 layers deviate from quiet background values by at least 20%, and the virtual height of the sporadic *Es* layer deviates by more than 30 km for epicentral distances of ~300 km.
2. Anomalous changes in these ionospheric parameters are detected on the same day at stations spaced apart by several hundred kilometers.
3. The lead-time of the possible SIEPs prior to underground impact is mainly determined by the energy of the upcoming earthquake and increases with a rise of its magnitude *M*. The maximum lead-time for the seismic impact in the specified range of magnitudes is 2.5 days.

4. For possible SIEPs, the inverse proportional dependence of lead-times ΔT on the epicentral distances R from the observational station is absent. At the same time, such dependence is typical for possible MIEPs. This is the main difference in the dependences of lead-times ΔT of possible IEPs of various immediacy.

The described features may eventually make it possible to distinguish SIEPs from other ionospheric anomalies not in epignosis but in real-time quiet geomagnetic conditions, which is important for prediction of seismic hazard in comprehensive forecasts of strong earthquakes with account of the ionospheric data.

References

1. Dobrovolsky, I.P., Zubkov, S.I., Myachkin, V.I.: Estimation of the size of earthquake preparation zones. Pure. appl. Geophys. **117**(5), 1025–1044 (1979)
2. Hegai, V.V., Legenka, A.D., Kim, V.P.: Unusual enhancement of ionospheric F2 layer critical frequency before the 23 August 2011 Virginia (USA) earthquake. Open Trans. Geosci. (GEOS) **1**(1), 39–43 (2014)
3. Heki, K.: Ionospheric electron enhancement preceding the 2011 Tohoku-Oki earthquake. Geophys. Res. Lett. **38**, L17312 (2011). https://doi.org/10.1029/2011GL047908
4. Hobara, Y., Parrot, M.: Ionospheric perturbations linked to a very powerful seismic event. J. Atmos. Sol.-Terr. Phys. **67**(7), 677–685 (2005). https://doi.org/10.1016/j.jastp.2005.02.006
5. Klotz, S., Johnson, N.L.: Encyclopedia of statistical sciences. Wiley, New Jersey (1983)
6. Korsunova, L.P., Khegai, V.V.: Medium-term ionospheric precursors to strong earthquakes. Int. J. Geomagn. Aeron. **6**, GI3005 (2006). https://doi.org/10.1029/2005gi000122
7. Korsunova, L.P., Khegai, V.V.: Analysis of seismoionospheric disturbances at the chain of Japanese stations for vertical sounding of the ionosphere. Geomagn. Aeron. (Engl. Transl.) 48(3), 392–399 (2008)
8. Korsunova, L.P., Hegai, V.V.: Ionospheric precursors of crustal earthquakes in the northwestern part of the Asia-Pacific seismic belt. Open Trans. Geosci. (GEOS) **1**(1), 25–33 (2014)
9. Korsunova, L.P., Hegai, V.V.: Effectiveness criteria for methods of identifying ionospheric earthquake precursors by parameters of a sporadic E-layer and regular F2-layer. J. Astron. Space Sci. **32**(2), 137–140 (2015)
10. Liu, J.Y., Chen, Y.I., Chuo, Y.J., Chen, C.S.: A statistical investigation of pre earthquake ionospheric anomaly. J. Geophys. Res. **111**, A05304 (2006). https://doi.org/10.1029/2005JA011333
11. Parrot, M., Berthelier, J.J., Lebreton, J.P., Sauvaud, J.A., Santolik, O., Blecki, J.: Examples of unusual ionospheric observations made by the DEMETER satellite over seismic regions. Phys. Chem. Earth. Parts A/B/C **31**(4), 486–495 (2006). https://doi.org/10.1016/j.pce.2006.02.011. 2006
12. Perevalova, N.P., Sankov, V.A., Astafyeva, E.I., Zhupityaeva, A.S.: Threshold magnitude for ionospheric TEC response to earthquakes. J. Atmos. Sol.-Terr. Phys. **108**, 77–90 (2014). https://doi.org/10.1016/j.jastp.2013.12.014
13. Perrone, L., Korsunova, L.P., Mikhailov, A.V.: Ionospheric precursors for crustal earthquakes in Italy. Ann. Geophys. **28**(4), 941–950 (2010)
14. Pulinets, S.A., Boyarchuk, K.A.: Ionospheric Precursors of Earthquakes, 315 p. Springer, Berlin (2004)
15. Sarkar, S., Gwal, A.K., Parrot, M.: Ionospheric variations observed by the DEMETER satellite in the midlatitude region during strong earthquakes. J. Atmos. Sol.-Terr. Phys. **69**(13), 1524–1540 (2007)
16. Saroso, S., Liu, J.Y., Hattori, K., Chen, C.H.: Ionospheric GPS TEC anomalies and M $\geq$ 5.9 earthquakes in Indonesia during 1993–2002. Terr. Atmos. Ocean. Sci. 19(5), 481–488 (2008). https://doi.org/10.3319/tao.2008.19.5.481(t)

17. Xia, C., Yang, S., Xu, G., Zhao, B., Yu, T.: Ionospheric anomalies observed by GPS TEC prior to the Qinghai Tibet region earthquakes. Terr. Atmos. Ocean. Sci. **22**(2), 177–185 (2011). https://doi.org/10.3319/TAO.2010.08.13.01(TibXS)
18. Zakharenkova, I.E., Shagimuratov, I.I., Tepenitzina, NYu., Krankowski, A.: Anomalous modification of the ionospheric total electron content prior to the 26 September 2005 Peru earthquake. J. Atmos. Sol.-Terr. Phys. **70**(15), 1919–1928 (2008). https://doi.org/10.1016/j.jastp.2008.06.003
19. Zakharenkova, I.E., Shagimuratov, I.I., Krankowski, A., Lagovsky, A.F.: Precursory phenomena observed in the total electron content measurements before great Hokkaido earthquake of September 25, 2003 (M = 8.3). Stud. Geophys. Geod. **51**(2), 267–278 (2007)
20. Zubkov, S.I.: Times of the occurrence of earthquake precursors. Izv. Akad. Nauk SSSR: Fiz. Zemli. (5), 87–91 (1987). (in Russian)

Seismicity and Crustal Structure of the Southern Crimea and Adjacent Northern Black Sea from Local Seismic Tomography

V. Gobarenko and T. Yegorova

Abstract The Main Caucasus Thrust, which marks the southern boundary of the Greater Caucasus orogen can be traced westward along the northern margin of the Black Sea and southern Crimea and coincides with a zone of seismicity called the Crimea Seismic Zone (CSZ). The CSZ is characterized by earthquakes of M = 3–5 with foci in the crust and uppermost mantle with abundant weak seismicity (M $\leq$ 3). Weak seismicity was used to recover the velocity structure of the crust of southern Crimea Peninsula and adjacent northern Black Sea employing local seismic tomographic techniques. Events were recorded during 1970–2013 by nine stations on the Crimea Peninsula and by one station (Anapa) on the Caucasus coast of the Black Sea. Data for the tomographic inversion were relocated for the P- and S-wave arrivals at all permanent stations of CSZ. The new local tomographic study documents significant P- and S-wave velocity heterogeneities in the depth range 10–30 km. Stable solutions have been obtained for depths of 15, 20 and 25 km. A distinctive feature of the crust of southern Crimea is its clear division on two domains: a western domain (Crimean Mountains) and an eastern one (Kerch-Taman zone). They are separated by a linear low-velocity zone of ~N-S strike (in the area between Sudak and Feodosiya) interpreted as a zone of crustal weakness that has been repeatedly reactivated. Interpretation of velocity anomalies suggests a complex 3-D crustal geometry that involved a change of underthrusting polarity in the western Crimea Mountains crust compared to eastern Crimea.

Keywords Seismicity · Local seismic tomography · Crimea

V. Gobarenko (✉)
Institute of Seismology and Geodynamics, Vernadsky Crimean Federal University, Simferopol 295007, Russia
e-mail: valja-gobar@mail.ru

T. Yegorova
Institute of Geophysics, National Academy of Sciences of Ukraine, Pr. Palladina 32, Kiev 03680, Ukraine
e-mail: egorova@igph.kiev.ua

T. B. Yanovskaya et al. (eds.), *Problems of Geocosmos—2018*, Springer Proceedings in Earth and Environmental Sciences, https://doi.org/10.1007/978-3-030-21788-4_18

1 Introduction

The Crimean Seismic Zone (CSZ) is located along the Main Caucasus Thrust, which controls the southern boundary of the Greater Caucasus orogen and traces westwards along the northern margin of the Black Sea and southern Crimea. Fold-and-thrust belt of Greater Caucasus and Crimean Mountains was originated at the southern margin of Eurasia as a result of Cenozoic collision between the Eurasian and Arabian plates. The CSZ is manifested by the earthquakes of $M = 3–5$ with foci in the crust and upper mantle and abundant weak seismicity ($M \leq 3$) [5, 17]. In addition to information on ongoing collision processes, the CSZ seismological data are used to restore the velocity structure of the southern Crimea and adjacent part of the Black Sea applying the local seismic tomography method [6, 19]. For this we used data on first arrivals of P- and S-waves from weak earthquakes recorded in the 1970–2013 by nine seismic stations located in the Crimea (Crimean Seismological Network) and one station (Anapa) on the Caucasus coast of the Black Sea (Fig. 1, Table 1).

The seismic tomography study, performed for the Southern Crimea and adjacent part of the Black Sea, for the first time have provided information on seismic velocities in the crust. And, what is especially important, using the both types of waves for seismic tomography provides information on velocity distribution not only from P-waves, but also from S-waves. That generally increases the possibilities and quality

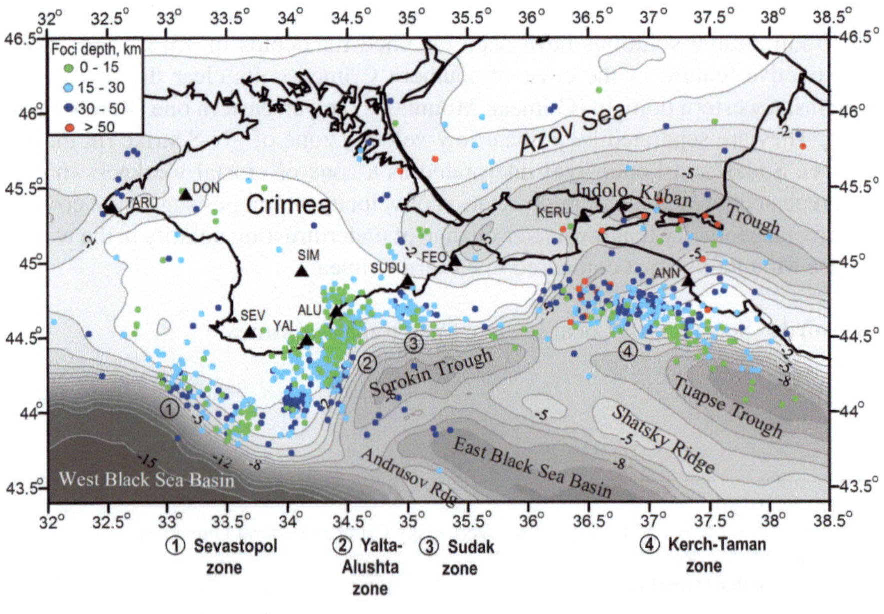

Fig. 1 Weak seismicity of the CSZ and seismic stations (black triangles) used for local seismic tomography study shown against the background of the Cretaceous basement depths (isolines in km) according to [16]

Table 1 Seismic stations of Crimea and Caucasus

No.	Seismic stations		Date of opening	Coordinates		Height
	Locality	Code		Lat (N)	Long (E)	m
1	Feodosiya	FEO	11.11.1927	45.02	35.39	40
2	Yalta	YAL	13.13.1928	44.48	34.15	23.6
3	Simferopol	SIM	14.05.1928	44.95	34.12	275
4	Sevastopol	SEV	28.06.1928	44.54	33.68	42
5	Alushta	ALU	03.10.1951	44.68	34.40	61
6	Sudak	SUDU	18.10.1988	44.89	35.00	108
7	Kerch	KERU	19.05.1997	45.31	36.46	50
8	Donuzlav	DON	05.12.1984	45.46	33.16	0
9	Tarhankut	TARU	11.07.2012	45.38	32.53	0
10	Anapa	ANN	09.05.1968	44.80	37.43	35

of interpretation by joint using the values of Vp and Vs and their ratios (Poisson's ratio). These parameters depend on SiO_2 content and on the composition of the rocks that constitute crust and upper mantle [2, 4, 14].

Our study for the first time represents the velocity structure of the crust and upper mantle of southern Crimea and adjoining part of the Black Sea at a depth of 10–40 km, and the resulting velocity distribution is interpreted on the basis of modern experimental data on distribution of the P- and S-wave velocities and their ratio in different conditions.

2 Seismicity of the Crimea Seismic Zone and Seismic Data

The Black Sea is located between two seismogenic zones - the North Anatolian zone on the south with strongest earthquakes in the vicinity of the Black Sea, commonly with magnitudes larger than 7 mainly associated with the North Anatolian transform fault in northern Turkey, and the CSZ lying along the northern margin of the Black Sea [17, 18].

The CSZ is characterized by earthquakes of $M = 3–5$ with foci in the crust and uppermost mantle with abundant weak seismicity ($M \leq 3$). The latter was used to recover the velocity structure of the crust of southern Crimea Peninsula and adjacent northern Black Sea employing local seismic tomography techniques. The data on arrivals of seismic waves (P- and S-wave phases) were taken from the bulletins of the regional network of Crimea seismic stations [12, 13]. Events were recorded during 1970–2013 by nine permanent seismic stations on the Crimea Peninsula (at Feodosiya (FEO), Yalta (YAL), Simferopol (SIM), Sevastopol (SEV), Alushta (ALU), Sudak (SUDU), Kerch (KERU), Tarhankut (TARU), Donuzlav (DON)) and by one station at Anapa (ANN) on the Caucasus coast of the eastern Black Sea (Fig. 1,

Table 1). Stations are equipped with the three-component short-period seismographs of Kharin system with high gain (10,000–50,000) for periods of 0.2–0.5 s [8].

Data for the tomographic inversion were relocated for the P- and S-wave arrivals at all permanent stations of CSN. Earthquake relocation was done via error minimization starting with a 1-D reference velocity model based on seismic surveys (active and passive) in the study area [6]. The number of phases, used for relocation, ranges from 8 to 16. In addition, use of both P- and S-waves for relocation led to an increased number of phases maximizing the number of earthquakes that could be relocated. As a result, over 1200 earthquakes with focal depths in the range 0–90 km were relocated. The deepest earthquakes (with focal depths in the range 70–90 km depth) occur below the Kerch Peninsula shelf and Caucasus coast with foci deepen northwards [6].

The obtained spatial distribution of relocated foci in the CSZ according to source depth is shown in Fig. 1. According to characteristic pattern of seismicity, within the CSZ from east to west, four major subzones are recognized: Kerch–Taman, Sudak, South Coast (or Yalta–Alushta), and Sevastopol ones [5]. In the Kerch–Taman subzone, located at the continental slope south of the Kerch and Taman peninsulas, the earthquakes are generated in the seismogenic zone with foci in the lower crust and upper mantle ($H = 29$–38 km), which plunge northward at about 30°. Thus, the deepest (70–90 km) hypocenters are located at the very north of the Kerch and Taman peninsulas (Fig. 1). In the Sudak subzone, a small number of earthquakes clustered at the continental slope to a depth of 35 km, is similar to the Kerch–Taman subzone tendency to plunge northward. The Yalta–Alushta (South Coast) subzone is known for generating a majority of earthquakes and is characterized by their maximum density. Their foci are located at 10–40 km depth within a zone running along the Black Sea continental slope ($\lambda \approx 34°$–$34.7°$) partly overlapping shelf and adjacent part of the Crimean Mountains (see Fig. 1). In the South Coast (Yalta-Alushta) subzone, there is a tendency of foci being uplifted northward (or to the northwest) toward the southern Crimea at an angle of 17°–18°. The Sevastopol subzone, which is orthogonal to the South Coast subzone, is characterized by scattered seismicity down to a depth of 40 km.

Thus, the main feature of the CSZ seismicity observed along the arc of the continental slope is the change in polarity of the seismic focal zone from the northern direction in the Kerch-Taman subzone, where the foci deepen landwards to depths of 80–90 km, at a gentle southeast direction (seawards) in the South Coast subzone with focal depths down to 50 km. Such phenomenon reflects the present seismotectonic process in the region and indicates a more complex collision environment in the area of Crimean Mountains compared to the Kerch-Taman zone [5].

The number of traces (source-receiver pairs) used for the seismic tomography calculations was 2430 for P-waves and 3015—for S-waves. In Fig. 2, showing the projections of seismic traces on the Earth's surface, one can see that Yalta-Alushta (South Coast) zone is distinguished by the highest density of traces. Time residuals (difference between observed arrival times and the arrival times predicted based on the new earthquake locations and the 1-D model) as a function of depth for P- and S-waves range from -2 to $+2$ s and from -3 to $+3$ s, respectively.

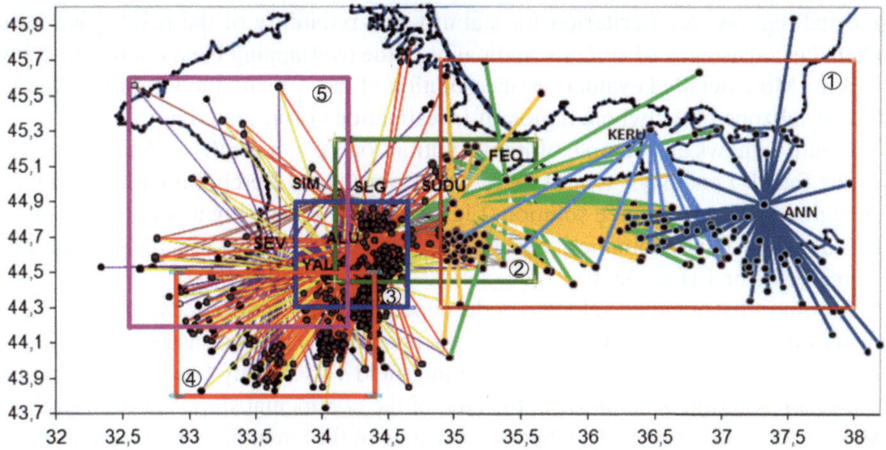

Fig. 2 Ray paths for P-waves used in the seismic tomography study. Rectangles of different colors indicate different regions (numbers in circles), for which seismic tomography solutions were obtained. Different colors of rays show seismic paths to a definite seismic station

3 Tomographic Inversion Method

To restore the velocity structure of the Earth's crust of the study region we used tomographic inversion method based on the Backus-Gilbert approach [1] for the 3-D case [7]. The method is intended to estimate the average velocity corrections relative to the starting model from a data base of travel times along the ray-passes traversing the study volume. In more detail the methodology of local seismic tomography and evaluation of resolution/accuracy of obtained solutions for the study region are presented in [6].

A certain disadvantage of applied method is assumed to be the dependence of obtained solution from coordinate system orientation being a sum of three functions, each of which is dependent on one coordinate. Minimization of this limitation is realized by choosing the optimal direction of the coordinate system along the strike of the main tectonic structures. Such an approximation only works in regions with relatively simple geology.

For the Crimea—Black Sea region, which is characterized by the presence of differently oriented units, it is very difficult to choose the preferred orientation of coordinate system. Therefore, the entire study area is divided into several overlapping regions (see Fig. 2), each of which determined by its own optimal orientation of the coordinate system. For that we used method of principal components based on allocation of linear combinations of travel times and on analysis of their covariance matrix [6]. The optimum direction of the coordinate system was chosen for reasons of obtaining the most contrasting solution.

The final solution for the entire area is obtained by alignment, joining and bringing together the results of tomographic inversion for different data sets within all the selected regions. As a criterion for stability and reliability of the results, we have taken the consistency of velocity anomalies in the overlapping bands of neighboring regions. More detailed evaluation of resolution of the velocity model for the Kerch-Taman subzone of the whole region of investigations is presented in [6].

Figure 3 shows an example of joint solution for Vp at a depth of 15 km, obtained for the Sudak-Feodosiya zone (Fig. 3a for region 2) and Kerch-Taman one (Fig. 3b for region 1). The combined solution (Fig. 3c) in the overlapping band of two regions does not show false anomalies or gradient zones that are absent in separate solutions for regions 2 or 1 (Fig. 3a, b).

Resolution of obtained velocity model depends on the number and mutual arrangement of the ray paths (see Fig. 2). Since the density of observations in the study area is extremely uneven, the sizes of smoothing field vary widely from point to point. The best resolution is obtained in the area of the continental slope between stations Alushta and Yalta (region 3), where dimensions of the smoothing area on horizontal coordinates do not exceed 10–20 km. The lowest resolution and the largest dimensions of the smoothing area (up to 40–50 km) are determined for the Kerch Peninsula region (region 1) and the plain Crimea region (regions 2 and 5) with a limited number of traces. The remaining areas are mainly characterized by resolution in the range

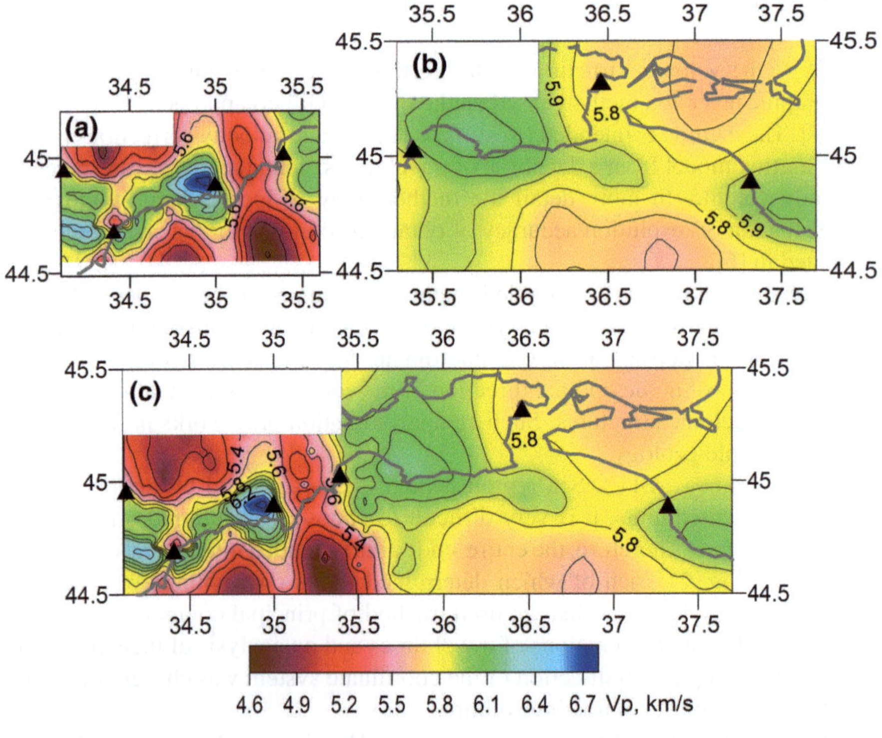

Fig. 3 Example of combined solution (**c**) for P-wave velocity at 15 km depth for Sudak-Feodosiya (**a**) and Kerch-Taman (**b**) zones

20–25 km horizontally, and 10–12 km in depth. More detailed evaluation of velocity model resolution for the region 1 is represented by Gobarenko et al. [6].

4 Results and Interpretation

The new local tomography results document significant P- and S-wave velocity heterogeneities in the depth range 10–30 km. Stable solutions have been obtained for the depths of 15, 20 and 25 km. Distribution of P-wave velocities and Vp/Vs ratio for the depth 15–25 km is shown in Fig. 4. Obtained distribution of P- and S-wave velocities and their relation (Vp/Vs) for the region of Crimean Mountains, Kerch and Taman Peninsulas and adjacent part of the Black Sea definitely shows two crustal domains—the Crimean Mountains (westwards of Sudak) of complicated structure with prevalence of high-velocity anomalies and the Kerch-Taman zone of more simple structure, which are separated by linear low-velocity zone of ~NS strike in the area between Sudak and Feodosiya (Fig. 4). In the Sudak-Feodosiya zone the Vp and Vs at the depth of 15–25 km are in the range 5.2–6.2 km/s and 3.3–3.5 km/s respectively. A consistent decrease in the velocities of both types of waves explains the lack of the Vp/Vs anomaly in this zone (Figs. 4 and 5). It is interpreted as a manifestation of crustal zone of weakness filled with felsic granites and gneisses. It could be associated with Korsaksko-Feodosiya Fault in Crimea located within the southern continuation of Late Proterozoic Orekhovo-Pavlograd fault zone, which separates two different domains of the Ukrainian Shield [19]. This zone, most probably, was reactivated during the Cenozoic time, when southern Crimea was subjected to tectonic processes related to evolution of the northern branch of the Tethys Ocean and Alpine stage of tectogenesis.

To the east of this zone there is located the Kerch-Taman zone that lacks notable velocity anomalies in the crust (Fig. 4). It should be noted here that in general the long wavelength velocity distribution in the Kerch-Taman zone in comparison with short wavelength in the crust of Crimean Mountains could be explained in part by distribution of seismic stations in the region with long distances in the Kerch-Taman zone (between stations FEO, KERU and ANN) and shorter ones in the Crimean Mountains (Fig. 1). That could produce the shorter ray paths (Fig. 2) and help to restore velocity anomalies in more detail in the western domain (Crimean Mountains) than in the eastern one (Kerch-Taman zone).

The velocity image and seismicity of the Kerch–Taman branch of the CSZ, where the majority of earthquakes are located on the continental slope and form a seismic focal zone that dips gently northwards, is explained by underthrusting of the East Black Sea microplate beneath the Scythian Plate [6]. In the southern segment of this part of the model, in the area of northern part of the East Black Sea, at the depths of 25–30 km, the uppermost mantle below the thin sub-oceanic crust of the Black Sea has rather low P-wave velocities (6.5–6.8 km/s) with high P- to S-wave velocity ratios (1.75–1.85). This is interpreted by Gobarenko et al. [6] as representing

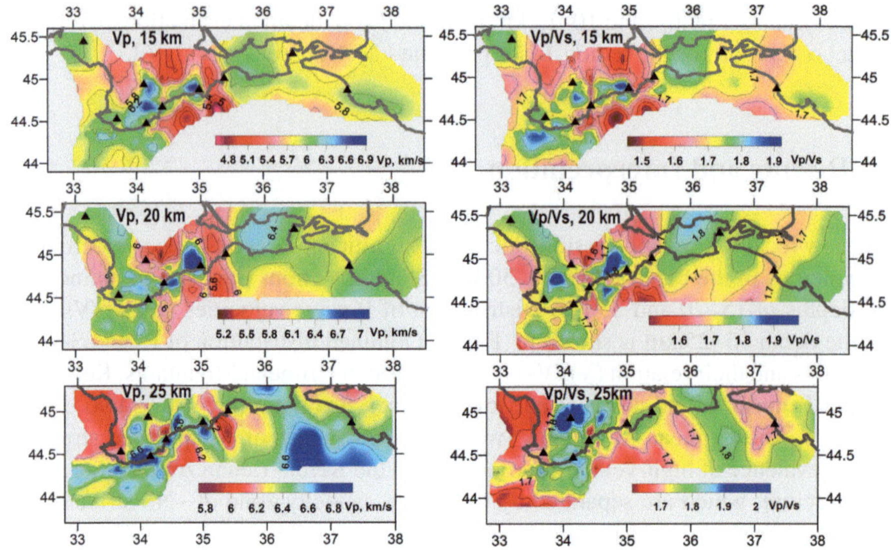

Fig. 4 P-wave velocities and Vp/Vs ratio at the depths of 15, 20 and 25 km according to the final tomographic model of the northeast Black Sea and southern Crimea—Taman peninsula study area

serpentinized upper mantle of continental lithosphere exhumed during Cretaceous rifting and lithospheric hyperextension of the eastern Black Sea.

The velocity model in Fig. 4 shows anomalous domain of complex configuration in the western part of the region, covering the area between the western Crimean Mountains (between the cities of Sevastopol and Alushta) and the plain Crimea (near the city of Simferopol), which continues offshore up to approximately 44°N (Fig. 4). This domain of general N-S strike comprises several anomalies distinguished in both types of waves, with the prevalence of anomalies with high and increased velocities, which at the depth of 15–25 km range from 6.2 to 6.8 km/s for Vp and from 3.45 to 3.8 km/s—for Vs. The Vp/Vs in this domain are in the range 1.75–1.9 (Fig. 4). Most clearly this anomalous domain is represented on the horizontal cross-section at 25 km (Fig. 4), which shows, in addition, the anomaly of Vp/Vs > 1.9 in the vicinity of Simferopol, which is caused by low S-wave velocity domain here.

Velocity distribution in the crust of the Southern Crimea is shown on the vertical cross-section of the seismic tomography model along the Sevastopol-Kerch line (Fig. 5), constructed along the deep seismic refraction profile of the same name.

The seismic profile Sevastopol-Kerch, acquired in 1974–1975, was recently reinterpreted using modern ray-tracing technique resulting in construction of velocity model for the crust [19]. The latter was supplemented by more detailed velocity distribution in the interval 10–35 km taken from present local seismic tomography study in Crimea. A significant difference in the structure of two major crustal blocks is revealed on the Sevastopol-Kerch seismic profile. One block characterized by higher velocities and located in the western and central Crimean Mountains, and the

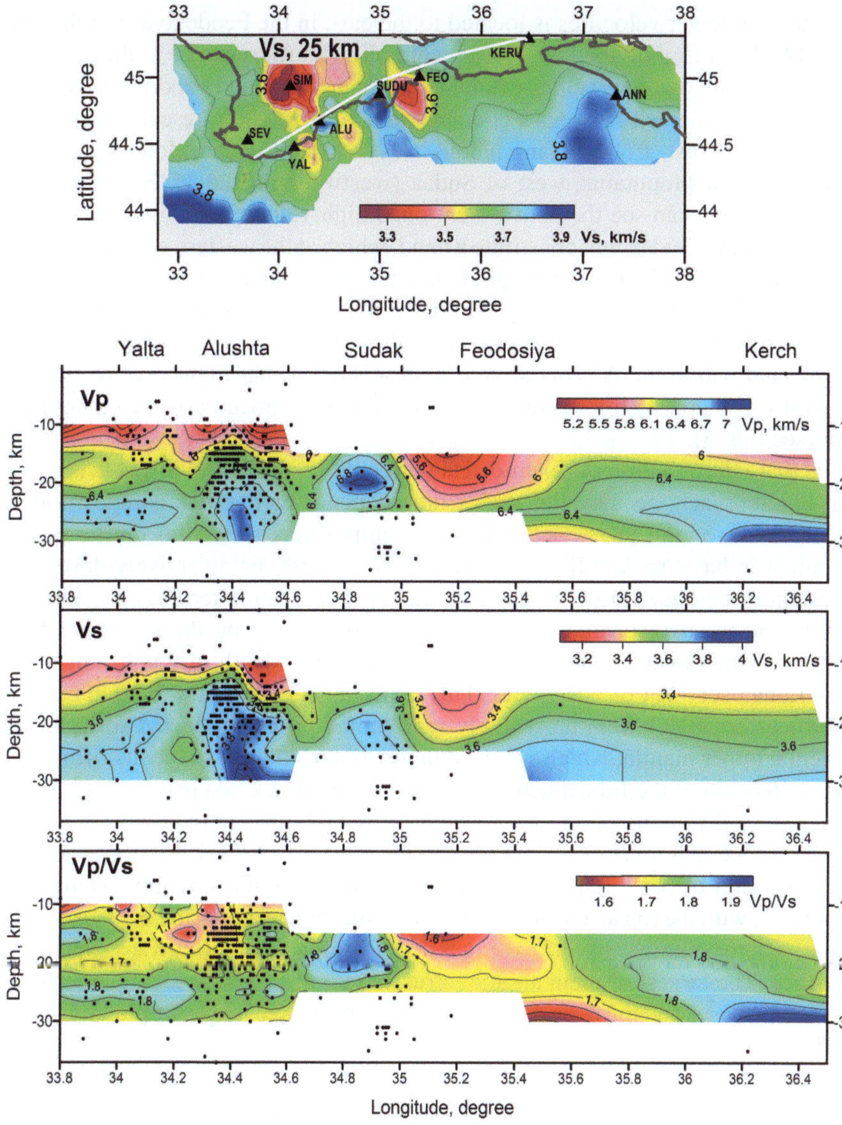

Fig. 5 Vertical cross-section of the seismic tomography model on Sevastopol-Kerch line constructed for P- and S-wave velocities, and for their ratio (Vp/Vs), with the foci location of weak earthquakes (black dots). The location of the section (white line) is shown on the horizontal cross-section of S-wave velocity at 25 km depth (upper panel)

other one with lower velocities is located to the east, in the Feodosiya–Kerch zone; the two blocks are subdivided by a basement uplift (Starokrymskoe Uplift) [19].

Figure 5 shows that low-velocity domain between Sudak and Feodosiya separates the eastern area located between Feodosiya and Kerch, with rare earthquakes and aseismic upper crust, from the anomalous crust of the Crimean Mountains. In the crust of Crimean Mountains, west of Sudak (westwards of the 35°E) at the depth of 10–30 km, one can see the above mentioned high-velocity bodies in the area of Yalta-Alushta and Sudak that constitute a band, which at the western end of the cross-section occurs at 25–30 km depth, gradually expanding and elevating eastwards to 15–25 km in the Sudak area (see Fig. 5). Three bodies are localized in this band with Vp = 6.6–6.8 km/s, Vs = 3.7–3.8 km/s, Vp/Vs = 1.8–1.85. Among them a central body, located within the Alushta area of high seismicity, has highest velocities and dimensions. It starts with a depth of 12 km and can be traced down to a depth of 30 km (see Fig. 5).

Note that interpretation of seismic data with joint use of P- and S-waves is a very promising approach, because the Vp/Vs ratio depends on the rock composition, or more precisely, on the SiO_2 content in bulk composition, as well as on the Fe and Mg content in basic rocks [2]. This approach is currently being actively developed for laboratory determination of velocities and is used when interpreting the results of active and passive seismic studies. For example, based on the results of local tomography, the upper mantle layer was identified at a depth of 25–40 km in the Kerch–Taman segment of our model. This upper mantle underlies the thin crust of the northeastern Black Sea, most likely represented by serpentinized rocks of the continental upper mantle, which were exhumed as a result of Cretaceous rifting and intense extension of the lithosphere in this part of the Black Sea [6].

The obtained velocity parameters of high-velocity bodies in the Crimean Mountains (Vp = 6.6–6.8 km/s, Vs = 3.6–3.8 km/s, Vp/Vs = 1.75–1.85) correspond to the composition of basic rocks [2] of deep intrusive massifs. The most significant crustal body, with the top at a depth of ~12 km, is located near the cities of Gurzuf and Alushta, where a number of intrusive massifs of Middle Jurassic age are exposed at the surface. The largest of them are the intrusive massif of Mt. Ayu-Dag, composed of gabbro-diabases and the massif of Mt. Kastel, represented by garnet plagiogranites [9, 15].

We interpret the high-velocity anomalies identified by seismic tomography as deep sources of intrusive massifs outcropping near Gurzuf and Alushta. Such a powerful emplacement of numerous Middle Jurassic intrusions related to the island-arc formation [10, 15] is associated with the Early-Middle Jurassic stage of the Cimmerian orogeny [11]. These include the intrusive massifs and bodies on the southern coast of Crimea (Ayu-Dag, Chamly-Burun, etc.), which constitute the Pervomaisko-Ayu-Dag dolerit-gabbro-diorite complex and are located within the band of ~N-S strike extended from Yalta to Simferopol [15].

Intensive magmatism reworked almost the entire crust of the Crimean Mountains causing a significant gravity maximum (Fig. 6) and magnetic anomaly. A distinctive feature of Southern Crimea is strong Bouguer gravity anomaly reaching 100 mGal in amplitude [3]. Good agreement of velocity image of our tomographic model and

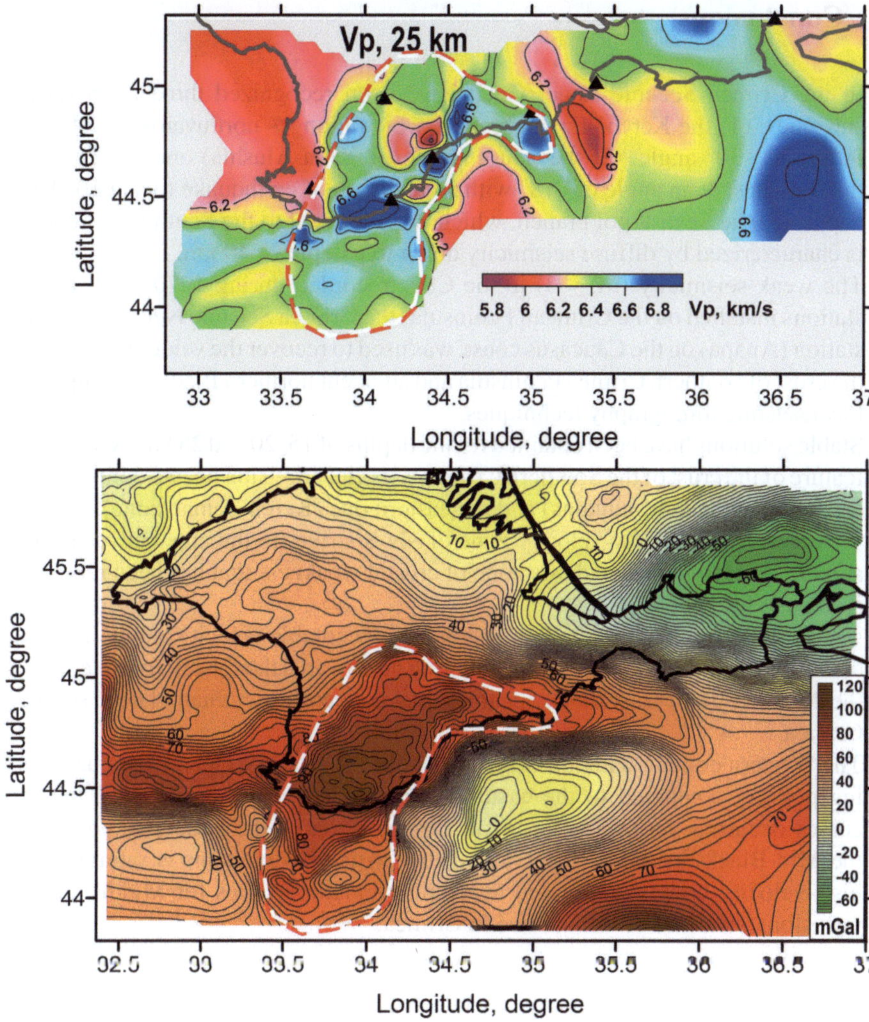

Fig. 6 Comparison of the P-wave velocity anomaly at 25 km depth (upper panel) and Bouguer gravity anomaly (lower panel) for the southern Crimea area. The dotted contour indicated the anomalous domain of high velocity and density

the gravity anomaly in Crimea (Fig. 6) could be explained by the fact that both anomalies (velocity and density) have common sources caused by reworking of the crust of southern Crimea by severe Middle Jurassic magmatism, combined with recent and ongoing collision processes (wedging and underthrusting) occurring in the environment of regional compression and transpression at the Alpine stage of tectogenesis.

5 Conclusions

- In the Crimea Seismic Zone (CSZ) have been recognized three main seismic subzones: (1) the Kerch–Taman branch, which plunges northwards to a depth of 90 km at ~30° angle; (2) the South Coast (or Yalta-Alushta) one, which tilts to the southeast at an angle of ~18° with localization of earthquake foci at 10–40 km depth; (3) the Sevastopol branch, which is orthogonal to the South Coast one and is characterized by diffuse seismicity down to a depth of 40 km.
- The weak seismicity ($M \leq 3$) in the CSZ, recorded during 1970–2013 by nine stations installed on the Crimean Peninsula (the Crimea Seismic Network) and one station (Anapa) on the Caucasus coast, was used to recover the velocity structure of the crust of southern Crimea Peninsula and adjacent northern Black Sea employing local seismic tomography techniques.
- Stable solutions have been obtained for the depths of 15, 20 and 25 km. A distinctive feature of the crust of the Southern Crimea is its clear division on two domains—the western (Crimean Mountains) and the eastern one (Kerch-Taman zone). They are separated by linear low-velocity zone of ~N-S strike (in the area between Sudak and Feodosiya) interpreted as a zone of crustal weakness that has been repeatedly reactivated.
- The Eastern domain (Kerch-Taman zone) lacks notable velocity anomalies in the crust. The velocity image here and plunging northwards the seismicity zone suggest the thrusting of the East Black Sea microplate under the Scythian Plate in southern Crimea.
- The Western domain with the Crimean Mountains shows a change in underthrusting polarity compared to the eastern Crimea. This may be related to the processes occurred during the Cenozoic compression of the structures controlling the Mid-Black Sea Rise. The crust of the Crimean Mountains is saturated with high-velocity bodies interpreted as deep magmatic sources for Middle-Jurassic intrusions observed on the surface in Crimea.

References

1. Backus, G., Gilbert, F.: The resolving power of gross earth data. Geophys. J. R. Astron. Soc. **16**, 169–205 (1968)
2. Christensen, N. I.: Poisson's ratio and crustal seismology. J. Geophys. Res.: Solid Earth **101**, 3139–3156 (1996)
3. Entin, V. A., Gintov, O. B., Guskov, S. I.: Once more on the nature of the Crimean gravitational anomaly. Geophys. J. **32**(6), 119–134 (2010) (in Russian)
4. Fernández-Viejo, G., Clowes, R.M., Welford, J.K.: Constraints on the composition of the crust and uppermost mantle in northwestern Canada: Vp/Vs variations along Lithoprobe's SNorCLE transect. Can. J. Earth Sci. **42**, 1205–1222 (2005)
5. Gobarenko, V.S., Murovskaya, A.V., Yegorova, T.P., Sheremet, E.E.: Collision processes at the northern margin of the Black Sea. Geotectonics **50**, 407–424 (2016)

6. Gobarenko, V., Yegorova, T, Stephenson, R.: Local tomography model of the northeast Black Sea: intraplate crustal underthrusting. In: Sosson, M, Stephenson, R.A., Adamia, S.A. (eds) Tectonic Evolution of the Eastern Black Sea and Caucasus. Geological Society, London, Special Publications, vol. 428, pp. 221–239, London (2017). https://doi.org/10.1144/sp428.2
7. Gobarenko, V.S., Nikolova, S.B., Yanovskaya, T.B.: 2-D and 3-D velocity patterns in southeastern Europe, Asia Minor and the Eastern Mediterranean from seismological data. Geophys. J. Roy. Astron. Soc. **90**, 473–484 (1987)
8. Kulchitsky, V.E., Pustovitenko, B.G.: 80 years of instrumental seismic observations in Crimea: history, results and prospects. Geophys. J. **30**, 9–49 (2008). (in Russian)
9. Lebedinskii, V.I., Solov'ev, I.V.: Bajocian volcanic structures of Gorny Crimea. Geol. J. **4**, 85–93 (1988)
10. Meijers, M.J.M., Vrouwe, B., van Hinsbergen, D.J.J., Kuiper, K.F., Wijbrans, J., Davies, G.R., Stephenson, R.A., Kaymakci, N., Matenco, L., Saintot, A.: Jurassic arc volcanism on Crimea (Ukraine): implications for the paleo-subduction zone configuration of the Black Sea region. Lithos **119**, 412–426 (2010). https://doi.org/10.1016/j.lithos.2010.07.017
11. Mileev, V.S., Rosanov, S.B., Baraboshkin, E.Yu., Rogov, M.A.: Cimmerian and alpine tectonics of Gorny Crimea. Byull. Mosk. O-va. Ispyt. Prir., Otd. Geol. **81**(3), 22–33 (2006) (in Russian)
12. Pustovitenko, B.G. (ed). Seismological Bulletin for the Western ESSN Zone (Crimea-Carpathians), 1970–1990. Naukova Dumka, Kiev (1970–1990) (in Russian)
13. Pustovitenko, B.G. (ed.) Seismological Bulletin of Ukraine, 1991–2013. EKOSI-Gydrofizika, Sevastopol (1991–2013) (in Russian)
14. Rudnick, R.L., Fountain, D.M.: Nature and composition of the continental crust: a lower crustal perspective. Rev. Geophys. **33**, 267–309 (1995)
15. Spiridonov, E.M., Fedorov, T.O., Ryakhovskii, V.M.: Magmatic rocks of Gorny Crimea. Part 1. Byull. Mosk. O-va. Ispyt. Prir., Otd. Geol. **65**(4), 119–133 (1990) (in Russian)
16. Tugolesov, D.A., Gorshkov, A.S., Meisner, L.B., Solov'ev, V.V., Khakhalev, V.I.: Tectonics of Mesozoic–Cenozoic Deposits of the Black Sea Basin. Nedra, Moscow, 215 p. (1985) (in Russian)
17. Yanovskaya, T.B., Gobarenko, V.S., Yegorova, T.P.: Subcrustal structure of the Black Sea Basin from seismological data. Izv., Phys. Solid Earth **52**, 14–28 (2016)
18. Yegorova, T., Gobarenko, V., Yanovskaya, T.: Lithosphere structure of the Black Sea from 3-D gravity analysis and seismic tomography. Geophys. J. Int. **193**, 287–303 (2013). https://doi.org/10.1093/gji/ggs098
19. Yegorova T.P., Baranova E.P., Gobarenko V.S., Murovskaya A.V.: Crustal structure of the Crimean mountains along the Sevastopol-Kerch profile from the results of DSS and local seismic tomography. Geotectonics **52**, 77–95 (2018)

Anisotropy of the Uppermost Mantle in Europe as Obtained from Surface Wave Data

T. B. Yanovskaya, E. L. Lyskova and T. Yu. Koroleva

Abstract A new approach is proposed to study radial anisotropy of the upper mantle under the European continent. Unlike the methods used so far, we propose to interchange the order of inversion procedures. At the first stage the path-averaged velocity sections of the SH and SV waves are constructed on each path and the path-average anisotropy coefficient is calculated from these sections at different depths. At the second stage the anisotropy coefficients for individual paths are used as initial data in the tomography procedure. This approach excludes different smoothness of Rayleigh and love wave velocities, which leads to significant errors in the estimation of the anisotropy coefficient obtained from the results of SV and SH velocity tomography. The method was applied to the group velocity dispersion data of Rayleigh and Love waves from both earthquake records and ambient seismic noise along the paths crossing the European continent. The results are presented as maps of anisotropy coefficient in three 20 km layers under the Moho boundary. It was found that in the central part of the study area the anisotropy coefficient is either close to zero or characterized by small negative values (about −1%). The values of the anisotropy coefficient decrease with depth. In the sea and coastal areas, the anisotropy coefficient is positive and its values are close to those of the oceanic mantle. The causes for exceeding the velocity of SV waves with respect to the velocities of SH in the continental mantle are discussed.

Keywords Anisotropy · Upper mantle · Europe · Surface wave tomography

T. B. Yanovskaya (✉) · E. L. Lyskova · T. Yu. Koroleva
St. Petersburg State University, 7-9 Universitetskaya Emb., St. Petersburg 199034, Russia
e-mail: t.yanovskaya@spbu.ru

E. L. Lyskova
e-mail: e.lyskova@spbu.ru

T. Yu. Koroleva
e-mail: t.koroleva@spbu.ru

T. B. Yanovskaya et al. (eds.), *Problems of Geocosmos—2018*, Springer Proceedings in Earth and Environmental Sciences, https://doi.org/10.1007/978-3-030-21788-4_19

1 Introduction

In the early sixties it was found that Love and Rayleigh dispersion curves along oceanic paths cannot be satisfied by a single S-wave velocity-depth section [1, 9, 11, 13]. The dispersion curves of Love waves result in overestimated values of the velocities at depths up to ~200 km in comparison with those obtained from Rayleigh wave observations. This discrepancy was explained by radial anisotropy of seismic wave velocities in the mantle. A difference between SH and SV velocities resulted from radial anisotropy as a property of the Earth's material was included in the PREM model [6], though this information was based mainly on data obtained along oceanic paths—on the one hand, because the oceanic structure is the most homogeneous, and on the other—because the oceans cover most of the globe [7, 16, 25]. Since the PREM model was originally built as spherically symmetric, it is clear that the anisotropy model could only be transversally isotropic (radial-anisotropic). The true anisotropy of the earth's material is certainly much more complicated, and, of course, the model parameters must depend on lateral coordinates.

In particular, observations of surface waves show the existence of azimuthal anisotropy [12, 17, 19], although it is considered to be less pronounced. In the PREM model, based on the data obtained by that time, the difference in the velocities of SH and SV waves under the Moho boundary reaches ~4%, decreases with depth and disappears below the boundary of about 200 km. Anisotropy is varied over the globe, it characterizes the properties of the earth's material, as well as the history of the tectonic structures and modern dynamic processes.

Development of tomography methods in the last two decades, as well as accumulation of the surface wave data led to possibility to regionalize the velocity sections obtained from the data of Love and Rayleigh waves and, accordingly, to conclude about lateral variations of anisotropy. In [27], the authors constructed a three-dimensional model of radial anisotropy for the Earth using the data on phase delays of Rayleigh and Love waves and applying the tomography method based on the Born approximation [26]. In their global anisotropy model SH velocity exceeds the velocity of SV waves to 220 km depth, as in the PREM model, though at lower depths, in contrast to PREM, V_{SV} turns out to be larger than V_{SH}. Lateral variations of the anisotropy coefficient were also detected. Later such an approach to tomographic reconstruction was not widespread. In most of studies the data on group velocities of Rayleigh and Love waves were used to estimate lateral anisotropy variations.

Information on anisotropy under continental areas is extremely scarce that is probably due to significant variations in the structure of the crust and upper mantle of the continents as compared to the oceanic areas. Yanovskaya and Kozhevnikov [22] used the data on the group velocity dispersion of Rayleigh and Love waves along the paths crossing the Asian continent, and obtained lateral velocity variations of SV and SH waves at different depths. It was found that SH velocities exceed the velocities of SV up to 6% at the depths of 100–200 km in Baikal rift region and in Tibet, but under Siberian Platform the anisotropy was not practically revealed.

Radial anisotropy in Tibet was studied using Rayleigh and Love wave data [5]. It was found that both proportions between SV and SH velocities may exist, i.e. $V_{SH} > V_{SV}$ and $V_{SV} > V_{SH}$, the anisotropy coefficient varying from -10 to $+10\%$. A similar result concerning the radial anisotropy of the crust in the Himalayas and Tibet was obtained in [8].

Estimations of the radial anisotropy in the European region are rather discrepant [3, 4, 10, 14, 15, 18]. In particular, in the model proposed in [4] some zones with $V_{SV} > V_{SH}$ were revealed, whereas in [15] $V_{SH} > V_{SV}$ was found everywhere. This discrepancy was explained (1) by a strong variability of the upper mantle structure under the continents as compared to the oceanic regions, (2) by different averaging areas for Rayleigh and Love wave velocities due to different data sets that leads to errors in the velocity-depth sections of SV and SH waves. And since these velocities are rather close, the errors in their differences may be of the order of these differences. Therefore to reduce errors in the final results is necessary to use uniform data sets in tomographic inversion.

In the present paper we applied an alternative approach for analysis of the anisotropy coefficient in the upper mantle below European continent. In this study we used the data from records both of earthquakes and seismic noise.

2 Method

As mentioned above, the anisotropy of the mantle in the oceanic regions was estimated from dispersion curves of Love and Rayleigh waves along the individual paths [2, 11, 13, etc.]. For each path the velocity sections of SH and SV waves were obtained from the group velocity dispersion curves, and the anisotropy coefficients $\alpha = \frac{V_{SH}-V_{SV}}{V_{SH}}$ at different depths were determined from their difference. Since there are no noticeable differences in the upper mantle structure under oceans, variations of the SH and SV velocities with depth obtained along different paths does not differ notably, and therefore could be averaged. The averaged values of the anisotropy coefficient, obtained from the data on the oceanic paths, were taken in the PREM model as the average for the whole Earth. The coefficient α varies from 4.8% under the Moho boundary and becomes negligible at the depth of 220 km.

Studies of anisotropy on the continents appeared in the beginning of this century, due to the widespread use of tomography methods. The traditional method for constructing 3D velocity sections of SH and SV waves from dispersion curves of Love and Rayleigh waves and subsequent estimate of the anisotropy coefficient consists of two stages. At the 1st stage 2D tomography problem is solved for determining so-called 'local' (more correctly a locally-averaged) dispersion curves from Rayleigh and Love wave velocities obtained along different paths for a set of periods. At the 2nd stage 1D inversion problem is solved for determining S-wave velocity section in separate points from local dispersion curves obtained at the 1st stage. The 'local' SH and SV velocity sections are used to estimate anisotropy coefficient at different points and at different depths.

Only reliable conclusion that have been obtained from such studies is that the anisotropy coefficients in the continental upper mantle can be both positive and negative. Lateral variations of this coefficient differ greatly in different studies. In particular, Chang et al. [4] revealed the zones in which $V_{SV} > V_{SH}$, whereas in [15] $V_{SH} > V_{SV}$ was obtained everywhere. As mentioned in the Sect. 1, this discrepancy is explained by a strong variability of the upper mantle structure under the continents in comparison with oceanic regions, and by different degree of smoothness of Rayleigh and Love wave velocities due to usage of the dispersion curves not along the same paths, and sometimes not in the same intervals of periods. This results in a variable degree of smoothness of SV and SH local velocity sections and, accordingly, to errors in anisotropy coefficient. Therefore, to obtain more reliable conclusions we modified the method for obtaining lateral variations of the anisotropy coefficient.

At the 1st stage we propose to solve 1D inverse problem to determine SV and SH velocity sections averaged along each paths. This allows us to estimate the path-averaged anisotropy coefficients at different depths. At the 2nd stage the 2D tomographic technique is applied to the data of 'path' anisotropy coefficients at different depths. Such procedure leads to a decrease in the amount of the data for tomography, because it requires to use the data only of those paths, for which it is possible to obtain sufficiently confident dispersion curves of both Rayleigh and Love waves in the same interval of periods.

Due to the ambiguity in constructing the velocity-depth sections from dispersion curves, the anisotropy coefficients are calculated not at individual depths, but as averaged in some selected depth intervals. Then, the tomography technique is applied to the data on the path averaged anisotropy coefficients in the selected depth intervals in order to obtain their lateral variations. Of course, this procedure is rather laborious in case of large data sets, but even when the amount of the data is limited, it gives steady though smoothed results.

3 Data

In this study we used the data on group velocity dispersion curve of Love and Rayleigh waves in the interval of periods 10–100 s obtained both from earthquake records and from cross-correlation functions of seismic noise (CCF). The advantage of earthquake data is high signal-to-noise ratio, and disadvantage is that sources are localized to confined seismic zones, so that it is impossible to cover the region under study by paths in wide range of azimuths. On the other hand, station-to-station paths cover the territory more or less uniformly in different directions but cross-correlation functions are often distorted, especially at large periods. This results from unevenness of noise sources: as was shown in [24] that the sources of the ambient noise at the periods larger than 40–50 s are distant earthquakes. To achieve more or less uniformity in the spatial distribution of the noise sources it was proposed to calculate the CCF by summation for a long period of time.

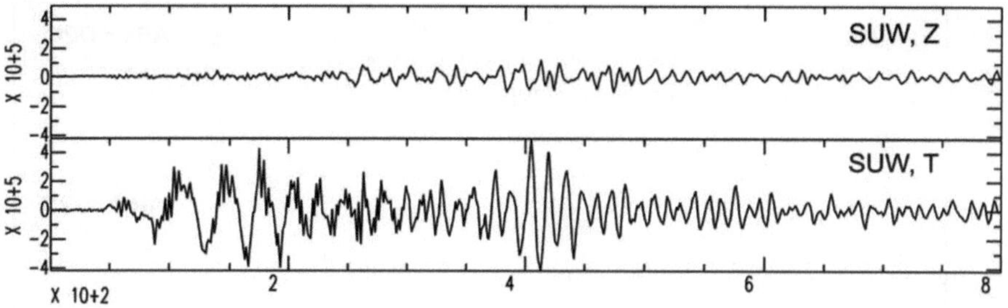

Fig. 1 Earthquake in Turkey (12-11-1999, Mw7.2). At the SUW station (Poland) a strong Love wave (T-component) was recorded, while the Rayleigh wave (Z-component) turned out to be much weaker

These reasons led to impossibility to select a large body of data on dispersion curves of Love and Rayleigh waves at the same paths and in the same interval of periods. In particular it was difficult to get necessary data at the European stations from earthquakes in Minor Asia, though many strong earthquakes occur there. It was easy to obtain dispersion curves of Love waves but practically impossible to do this for Rayleigh waves. Source mechanisms of these earthquakes are mostly horizontal strike-slips that practically do not excite Rayleigh waves. An example of seismogram from an earthquake in Turkey recorded at the station SUW is shown in Fig. 1.

As mentioned above, determination of the anisotropy coefficient from the CCF of the ambient noise suffers from some uncertainty. This is due to the fact that at periods above 40–50 s the main source of seismic noise is the earthquakes, and their distribution over the surface is uneven due to the occurrence of earthquakes in localized seismic zones. This leads in some cases to an overestimation of surface wave velocity due to influence of sources on the off-line paths between the stations: the difference in arrival times of the waves from such sources is less than in the case when the sources are located on the station-to-station line. If this were the only cause, such an understatement would take place approximately equally for the velocities of Love and Rayleigh waves and would have a little effect on the magnitude of the anisotropy coefficient. But there is an additional reason for a distortion of the transverse-transverse (T-T) cross-correlation. If the source is outside the station-to-station line, the transverse component (projection of the horizontal displacement to a perpendicular to the station-station line) will contain a contribution not only from Love waves, but also partly from Rayleigh wave. Thus, the T-T cross-correlation will include a contribution from Rayleigh wave so that the Love wave velocity determined from it will be underestimated. Moreover in many cases the T-T cross-correlations are distorted so strongly that it is impossible to determine Love wave dispersion from them. Figure 2 shows such an example of cross-correlation functions Z-Z and T-T between the stations ARU and DPC.

Necessity to obtain high-quality data for both Love and Rayleigh dispersion curves at the same paths leads to considerable reduction of the data. However this drawback is compensated by selection of the high-quality data.

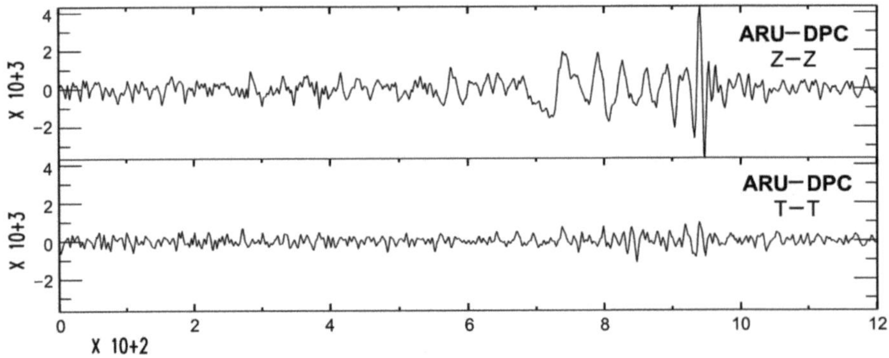

Fig. 2 Cross-correlation functions between the stations ARU and DPC: above Z-Z, below T-T

Inversion of Rayleigh and Love wave data was carried out by successive approximations. At the first step a joint inversion is carried out to obtain a common solution from both Rayleigh and Love wave dispersion curves in the framework of the isotropic model. The solution obtained at this stage is accepted as a starting model for separate inversion—for $V_{SV}(z)$ from Rayleigh wave and for $V_{SH}(z)$ from Love wave dispersion curves. The two solutions are averaged, and the averaged velocity-depth curve is accepted as a new starting model for the joint inversion, etc. The process stops when the successive solutions are stabilized.

Examples of getting the data for tomography are demonstrated in Fig. 3.

Anisotropy coefficients α averaged in the selected depth intervals are indicated at the right panels. As is seen from Fig. 3, values of α may be positive and negative, though the most of α values turn out to be positive or close to zero. The negative anisotropy coefficients are obtained at about 20% paths. Therefore we may confidentially assert that the zones in the European upper mantle where $V_{SV} > V_{SH}$ really exist.

Total number of paths on which we succeeded to determine the anisotropy coefficients amounts to 250 in the in three uppermost layers below Moho (35–55, 55–75, 75–95 km). Values of the anisotropy coefficients in these layers were used for tomographic inversion. A pattern of paths along which the anisotropy coefficients in three layers below Moho have been reliably obtained is shown in Fig. 4. For tomographic inversion we used the method based on Backus-Gilbert approach [21]. Resolving power of the solution was estimated by a linear size of the averaging area [20].

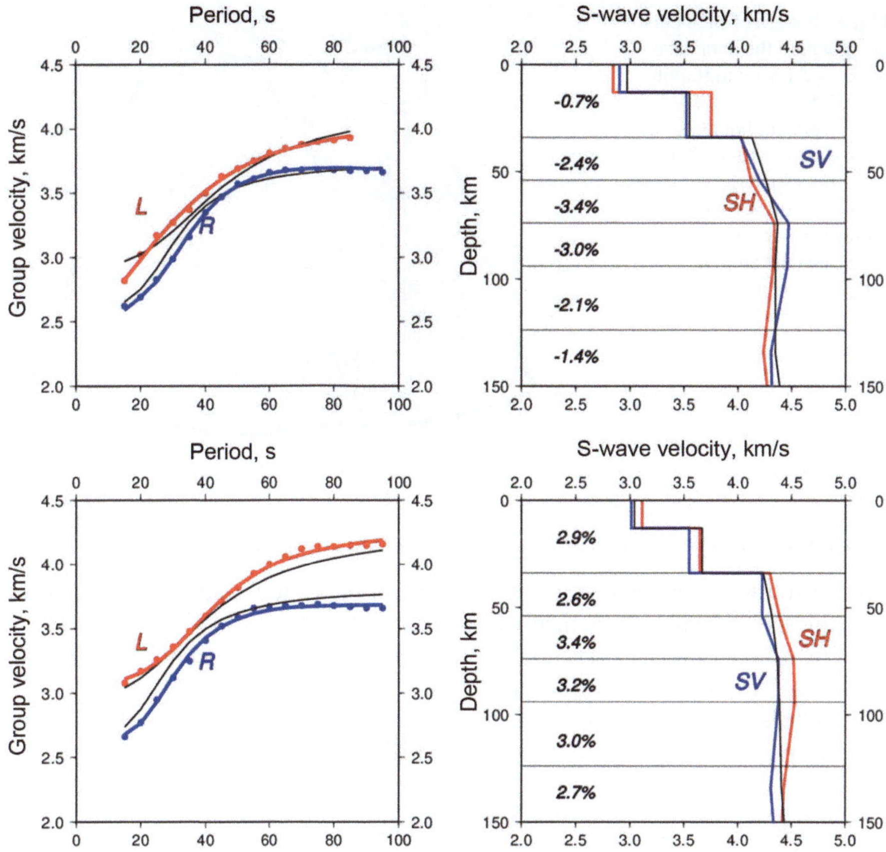

Fig. 3 Examples of determination of the path anisotropy coefficients. At left: points are group velocity data, black lines—calculated dispersion curves for common velocity-depth curve satisfied simultaneously to both data for Love and Rayleigh waves, color lines—dispersion curves calculated for velocity-depth solutions satisfied to Love and Rayleigh wave data separately. At right: black lines—velocity-depth solutions from joint inversion of both Rayleigh and Love wave data, color lines—from separate inversion. Horizontal lines indicate boundaries of the layers in which the anisotropy coefficients in percentage were determined. Above—at the paths between source in South Greece ($\varphi = 22.10°$N, $\lambda = 36.35°$E) and the station DPC ($\varphi = 16.41°$N, $\lambda = 50.3°$E), $V_{SV} > V_{SH}$. Below—at the paths between source in Northern Iran ($\varphi = 33.88°$N, $\lambda = 59.82°$E) and the station TAM ($\varphi = 22.79°$N, $\lambda = 5.52°$E), $V_{SH} > V_{SV}$

4 Results

Blue line in Fig. 4 delineates the area of acceptable resolution for three upper mantle layers: linear size of the averaging area within this contour is less than 1200 km and a ratio of the largest to the smallest linear dimensions does not exceed 2. Lateral variations of the anisotropy coefficient within this area are represented in Fig. 5 for three upper layers below Moho boundary.

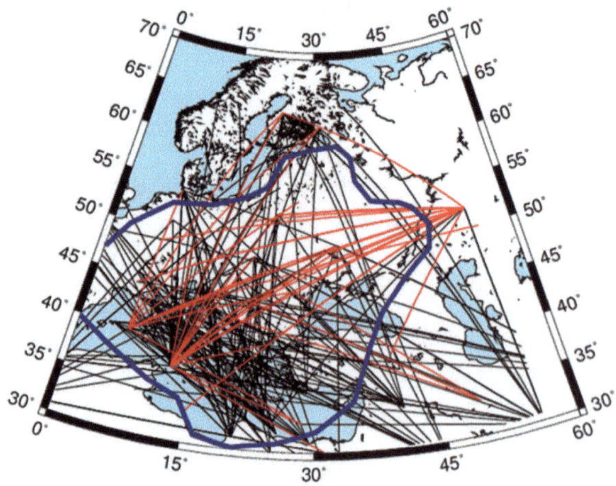

Fig. 4 Pattern of the paths along which the data have been used for tomographic inversion. Black lines—the paths from earthquakes, red lines—station-to-station paths for which the data were obtained from CCFs of ambient noise. Blue line delineates the area of acceptable resolution

Main features of the lateral variations of the anisotropy are as follows:

(1) anisotropy coefficient α is close to zero in the central part of the region, i.e. European mantle is practically isotropic;
(2) in the East-European Platform α is weakly negative (about −1%), the difference V_{SV}–V_{SH} increasing with depth;
(3) in the sea and coastal areas α is positive and decreases with depth as in oceanic regions, values being close to those in PREM—about 3–4%.

5 Discussion and Conclusions

Preliminary results on lateral variations of the anisotropy coefficient obtained from smaller amount of data (~170 traces) are published in [23]. The main features of the anisotropy pattern given there are consistent with those represented in Fig. 5

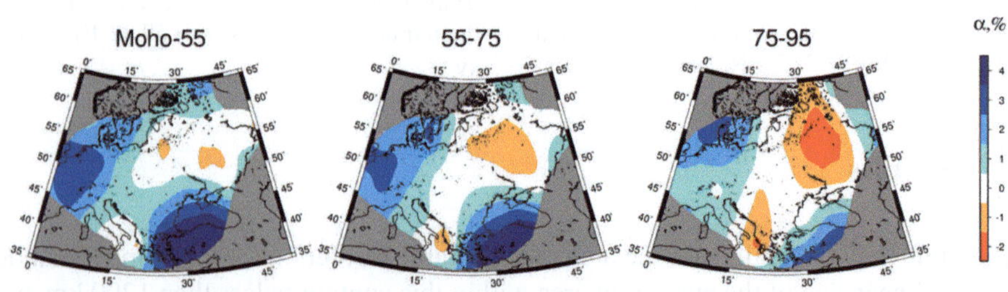

Fig. 5 The lateral distributions of the anisotropy coefficient α (%) in the region of acceptable resolution in the three depth intervals. They were obtained both from earthquake records and from cross-correlation functions of seismic noise

Fig. 6 Pattern of the paths at which the anisotropy coefficient turns out to be negative

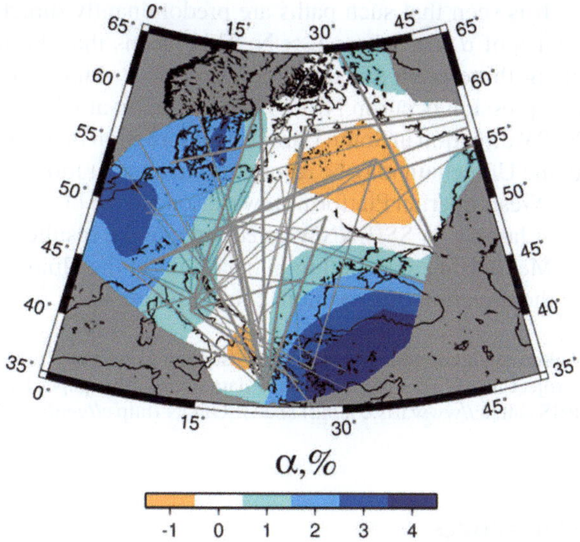

though being more smoothed. The area in the uppermost mantle characterized by negative anisotropy coefficient was also stretched from Central Ural to Apennines. This confirms the reliability of the results since the increase in the data sample by half did not change the main conclusions about the lateral distribution of the anisotropy coefficient in the European upper mantle but made it more detailed.

A possible explanation of the obtained anisotropy behavior may be as follows. In the framework of the transversely isotropic model the case $V_{SH} > V_{SV}$ that is typical for oceanic regions corresponds to a relation between the elastic modulus $L < N$. On the contrary $V_{SV} > V_{SH}$ corresponds to $L > N$. The first case means that the shear deformation in all horizontal directions should be larger comparatively with the vertical one. This makes the horizontal flow of masses easier in the oceanic upper mantle. It is completely in agreement with the concept of plate tectonics. The case $L > N$ (negative anisotropy coefficient) means that a medium is subjected to a horizontal pressure. In the framework of transversely isotropic model the modulus L should not depend on azimuth in a horizontal plane, so that the rocks should be in a state of uniform horizontal compression. But this is difficult to explain from tectonic point of view: tectonic movements in continental regions happen in some preferred direction that leads to non-uniform compression forming mountain chains. In such a case the modulus L should depend on azimuth, so that the model of transverse isotropy becomes unfounded. In other words, the azimuthal anisotropy cannot be neglected.

Dependence of the anisotropy coefficient on azimuth is confirmed by a pattern of paths on which negative values of α have been determined. The paths on which the anisotropy coefficient is strongly negative (to −4%) are shown Fig. 6 in comparison with results of tomography for a depth interval 55–75 km (from Fig. 5).

It is seen that such paths are predominantly directed along the area of negative values of α, i.e. where L > N. This means that the largest values of L take place along these predominant directions, i.e. L values, and correspondingly, maximum compression occur in NEE-SWW direction at East-European Platform and in SSE-NNW direction in South Europe. Compression in the former area being perpendicular to the Ural chain may be resulted from ancient movements of the East European and the West Siberian Platforms towards one to the other. The southern branch of the area with large L in SSE-NNW direction may be resulted from the present-day opening of Mediterranean resulting in the origin of the alpine folded structure perpendicular to the direction of maximum compression.

Acknowledgements The study was supported by the Russian Foundation for Basic Research (project no. 17-05-00522). Seismic data used in this paper have been obtained from data centers IRIS (https://www.iris.edu/hq) and GEOFON (https://geofon.gfz-potsdam.de).

References

1. Anderson, D.L.: Recent evidence concerning the structure and composition of the Earth's mantle. In: Physics and Chemistry of the Earth. Pergamon Press, Oxford, pp 1–131 (1966)
2. Anderson, D.L., Regan, J.: Upper mantle anisotropy and the oceanic lithosphere. Geophys. Res. Lett. **10**(9), 841–844 (1983)
3. Boschi, L., Fry, B., Ekström, G., Giardini, D.: The European upper mantle as seen by surface waves. Surv. Geophys. **30**, 463–501 (2009)
4. Chang, S.-J., van der Lee, S., Matzel, E., Bedle, H.: Radial anisotropy along the Tethyan margin. Geophys. J. Int. **182**(2), 1013–1024 (2010). https://doi.org/10.1111/j.1365-246X.2010.04662.x
5. Chen, Y., Badal, José, Zhang, Zh: Radial anisotropy in the crust and upper mantle beneath the Qinghai-Tibet Plateau and surrounding regions. J. Asian Earth Sci. **36**, 289–302 (2009)
6. Dziewonski, A.M., Anderson, D.L.: Preliminary reference Earth model. Phys. Earth Planet. Inter. **25**, 297–356 (1981)
7. Forsyth, D.W.: The early structural evolution and anisotropy of the oceanic upper mantle. Geophys J. R. Astr. Soc. **43**, 103–162 (1975)
8. Guo, Z., Gao, X., Wang, W., Yao, Z.: Upper- and mid-crustal radial anisotropy beneath the central Himalaya and southern Tibet from seismic ambient noise tomography. Geophys. J. Int. **189**:1169–1182 (2012)
9. Harkrider D.G., Anderson D.L.: Computation of surface wave dispersion for multilayered anisotropic media. Bull. Seismol. Soc. Am. **52**(2), 321–332 (1962)
10. Kustowski, B., Ekström, G., Dziewonski, A.M.: The shear wave velocity structure in the upper mantle beneath Eurasia. Geophys. J. Int. **174**, 978–992 (2008)
11. Mitchell, B.J.: On the inversion of Love- and Rayleigh-wave dispersion and implications for the Earth structure and anisotropy. Geophys. J. R. Astr. Soc. **76**, 233–341 (1984)
12. Montagner, J.-P., Nataf, H.C.: On the inversion of the azimuthal anisotropy of surface waves. J. Geophys. Res. **91**, 511–520 (1986)
13. Nishimura, C.E., Forsyth, D.W.: The anisotropic structure of the upper mantle in the Pacific. Geophys. J. Int. **96**, 203–229 (1989)
14. Schaefer, J.F., Boschi, L., Becker, T.W., Kissling, E.: Radial anisotropy in the European mantle: tomographic studies explored in terms of mantle flow. Geophys. Res. Lett. **38**, L23304 (2011). https://doi.org/10.1029/2011GL049687
15. Schivardi, R., Morelli, A.: EPmantle: a 3-D transversely isotropic model of the upper mantle under the European Plate. Geophys. J. Int. **185**, 469–484 (2011)

16. Schlue, J., Knopoff, L.: Shear-wave polarization anisotropy in the Pacific basin. Geophys. J. R. Astr. Soc. **49**, 145–165 (1977)
17. Shaeffer, A.J., Lebedev, S., Becker, T.W.: Azimuthal seismic anisotropy in the Earth's uppermantle and the thickness of tectonic plates. Geophys. J. Int. **207**(2), 901–933 (2016)
18. Shapiro, N.M., Ritzwoller, M.H.: Monte-Carlo inversion for a global shear velocity model of the crust and upper mantle. Geophys. J. Int. **151**, 88–105 (2002)
19. Smith, M.L., Dahlen, F.A.: The azimuthal dependence of Love and Rayleigh wave propagation in a slightly anisotropic medium. J. Geophys. Res. **78**, 3321–3333 (1973)
20. Yanovskaya, T.B.: Resolution estimation in the problems of seismic ray tomography. Izv. Phys. Solid Earth **33**(9), 762–765 (1997)
21. Yanovskaya, T.B., Ditmar, P.G.: Smoothness criteria in surface wave tomography. Geophys. J. Int. **102**(1), 63–72 (1990)
22. Yanovskaya, T.B., Kozhevnikov, V.M.: Upper mantle anisotropy beneath the Asian continent from group velocities of Rayleigh and Love waves. Russ. Geol. Geophys. **47**(5), 618–625 (2006)
23. Yanovskaya, T.B., Lyskova, E.L., Koroleva, T.Yu.: Radial anisotropy in the European upper mantle from surface wave data. Izv. Phys. Solid Earth **55**(2), 195–204 (2019)
24. Yanovskaya, T., Koroleva, T., Lyskova, E.: Effect of earthquakes on ambient noise surface wave tomography in upper-mantle studies. Geophys. J. Int. **205**, 1208–1220 (2016). https://doi.org/10.1093/gji/ggw083
25. Yu, G., Mitchell, B.J.: Regionalized shear velocity models of the Pacific upper mantle from observed Love and Rayleigh wave dispersion. Geophys. J. R. Astr. Soc. **57**, 311–341 (1979)
26. Zhou, Y., Dahlen, F.A., Nolet, G.: Three-dimensional sensitivity kernels for surface wave observables. Geophys. J. Int. **158**, 142–168 (2004)
27. Zhou, Y., Nolet, G., Dahlen, F.A., Laske, G.: Global upper-mantle structure from finite-frequency surface-wave tomography. J. Geophys. Res. **111**, B04304 (2006). https://doi.org/10.1029/2005JB003677

Heterogeneities of the Earth's Inner Core Boundary from Differential Measurements of PKiKP and PcP Seismic Phases

Dmitry Krasnoshchekov and Vladimir Ovtchinnikov

Abstract The Earth's crystalline inner core (IC) solidifies from the liquid Fe alloy of the outer core (OC), which releases latent heat and light elements sustaining the geodynamo. Variability in solidification regime at the inner core boundary (ICB) may result in compositional and thermal multi-scale mosaic of the IC surface and dissimilarity of its hemispheres. Both the mosaic and hemisphericity are poorly constrained, not least due to a lack of available sampling by short-period reflected waves. Measured amplitude ratio of seismic phases of PKiKP and PcP reflected, respectively, off the inner and outer boundary of the liquid core, yields direct estimate of the ICB density jump. This parameter is capable of constraining the inner–outer core compositional difference and latent energy release, but is not well known (0.2–1.2 g/cub. cm), and its distribution is obscure. Travel time measurements of PKiKP and PcP waveforms can be useful in terms of getting an insight into fine structure of ICB and its topography. We analyse a new representative sample of pre-critical PKiKP/PcP differential travel times and amplitude ratios that probe the core's spots under Southeastern Asia and South America. We observe a statistically significant systematic bias between the Asian and American measurements, and carefully examine its origin. Separating the effects of core-mantle boundary and ICB on the measured differentials is particularly challenging and we note that a whole class of physically valid models involving D″ heterogeneities and lateral variation in lower mantle attenuation can be employed to account for the observed bias. However, we find that variance in PKiKP-PcP differential travel times measured above the epicentral distance of 16° is essentially due to mantle heterogeneities. Analysis of data below this distance indicates the ICB density jump under Southeastern Asia can be about 0.3 g/cub. cm, which is three times as small as under South America where also the thickness of the above liquid core can be by 1–3 km in excess of the one in the East. The findings preclude neither IC hemispherical asymmetry (whereby crystallization dominates in the West and melting in the East) nor patchy IC surface, but provide an improved and robust estimate of the ICB density jump in two probed locations.

D. Krasnoshchekov (✉) · V. Ovtchinnikov
Institute of Dynamics of Geospheres, Russian Academy of Sciences, Leninsky pr. 38 korp. 1, 119334 Moscow, Russia
e-mail: krasnd@idg.chph.ras.ru

T. B. Yanovskaya et al. (eds.), *Problems of Geocosmos—2018*, Springer Proceedings in Earth and Environmental Sciences, https://doi.org/10.1007/978-3-030-21788-4_20

Keywords Inner core boundary · Reflected waves · Density jump

1 Motivation

The key region to understanding the Earth's crystalline iron inner core (IC) is the transition from liquid to solid core—the region where it grows by freezing [12]. Many details of this crystallisation are currently unknown, but there have been developed a number of scenarios that result in formation of inhomogeneous structures such as slurry, dendritic and non-dendritic morphologies, etc. (e.g., [9, 18, 27]. These heterogeneities are essentially due to the fact that solidification regime and the cooling rate during phase transition are not uniform throughout the inner core boundary (ICB) (e.g., [2, 11]. In seismic data such multi-scale heterogeneities may look like compositional and thermal mosaic of the IC exterior [15] and even dissimilar hemispheres [1, 2, 19].

A lot of seismological constraints on ICB structure and parameters have been obtained using the body waves reflected from it—the seismic phase of PKiKP. Studies of ICB and its vicinity by short-period pre-critical PKiKP waves normally involve a reference seismic phase whose path (in an ideal case) coincides with the one of PKiKP in crust and mantle. Differential travel times and amplitudes of PKiKP and the reference phase measured on a record of one and the same event are less contaminated with out-of-core effects and, consequently, yield insights into the ICB structure. For steeply reflected PKiKP waveforms, the seismic phase of PcP reflected off the core-mantle boundary (CMB) is the best reference, since this pair obviously yields adequate resolution, and close proximity of their crust and mantle paths cancels the influence of heterogeneities localized outside the Earth's core. It is these waveforms that produce the main evidence that the IC surface is mosaic [15] or rough [5], and enable localized estimates of the ICB density jump [3]. On the other hand, part of the results of joint analysis of PKiKP and PcP cannot be easily reconciled with other studies. For instance, the relevant ICB density jump estimates are usually larger (up to 1.8 g/cm^3) than normal mode outputs (0.9 g/cm^3 and below). Another controversy may arise from evidence for hemispherical pattern in ICB properties—it can be seen in teleseismic PKiKP data [19], but is not confirmed [31] by pre-critical reflections.

Lack of differential measurements of steeply reflected PKiKP and PcP is one of the reasons for controversies. Detection of these waveforms is tricky because they are subtle and always hidden in seismic coda formed by intensive reverberations in crust and mantle. In practice, PKiKP–PcP differential measurements are usually too few in number, yield sparse sampling, and exhibit large differences over small lateral length scales. For instance, on average, the observed PKiKP/PcP amplitude ratios should scatter around the 'true' amplitude ratio, and the best way to resolve concerns as to usability [4] and uncertainties [28] of such data and related estimates is to increase the number of measurements. This, however, may be difficult because the chief factor governing observability of steeply reflected PKiKP and PcP waveforms on daylight surface hasn't been established yet, despite efforts to examine various effects of seismic source, CMB [29] and ICB [15]. Thus, in the problem of sampling the ICB with PKiKP, where the coverage and number of samples are extremely limited, enhancing the dataset is of chief importance to improve the level of accuracy and confidence in the observations. In this paper we revisit heterogeneities of ICB using the newly presented in [16] database of a total of more than 1300 new differential travel times and amplitude ratios of PKiKP and PcP measured at 3.2°–35.2° and reflected off two spots in the western and eastern hemispheres of the Earth's core.

2 Data and Measurements

Pre-critical PKiKP and PcP waveforms were successfully detected on broadband and short-period records of four deep earthquakes (Table 1). The reflection points of the analysed dataset provide good sampling of two IC spots of about 125×240 km^2 under Bolivia and to the southeast of Sakhalin Island (Fig. 1). Each spot is scanned by hundreds of ray traces with incident angles from 2° to 20°. The number of rays and density of scanning are unprecedented and enable statistically significant and robust estimates of possible heterogeneities localized within the probed spots.

In general, observability of pre-critical PKiKP and PcP phases can be associated with favourable seismic energy radiation pattern at source and focusing along the propagation paths. According to the source solution of global CMT catalogue [7, 8], all four focal mechanisms of the analysed events including the one in Okhotsk Sea feature domination of vertical forces and faulting. The PKiKP and PcP incidence angles in such mechanisms are not far from maxima in the P-wave radiation pattern, which encourages observation of PKiKP and PcP waveforms. On the other hand, the

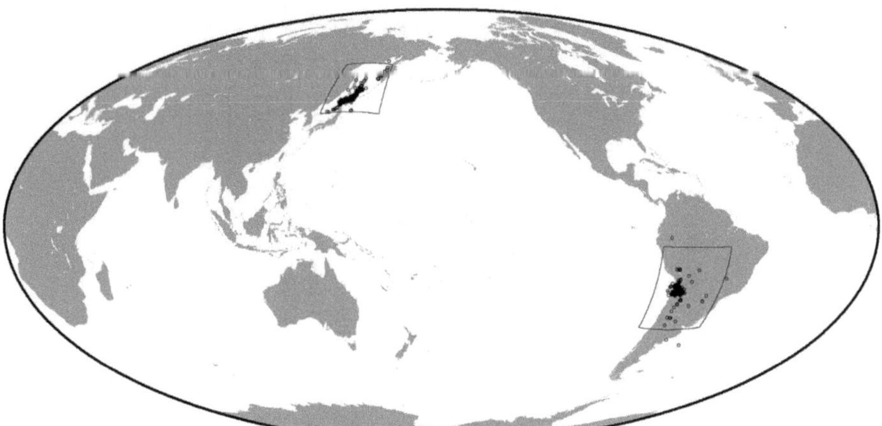

Fig. 1 Map with daylight surface projections of PKiKP reflection points below South America and Southeastern Asia. Surface rectangles roughly depict the ICB scanned spots

Table 1 Parameters of seismic sources and arrays/networks

Date (dd.mm.yyyy)	Origin time (hh:mm:ss.00)	Latitude (°)	Longitude (°)	Depth (km)	M_b	Δ (°)	Networks
Events under Southeastern America[a]							
12.07.2009	01:19:21.31	−15.0411	−70.5354	198.7	5.9	7.7–18.2	CX, TO, X6, XH, YS, ZL
24.05.2010	16:18:28.81	−8.1152	−71.6412	582.1	6.0	14.7–32.5	3A, XH, XP, XS
05.03.2012	07:46:09.23	−28.2579	−63.2916	551.9	6.0	5.4–34.2	ZD, ZG, ZV, XP
Event in Southeastern Asia (Okhotsk Sea)							
24.05.2013	14:56:31.60	52.1357	151.5688	632.0	6.8	3.2–35.2	SAGSR⁺, KAGSR⁺, J-array⁺⁺, Hi-net§

[a]DOI of networks contributed into the American dataset: ZL (https://doi.org/10.7914/sn/zl_2007), X6 (https://doi.org/10.7914/sn/x6_2007), XH (https://doi.org/10.7914/sn/xh_2008), TO (https://doi.org/10.7909/c3rn35sp), YS (https://doi.org/10.7914/sn/ys_2009), CX (https://doi.org/10.14470/pk615318), XS (https://doi.org/10.15778/resif.xs2010), XP (https://doi.org/10.7914/sn/xp_2010), ZG (https://doi.org/10.7914/sn/zg_2010), ZD (https://doi.org/10.7914/sn/zd_2010), ZV (https://doi.org/10.7914/sn/zv_2012). No DOI has the 3A network (Maule Aftershock Deployment (UK))

+Regional Network of the Russian Geophysical Survey http://www.ceme.gsras.ru/new/struct/

++Earthquake Research Institute, the University of Tokyo, Japan, http://www.eri.u-tokyo.ac.jp

§ [20, 21]

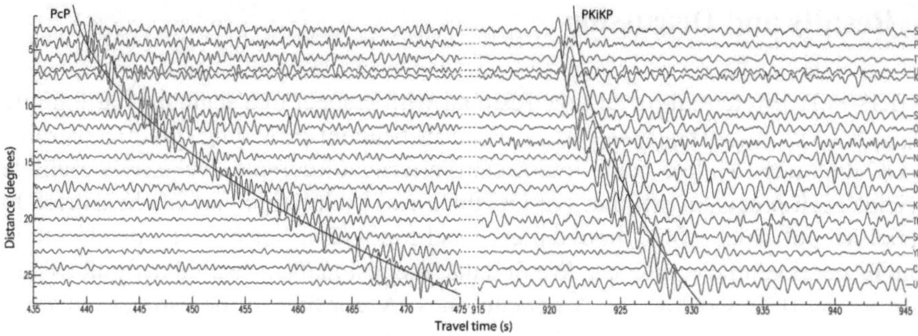

Fig. 2 Record section plot of 18 frequency-filtered vertical components (station names are on the right). Left pane: passage corresponding to arrival of PcP; right pane: to PKiKP. Standard travel time curves of PcP and PKiKP are computed for *ak135*

profound influence of seismic source can be questioned, because no PKiKP or PcP waveforms were observed in the records of the same stations after a nearby preceding event. It occurred 9 h before the analysed event in Okhotsk Sea, 300 km away to the North, and had the M_b magnitude of 7.3, similar focal depth and mechanism. Same considerations hold true and raise objections to the influence of focusing [31]. As a matter of fact, PKiKP and PcP from these two nearby events propagate in essentially the same geological settings, but are distinct after the analysed one, and not observable in the records of the preceding event in the whole distance range of 3°–35°.

To uniform the analysed dataset and increase signal-to-noise ratio of PKiKP and PcP waveforms, raw digital traces were frequency-filtered between 1.1 and 7 Hz. The filtering removed intensive crust and mantle reverberations and accentuated the detected pulse-shaped waveforms of PcP and PKiKP on vertical components. The revealed PKiKP and PcP waveforms build hyperbolic travel time curves, characteristic for the reflected phases (Fig. 2). We also note that each core-reflected arrival is followed by its own coda. It comes as continuous wavetrains with increased amplitudes observable not only on Hilbert transformed beams [30], but already on raw vertical and horizontal record section plots too. After filtering, both PKiKP and PcP waveforms dominate on the time interval tens of seconds long around the predicted arrival time and exhibit signal-to-noise ratios well above 2.5. Absolute travel times and amplitudes of pre-critical PKiKP and PcP are highly variable due to crust and mantle heterogeneities, but their influence can be mitigated by using PKiKP–PcP differential measurements. Picking was performed automatically in Seismic Analysis Code [10] by cross-correlation of PKiKP and PcP waveforms.

3 Results and Discussion

PKiKP–PcP differential travel time residual with respect to a standard Earth model $(t_{PKiKP}-t_{PcP})_{measured}-(t_{PKiKP}-t_{PcP})_{model}$ is an instrument providing direct constraints on OC thickness. Unless being interpreted in terms of complex velocity variations, it yields high precision estimates of OC thickness undulation that can be caused by spatial irregularities of OC boundaries, local topography of ICB, IC displacement from its geopotential centre, etc. We find that the residuals measured in Southeastern Asia are by 0.72 s below the American ones. According to the estimate given in [23], this indicates the OC under the Americas can be about 3 km thicker than under Southeastern Asia. Concretely, the mean PKiKP–PcP differential travel time residuals calculated with respect to *ak135* [13] and *PREM* [6] over 1016 Japanese records were, respectively, −1.79 and −0.41 s with s.d. of 0.51; same estimates for the American data made −1.07 ± 0.45 s and 0.31 ± 0.45 s. This estimate of OC thickness variability is rather an upper bound subject to allowance for ellipticity and possible influence of strongly heterogeneous lower mantle. We expect the elliptic corrections are almost one order of magnitude smaller than the observed 0.7 s, but mantle corrections can be quite large and variable depending on the model used. To get more accurate estimate, we invoke the up-to-date 3-D high-resolution tomographic model [24] of LLNL-Earth3D that takes into account Earth's ellipticity, undulating discontinuity surfaces and heterogeneities in mantle and crust. Also, it yields a credible result reconcilable with basic physical notions. Specifically, qualitative analysis predicts that crust and mantle heterogeneities have essentially similar influence on almost vertically propagated PKiKP and PcP (e.g. at the epicentral distance of 3.2°, where the respective Fresnel zones are larger than separation between the PcP and PKiKP pierce/reflection points). Farther away (at larger epicentre distances), where PKiKP and PcP paths start to diverge in lower mantle, mantle heterogeneities may act differently. This pattern can be observed in Fig. 3. It plots modelled and measured residuals overlaid by 0.5° binned averages for 1016 records of the Okhotsk Sea event (Table 1) by Japanese stations. Above the epicentral distance of 16.5°, the measured residuals completely coincide with predictions by LLNL-Earth3D model and therefore mostly contain information on heterogeneities outside the Earth's core. Thus, to decrease data contamination with non-core effects, we examined the statistics on residuals obtained only under 16.5° (data in the remainder of this paragraph are given with respect to PREM, because it has the IC radius of 1221.5 km, equal to that of LLNL-Earth3D). In this way, the mean of residuals calculated over 330 Asian reflections and 181 American ones were −0.45 ± 0.55 s and 0.27 ± 0.44 s, accordingly, while the relevant modelled residuals made −0.27 ± 0.07 s and 0.12 ± 0.10 s. These averages indicate that up to a half of the systematic bias between PKiKP–PcP differential travel time residuals measured in Asia and America can be accounted for by out-of-core structures, still the rest of the bias is induced by the Earth's core, statistically significant and equivalent to hemispherical disparity in OC thickness of about 1–3 km.

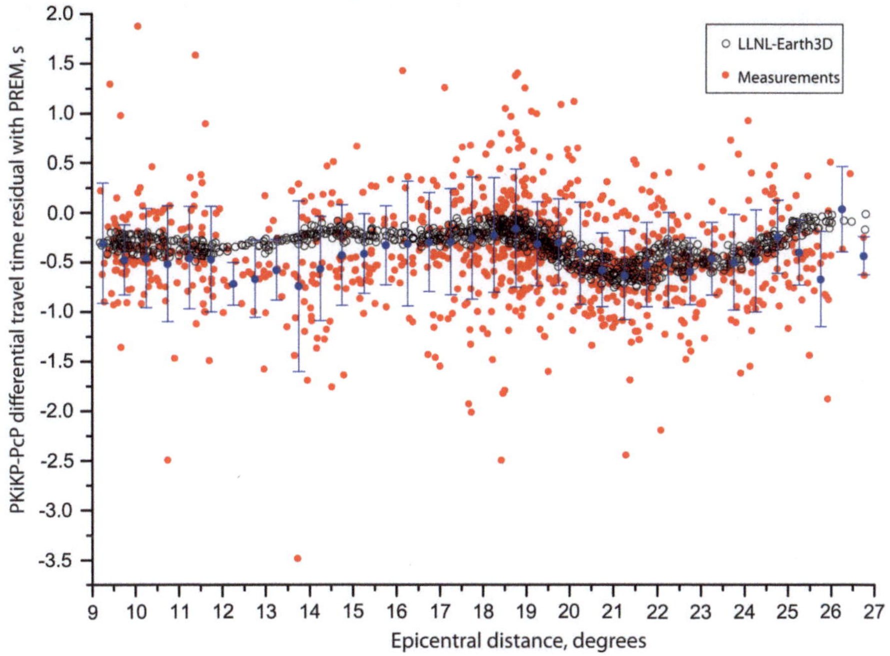

Fig. 3 Differential travel time residuals for 1016 records collected in Japan. Blue dots with standard deviation bars are 0.5° binned averages of measured residuals. Datapoint legend is in the upper right corner

Comparison of measured and model-predicted PKiKP/PcP amplitude ratios yields [3] direct estimates of the ICB density jump provided the CMB density and velocity contrasts are known and shear wave velocity in the top of the IC is fixed. Theoretical curves of various ICB density jump models are not far apart (especially above 10°), whereas the scatter of measurements is large—it can be seen in Fig. 4 that plots the whole dataset of measured PKiKP/PcP amplitude ratios. In addition, as argued above, it's reasonable to include only data under 16.5° (since above this limit the ratios may suffer from influence of heterogeneities outside the Earth's core). Figure 5 shows that Asian and American measurements up to epicentral distances of about 16° are consistently divided by a gap equivalent to the ICB density jump of about 0.6 g cm^{-3}. It is confirmed by binned averages. Given the notorious [14, 28] trade-off between variation of acoustic impedance contrast at ICB and CMB, and the resulting ICB density jump estimate, an alternative interpretation in terms of CMB density jump has to be examined too. The interpretation would assume 10–15% density variation between the probed American and Asian spots of the mantle bottom (Fig. 5). However, such variation can be controversial in geodynamical context because the sampled mantle sides of CMB feature essentially similar shear velocities [22] and material properties specific to regions outside the Pacific large low-shear-velocity province [17]. Strong density variations on the core side of CMB are hardly possible

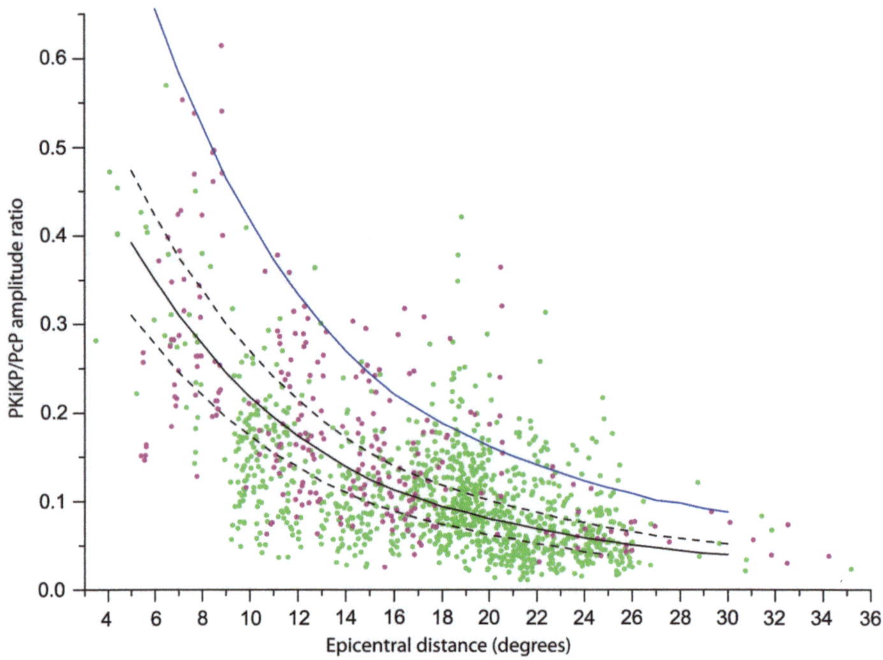

Fig. 4 Measured PKiKP/PcP amplitude ratios and their theoretical estimates for *ak135*. Green and violet dots—measured ratios from Asia and America, respectively. Theoretical curves are for ICB density jumps of 0.3 g/cm^3 (lower dash), 0.6 g/cm^3 (black solid), 0.9 g/cm^3 (upper dash), 1.8 g/cm^3 (blue solid)

too [25]. Still, the 3-D density variation in D$''$ has yet to be mapped, and thus the model with variable CMB density jump cannot be entirely ruled out, as well as, for example, a complex model with lateral variation of P velocity near CMB combined with lateral variation in either viscoelastic or scattering attenuation in the lowest 150 km of mantle. The latter model enables strong variation in PKiKP/PcP amplitude ratios [29], but can be a bit farfetched, especially if compared to a simple hemispheric scenario where ICB density jump in the eastern hemisphere is about 0.3 g cm^{-3}, and in the western—about 0.9 g cm^{-3}.

4 Conclusions

The analysed reflected data indicate that the sampled spots of the Earth's core below Southeastern Asia and South America feature dissimilar properties. The observations can be accounted for by a class of models assuming multifactorial contributions of out-of-core inhomogeneities, yet the most credible is a model with variable ICB density jump. We estimate it to be about 0.3 g/cm^3 under Southeastern Asia, and

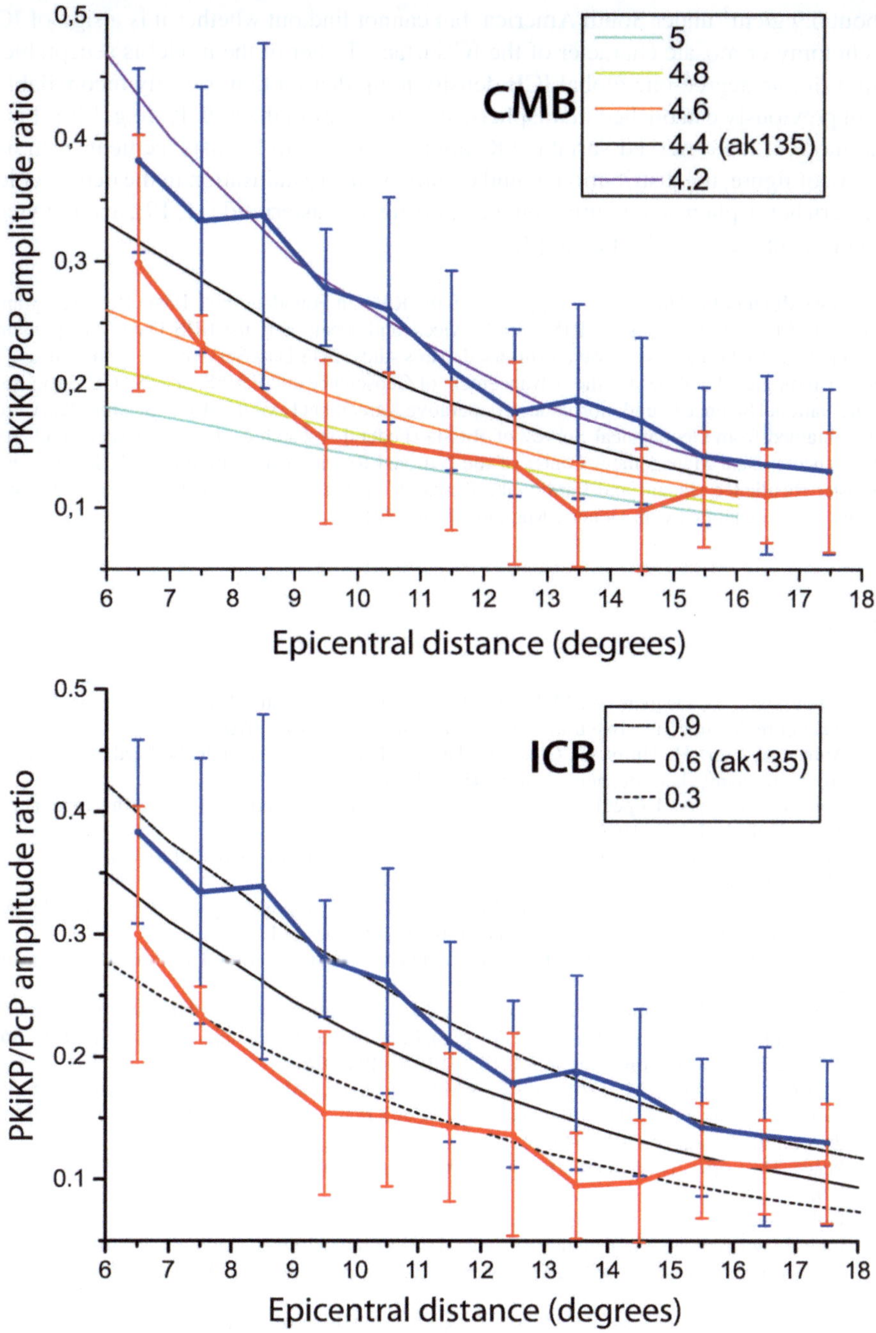

Fig. 5 Theoretical and observed dependencies of PKiKP/PcP amplitude ratio on distance. Theoretical curves on the base of *ak135* are for varying ICB and CMB density jumps given in the legends in g/cm^3. Thick blue and red lines are the polygon curves formed out of 1° binned amplitude ratios of the American and Asian subsets, respectively

about 0.9 g/cm^3 under South America, but cannot find out whether it is a sign of IC dichotomy or mosaic character of the IC surface. Either of the model is acceptable, but a simple degree-one global ICB density jump distribution is easily reconcilable with previously established hemispherical differences in the bulk IC (e.g. [26]. Furthermore, if the observed variable OC thickness is due to IC displacement from its centre of figure, the distribution would comply with crystallisation in the denser cold western hemisphere and melting on the opposite hot eastern side [1, 19], and not vice versa as argued by Aubert et al. [2].

Acknowledgements This work is supported by the Russian Foundation for Basic Research grant #18-05-00619. The facilities of IRIS Data Services, and specifically the IRIS Data Management Center, were used for access to waveforms used in this study. IRIS Data Services are funded through the Seismological Facilities for the Advancement of Geoscience and EarthScope (SAGE) Proposal of the National Science Foundation under Cooperative Agreement EAR-1261681. Seismic data were also obtained from Geophysical Survey of Russia (Federal research center of Russian Academy of Sciences), Data Management Center of the National Research Institute for Earth Science and Disaster Resilience (Japan) and Earthquake Research Institute, the University of Tokyo (Japan). Figure 1 was generated with Generic Mapping Tools [32].

References

1. Alboussiére, T., Deguen, R., Melzani, M.: Melting-induced stratification above the Earth's inner core due to convective translation. Nature **466**, 744–747 (2010)
2. Aubert, J., Amit, H., Hulot, G., Olson, P.: Thermochemical flows couple the Earth's inner core growth to mantle heterogeneity. Nature **454**, 758–761 (2008)
3. Bolt, B., Qamar, A.: Upper bound to the density jump at the boundary of the Earth's inner core. Nature **228**, 148–150 (1970)
4. Buchbinder, G.G.R., Wright, C., Poupinet, G.: Observations of PKiKP at distances less than 110°. Bull. Seismol. Soc. Am. **63**, 1699–1707 (1973)
5. de Silva, S., Cormier, V.F., Zheng, Y.: Inner core boundary topography explored with reflected and diffracted P waves. Phys. Earth Planet. Inter. **276**, 202–214 (2017)
6. Dziewonski, A.M., Anderson, D.L.: Preliminary reference Earth model. Phys. Earth Planet. Inter. **25**, 297–356 (1981)
7. Dziewonski, A.M., Chou, T.-A., Woodhouse, J.H.: Determination of earthquake source parameters from waveform data for studies of global and regional seismicity. J. Geophys. Res. **86**, 2825–2852 (1981). https://doi.org/10.1029/JB086iB04p02825
8. Ekström, G., Nettles, M., Dziewonski, A.M.: The global CMT project 2004-2010: Centroid-moment tensors for 13,017 earthquakes. Phys. Earth Planet. Inter. **200–201**, 1–9 (2012). https://doi.org/10.1016/j.pepi.2012.04.002
9. Fearn, D., Loper, D., Roberts, P.: Structure of the earth's inner core. Nature **292**, 232–233 (1981)
10. Goldstein, P., Dodge, D., Firpo, M., Minner, L.: SAC20.0: Signal processing and analysis tools for seismologists and engineers, in the IASPEI international handbook of earthquake and engineering seismology. In: Lee W.H.K. et al. (eds.), pp. 1613–1614. Academic Press, London (2003)
11. Gubbins, D., Sreenivasan, B., Mound, J., Rost, S.: Melting of the Earth's inner core. Nature **473**, 361–363 (2011)
12. Jacobs, J.A.: The Earth's inner core. Nature **172**, 297–298 (1953)
13. Kennett, B., Engdahl, E., Buland, R.: Constraints on seismic velocities in the Earth from travel times. Geophys. J. Int. **122**, 108–124 (1995)

14. Koper, D.K., Pyle, M.L.: Observations of PKiKP/PcP amplitude ratios and implications for Earth structure at the boundaries of the liquid core. J. Geophys. Res. **109**, B03301 (2004). https://doi.org/10.1029/2003JB002750
15. Krasnoshchekov, D.N., Kaazik, P.B., Ovtchinnikov, V.M.: Seismological evidence for mosaic structure of the surface of the Earth's inner core. Nature **435**, 483–487 (2005)
16. Krasnoshchekov, D.N., Ovtchinnikov, V.M.: The density jump at the inner core boundary in the eastern and western hemispheres. Doklady Earth Sci. 478, Part 2, 219–223 (2018)
17. Li, M., McNamara, A.K., Garnero, E.J., Yu, S.: Compositionally-distinct ultra-low velocity zones on Earth's core-mantle boundary. Nat. Commun. **8**, 177 (2017). https://doi.org/10.1038/s41467-017-00219-x
18. Loper, D.E., Roberts, P.H.: On the motion of an iron-alloy core containing a slurry. Geophys. Astrophys. Fluid Dyn. **9**(3–4), 289–321 (1978)
19. Monnereau, M., Calvet, M., Margerin, L., Souriau, A.: Lopsided growth of Earth's inner core. Science **328**, 1014–1017 (2010)
20. Obara, K., Kasahara, K., Hori, S., Okada, Y.: A densely distributed high-sensitivity seismograph network in Japan: Hi-net by National Research Institute for Earth science and disaster prevention. Rev. Sci. Instrum. **76**, 021301 (2005)
21. Okada, Y., Kasahara, K., Hori, S., Obara, K., Sekiguchi, S., Fujiwara, H., Yamamoto, A.: Recent progress of seismic observation networks in Japan -Hi-net, F-net, K-NET and KiK-net. Earth Planets Space **56**, xv–xxviii (2004)
22. Ritsema, J., van Heijst, H.J., Deuss, A., Woodhouse, J.H.: S40RTS: a degree-40 shear-velocity model for the mantle from new Rayleigh wave dispersion, teleseismic traveltimes, and normal-mode splitting function measurements. Geophys. J. Int. **184**, 1223–1236 (2011)
23. Shen, Z., Ai, Y., He, Y., Jiang, M.: Using pre-critical PKiKP–PcP phases to constrain the regional structures of the inner core boundary beneath East Asia. Phys. Earth Planet. Inter. **252**, 37–48 (2016)
24. Simmons, N.A., Myers, S.C., Johannesson, G., Matzel, E.: LLNL-G3Dv3: Global P wave tomography model for improved regional and teleseismic travel time prediction. J. Geophys. Res. **117**, B10302 (2012). https://doi.org/10.1029/2012JB009525
25. Stevenson, D.J.: Limits of lateral density and velocity variation in the Earth's outer core. Geoph. J. R. Astron. Soc. **88**, 311–319 (1987)
26. Tanaka, S., Hamaguchi, H.: Degree one heterogeneity and hemispherical variation of anisotropy in the inner core from PKP(BC)–PKP(DF) times. J. Geophys. Res. **102**(B2), 2925–2938 (1997)
27. Tian, D., Wen, L.: Seismological evidence for a localized mushy zone at the Earth inner core boundary. Nat. Commun. **8**, 165 (2017). https://doi.org/10.1038/s41467-017-00229-9
28. Tkalčić, H., Kennett, B., Cormier, V.: On the inner-outer core density contrast from PKiKP/PcP amplitude ratios and uncertainties caused by seismic noise. Geophys. J. Int. **179**, 425–443 (2009)
29. Tkalčić, H., Cormier, V., Kennett, B., He, K.: Steep reflections from the Earth's core reveal small-scale heterogeneity in the upper mantle. Phys. Earth Planet. Inter. **178**, 80–91 (2010)
30. Vidale, J.E., Earle, P.S.: Fine-scale heterogeneity in the Earth's inner core. Nature **404**, 273–275 (2000)
31. Waszek, L., Deuss, A.: Anomalously strong observations of *PKiKP/PcP* amplitude ratios on a global scale. J. Geophys. Res. Solid Earth **120**, 5175–5190 (2015)
32. Wessel, P., Smith, W.H.F.: New version of the generic mapping tools. EOS Trans. Am. Geophys. Union **76**, 329 (1995)

Superdeep Drilling and Its Impact on the Seismic Models of the Fennoscandian Shield

N. V. Sharov

Abstract Attempts to interpret the origin of seismic boundaries in the crystalline crust in the Fennoscandian Shield from deep drilling records and available geological and geophysical data were made by a number of authors. This article contains an overview of the most essential results reported so far. Deep drilling was conducted in several regions of the Fennoscandian Shield: Kola, Russia (SG-3), to a depth of 12,262 m (1970–1990); Gravberg (GR) and Stenberg-I, Central Sweden, to a depth of 6337 m (1986–1987); and 6529 m (1991–1992); Outokumpu (OU), SE Finland, to a depth of 2516 m (2004–2005); Pogranichnaya (P-1), Sredny Peninsula, Murmansk Region, Russia, to a depth of 5200 m (2004–2006); and Onega, Central Karelia, Russia (ON), to a depth of 3537 m (2007–2008). The results obtained were analyzed. They show considerable discrepancies between forecast seismic-geological models and the actual structure of the upper portion of the continental crust. Deep drilling has proved that heterogeneities in the upper portion of the crystalline crust, indicated by geophysical data, are due to changes in the composition and physical conditions of deep-seated rocks. Therefore, even approximate rock composition cannot be uniquely estimated from elastic wave velocities alone. Deep drilling showed the real pattern and origin of low-velocity zones in the upper portion of the crust and did not show that layers become more homogeneous with depth. The origin of intra-crustal seismic boundaries could be associated with various factors, such as variations in mineralogical composition, structural and textural differences between the rocks, and a thermodynamic setting.

Keywords Deep drilling · Fennoscandian shield · Seismic boundaries · Crystalline crust

N. V. Sharov (✉)
Karelian Research Centre, RAS, Institute of Geology, Petrozavodsk, Russia
e-mail: sharov@krc.karelia.ru

T. B. Yanovskaya et al. (eds.), *Problems of Geocosmos—2018*, Springer Proceedings in Earth and Environmental Sciences, https://doi.org/10.1007/978-3-030-21788-4_21

1 Introduction

Attempts to interpret the origin of seismic boundaries in the crystalline crust from deep drilling records and available geological and geophysical data in the Fennoscandian Shield were made by a number of authors. This article contains an overview of the most essential results reported so far.

Deep drilling was conducted in several regions of the Fennoscandian Shield: Kola, Russia (SG-3), to a depth of 12,262 m (1970–1990); Gravberg (GR) and Stenberg-I, Central Sweden, to a depth of 6337 m (1986–1987); and 6529 m (1991–1992); Outokumpu (OU), SE Finland, to a depth of 2516 m (2004–2005); Pogranichnaya (P-1), Sredny Peninsula, Murmansk Region, Russia, to a depth of 5200 m (2004–2006); and Onega, Central Karelia, Russia (ON), to a depth of 3537 m (2007–2008). The results obtained were analyzed (Fig. 1) [10]. They show considerable discrepancies between forecast seismic-geological models and the actual structure of the upper portion of the continental crust. Deep drilling has proved that heterogeneities in the upper portion of the crystalline crust, indicated by geophysical data, are due to changes in both the composition and physical conditions of deep-seated rocks. Therefore, even approximate rock composition cannot be uniquely estimated from elastic wave velocities alone. Deep drilling showed the real pattern and origin of low-velocity zones in the upper portion of the crust and did not show that layers become more homogeneous with depth. The origin of intra-crustal seismic boundaries could be associated with various factors, such as variations in mineralogical composition, structural and textural differences between the rocks, and a thermodynamic setting [8].

2 Results and Discussion

The Kola Superdeep Drillhole (SG-3) has opened a new epoch in the study of the continental crust [3, 4]. It has shown that our knowledge even of upper crustal layers is incomplete. The vertical thickness of the Pechenga complex in SG-3 turned out to be 6800 m, not 4700 m, as was estimated earlier by gravity and seismic methods [5]. Drilling of SG-3 showed that the inclined seismic boundary, considered at the drillhole design stage, reflects a large-scale interlayer fault, rather than the bottom of the Pechenga complex (Fig. 2). The vertical thickness of gneisses in the Kola series along the drill hole SG-3 was 4900 m (6800–11,700 m). The abundance of amphibolite in the gneiss increases from 15 to 40% from a depth of 7300 m throughout the entire sequence. An old basement or an Archean basement complex was revealed under the gneiss at a depth of 11,700–12,236 m. It is either the lower portion of a granitic-metamorphic layer or the upper portion of a new (diorite) layer [4]. The occurrence of ore mineralization and fluids to the maximum depth of the drill hole was confirmed.

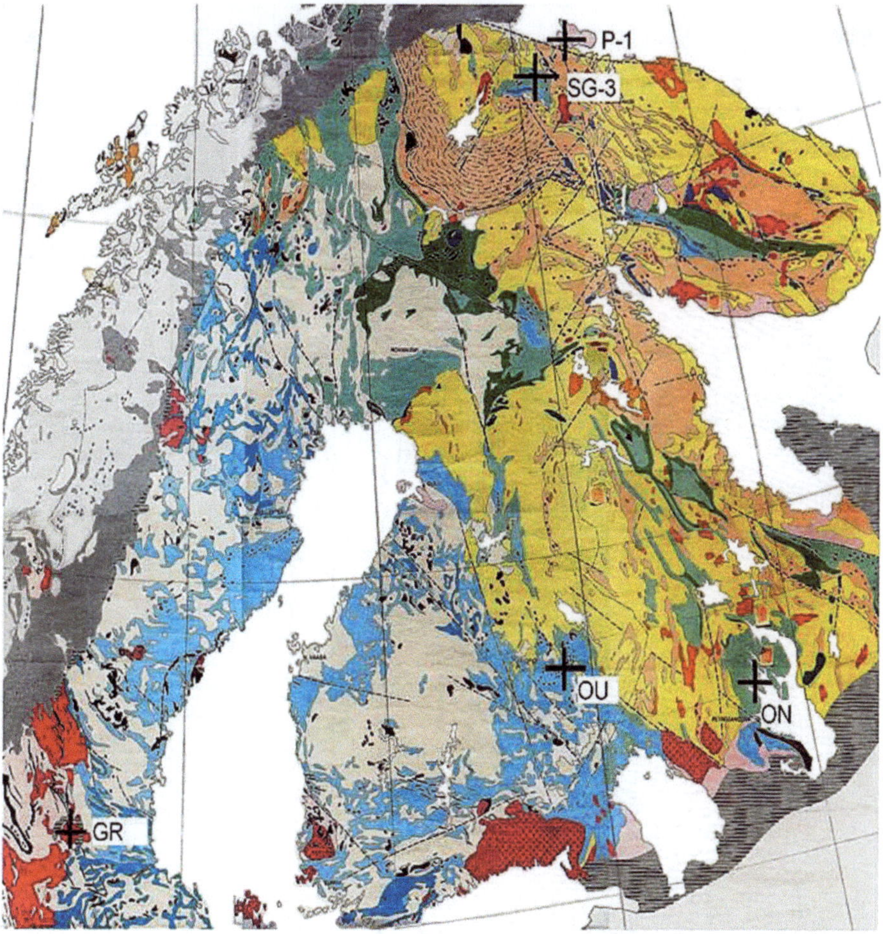

Fig. 1 Fragment of a geological map of the Fennoscandian Shield [10] and the location of deep drill holes

In summary, direct observations of the composition and structure of the old crystalline crust to a depth of 12,262 m were obtained for the first time. Thanks to that, our knowledge of the origin of geophysical boundaries has improved.

Pogranichnaya Drillhole (P-1) Four wells were drilled: the primary well P-1 to a depth of 5200 m and three additional wells; R-1 to a depth of 3001 m, R-2 to a depth of 800 m, and K-1 to a depth of 1000 m. Two prospecting drillholes R-2 and K-1 revealed tectonic breccias with a steady long-term gas (methane, ethane) inflow at a depth of 790 and 800 m.

The 5200 m deep P-1 drillhole intersected Proterozoic volcanic-sedimentary rocks and pierced Archean plagiogneiss at a depth of 1100 m. **Common Depth Point** (CDP)

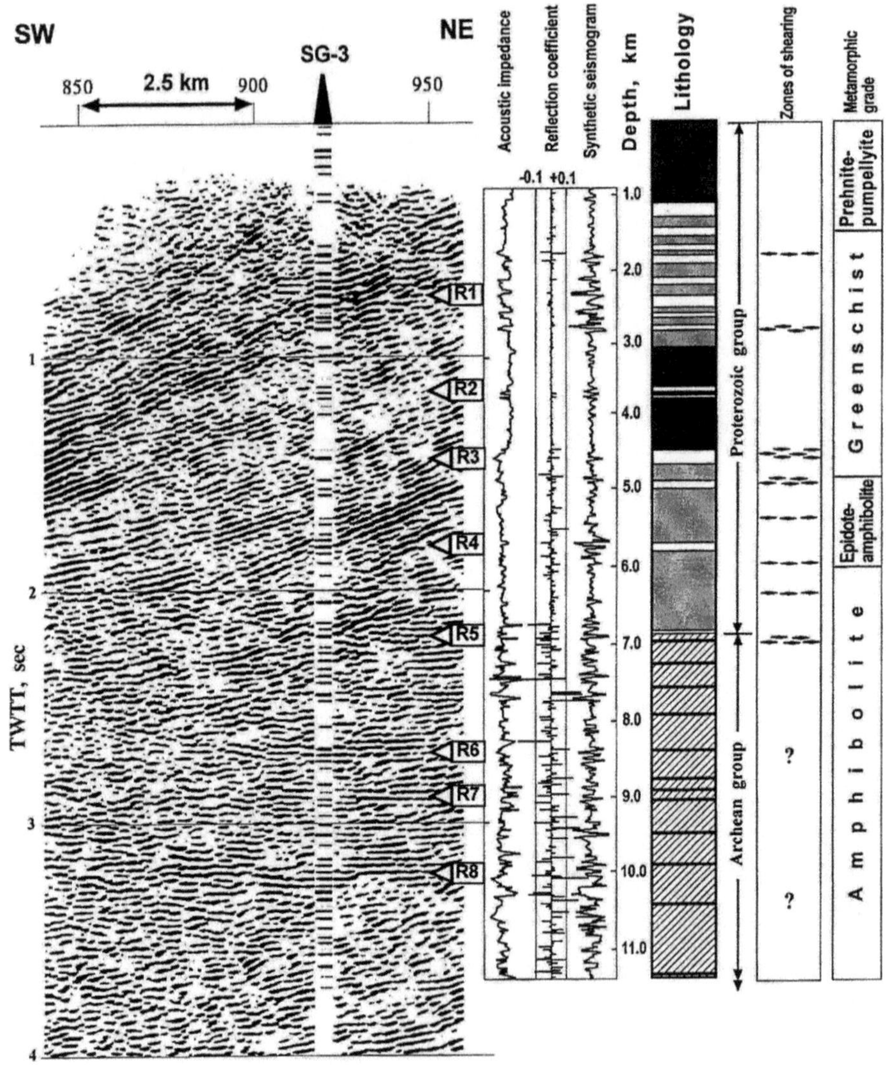

Fig. 2 Comparison of seismic data with the density and elasticity of rocks from the drillhole SG-3 [8]

seismic data indicated two reflecting boundaries at a depth of 1200 and 2300 m in the Pogranichnaya drillhole area (Fig. 3) [7]. There is a series of "strong" reflecting boundaries there. Their real existence raised no doubt. These boundaries must have been mentioned in the drilling record, although their depth position could differ by several percentages from the drilling data due to a seismic wave velocity estimation error. The structural pattern in the seismic section suggested the presence of a thick layered Upper Proterozoic sedimentary cover, at least to a depth of about 2.5 km.

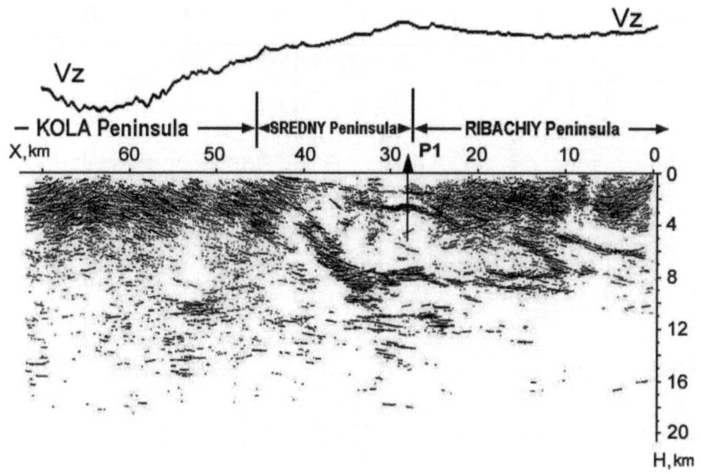

Fig. 3 CDP section and a plot of a gravity anomalies [7]

Downwards, the pattern becomes largely irregular (non-layered). However, Riphean sedimentary rocks were assumed to extend farther as a thick clinoform on the shield slope. The base of the clinoform in the Rybachy Peninsula (see P-I, Fig. 1) was assumed to be at a depth of 8 km, where the surface of the AR-PR_1 crystalline basement was presumably located [8].

Thus, analysis of geophysical data from the Fennoscandian Shield-Barents Plate transition zone shows that the reference reflecting horizons, identified in the seismic wave field of CDP reflections, are valid [7, 8]. Drilling records show, on the other hand, that the geological interpretation of these boundaries differs substantially from the original interpretation of the sequence [7] based on the initial concept of the essential contribution of thrust tectonics in the Sredny Peninsula-Rybachy Peninsula contact zone [8]. The tectonic data over the region, based on the study of gravity and magnetic fields, do not support such considerably vertical translocations either [1, 7]. The Riphean terrigenous complex in Sredny Peninsula is generally not more than 1.0–1.5 km thick and in Rybachy Peninsula not more than 4–5 km thick [1, 8].

Gravberg Drillhole (GR) Reflected seismic method studies along a series of profiles within the Silyan granite ring structure (Central Sweden) were conducted in 1985 and correlated with drilling records from two superdeep Gravberg drillholes (6337 m) [9] and Stenberg drillhole (6529 m) [8]. The seismic profiles intersected that structure in N–S and E–W directions. The most interesting result of the seismic study was revealing a series of sub-horizontal reflecting horizons in the northern portion (in the right of the Gravberg drillhole) of Silyan ring structure (Fig. 4). Three upper horizons were intersected by 6337 m deep Gravberg drill hole. The horizons were found to be caused by 20- and 60-m thick dolerite intrusions and a thinly-laminated zone, in which some of the layers are 3–9 m thick [9]. The intrusions penetrated

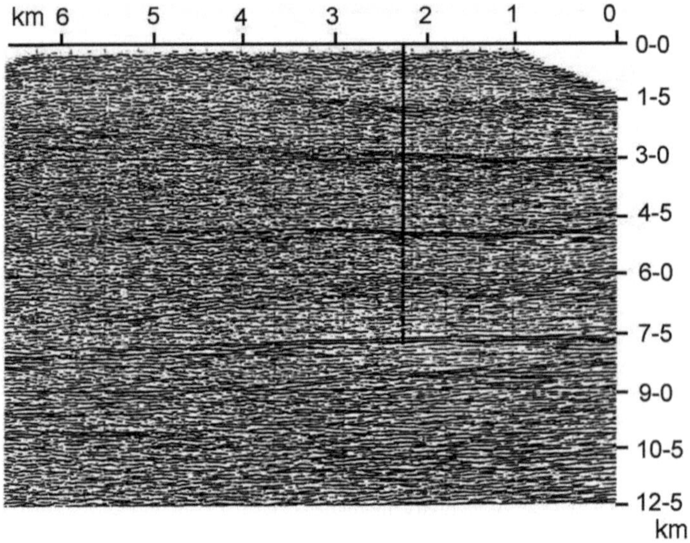

Fig. 4 Seismic CDP section and the position of Gravberg drillhole [9]

into the granite massif along the faults. The reflecting horizons are thus due to the contacts of different rocks and tectonic planes.

Outokumpu Drillhole The 2516 m deep well is located in SE Finland, near a worked-out base-metal deposit. The drillhole intersected an Archean-Proterozoic contact rock zone at a depth of 1310 m. Seismic reconstructions (Fig. 5a) suggested that mafic rocks, such as mica schist and black shale. Located at a depth of 1310 m is a boundary which separates Proterozoic rocks from the Archean basement. Metamorphic rocks were expected to occur at a depth of 1400–1900 m. The real sequence showed (Fig. 5b) mica schist in the upper portion of the well to a depth of 1310 m. The 1310–1515 m depth interval consists of alternating black shale, biotite gneiss and serpentinite layers. In the 1655–2516 m depth interval mica schist alternate with biotite gneiss and pegmatoid granite layers [11].

Onega Parametric Drill Hole (ON) A continuous 3537 m thick Early Precambrian sequence, which comprises basement granites of Late Archean age and the main strata of the early Karelides consisting of Jatulian, Ludicovian and Kalevian suprahorizons, was revealed. Core samples from the well were analyzed. Furthermore, the results of the geophysical study of the drill holes (GSD) and vertical seismic profiling (VSP) were interpreted.

A salt-bearing (halite) unit was identified in ON at a depth of 2750.8–2944.0 m. Halite rocks in the Onega structure were found for the first time. In view of the age of the salt (Early Precambrian, Jatulian superhorizon), it is the oldest of all known salts [6]. The total thickness of the salt-bearing unit is 193.2 m [2, 7].

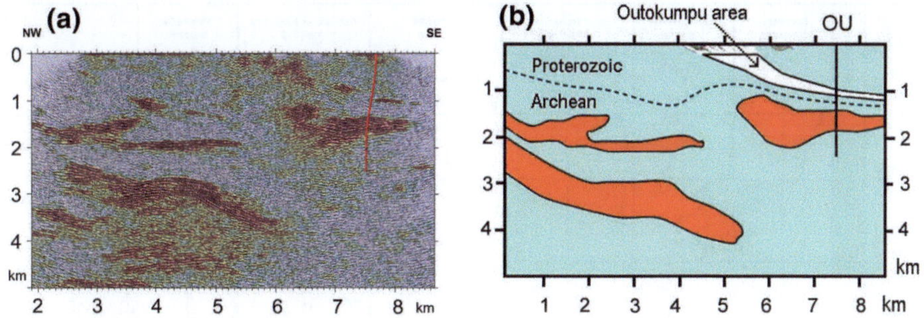

Fig. 5 Migrated CDP section, with the position of Outokumpu drillhole shown in red (**a**) and the geological-geophysical section (**b**) of the OU drillhole area. Modified after [8, 11] with permission of the author, Professor Ilmo T. Kukkonen

Fig. 6 Rock samples from the salt-bearing unit, OPD: 1—2868.5 m; 2—2897.4 m; 3—2905.6 m [2]

Analysis of the core samples (Fig. 6) showed that the composition and structure of the salt-bearing unit are rather homogeneous. The rocks have a bright-red color, which passes into dirty-grey with a slight pink tint in some depth intervals. The rock consists of medium-grained 3–6 mm thick halite (70–75%), partly dissolved anhydrite 2–3 mm thick crystals (12–20%) and 0.5–2.0 mm thick well-processed magnesite crystals (10–15%).

Figure 7 shows a litho-density model of the salt-bearing unit structure based on GSD and core analysis. The salt-bearing unit is a homogeneous halite rock sequence. The top and base of the sequence comprise transition zones formed as a result of tectonic, metasomatic and hydrodynamic processes that took place at the contact of the salt-bearing unit with underlying and overlying rocks. Elevated density intervals in the sequence, consisting dominantly of carbonate, terrigenous, sulphate and igneous rock veins or fragments, were identified. All these rocks are much denser than halite [2].

The formation of the salt-bearing unit was most probably affected by the spatial redistribution (outflow, lateral flow, injection) of rock masses under volumetric tectonic flow conditions at various depth levels of the lithosphere. This could have

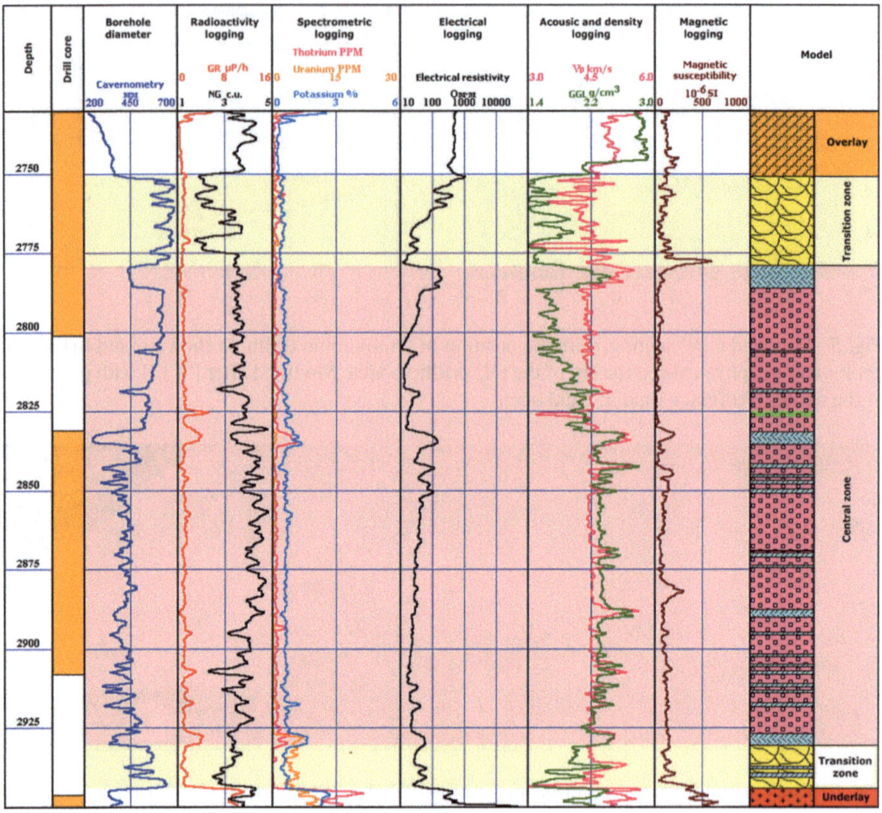

Fig. 7 Geophysical characteristics of the salt-bearing unit [2]

been invoked by the differentiation of accumulating volcanic-sedimentary rocks in thickness and composition, which led to changes in the vertical and horizontal components of geostatic pressure vectors. A greater number of tectonic stresses may have also occurred in the area where the survey to locate major deep faults was conducted [6].

Many questions regarding the formation of the salt-bearing unit may be answered by drilling one more borehole to estimate variations in the thickness and composition of the salt-bearing unit and to assess the local and regional scope of this phenomenon [2].

3 Conclusion

Deep drilling provided direct observations of the composition and structure of the upper portion of the crystalline crust in the Fennoscandian Shield down to a depth of 12 km. They showed that we know quite a bit about the continental crust and made us considerably revise our earlier concepts.

The results of the geophysical studies, conducted as part of deep drilling in the Fennoscandian Shield, make scientists re-interpret relevant geological and geophysical data. The mineralogical composition and petrophysical characteristics of rocks can be assessed only by deep drilling of the crystalline crust. The objective geological-geophysical sequence of the upper portion of the earth crust can then be constructed.

References

1. Glaznev, V.N., Sorokhtin, N.O., Rayevsky, A.B., Kozlov, N.E.: Identifying crustal reflections in the Baltic Shield-Barents Sea Plate transition zone as a basis for forecasting useful minerals. Vestnik VGU. Series: Geology, no. 1, pp. 95–102 (2014). (in Russian)
2. Yesipko, O.A., Neronova, I.V., Sharov, N.V.: Geophysical studies of the Onega Parametric drillhole, p. 62. OAO NPC Nedra (2014). (in Russian)
3. Kozlovsky, E.A. (ed.): Kola Superdeep Drill Hole. In: Study of the deep structure of the continental crust by boring the Kola Superdeep Drillhole, p. 490. Nedra, Moscow (1984). (in Russian)
4. Orlov, V.P., Laverov, N.P. (eds.): Kola Superdeep Drill Hole. In: Scientific results and research experience, p. 260. MF Tekhnoneftegaz, Moscow (1998). (in Russian)
5. Litvinenko I.V.: Seismic method for the study of the deep structure of the Baltic Shield, vol. 46, no. 2, pp. 3–13. Zap LGI (1963). (in Russian)
6. Glushanin, L.V., Sharov, N.V., Shchiptsov, V.V. (eds.): Paleoproterozoic Onega structure: geology, tectonics, deep structure and mineral genesis, p. 431. KarRC, RAS, Petrozavodsk (2011). (in Russian)
7. Sharov, N.V., Isanina, E.V., Pozhilenko, V.I., Stupak, V.M.: Increasing the validity of the geological models of the earth crust in the Sredny-Rybachy peninsula area by complexing CDP, DSS and PSM seismic methods. Uralski geofizicheski vestnik, no. 4, pp. 98–105. Yekaterinburg. Izd. IGUr, RAS (2007). (in Russian)
8. Sharov, N.V.: Lithosphere of North Europe based on seismic data, p. 173. Petrozavodsk KarRC, RAS (2017). (in Russian)
9. Beswick A.J.: Gravberg-1: Well design and drilling preparations. In: Boden, A., Eriksson, K.G. (eds.) Deep Drilling in Crystalline Bedrock. Volume 1: The Deep Gas Drilling in the Siljan Impact Structure, Sweden and Astroblemes, pp. 55–64. Springer, Berlin (1988)
10. Koistinen, T., Stephens, M.B., Bogachev, V., Nordgulen, O., Wennerström, M., Korchonen, J.: Geological map of the Fennoscandian Shield, scale 1:2000000. Geological Survey of Finland, Norway and Sweden and North-West Department of Natural Resources of Russia (2001)
11. Kukkonen, I.T., Rath, V., Kivekäs, L., Šafanda, J., Čermák, V.: Geothermal studies of the Outokumpu deep drill hole. In: Kukkonen, I.T. (ed.) Outokumpu Deep Drilling Project 2003–2010, Special Paper 51, pp. 181–198. Geological Survey of Finland, Espoo (2011)

Some Remarks on the Seismometric Experiments Taking into Account the Thickness of the Frozen Layer Soil

K. V. Kislov and V. V. Gravirov

Abstract Methodical studies of seismic wave propagation remain relevant in the shallow seismology, since the accuracy of the results depends on the appropriate analysis of the data. Monitoring of the thickness of the frozen ground is necessary for many scientific objectives and applications. The thickness of the frozen soil affects the accuracy of seismo-acoustic data in mining, seismo-acoustic logging and geological surveys' used to study the structure of the geological formations, coal-seam fractures, assessment of the geological environmental stress state and shallow tomography. While the effect of the soil temperature on the velocity of seismic waves has been investigated well the seismic waves attenuation affected by the thickness of the frozen Earth top layer needs to be investigated further. No sufficient data is available of the electrical properties of the frozen top layer of the Earth.

Keywords Freezemeter · Frozen soil layer thickness · Seismic survey · Monitoring · Seismic signal

1 Introduction

In Russia, about 11 million km^2 of the territory are covered with permafrost. Besides the fact that the thickness of the permafrost is uneven (due to different thickness, lenses and interlayers, taliks), the permafrost is also a dynamic system that follows the seasonal and climatic changes. Almost throughout the rest of the country, the active layer of the soil undergoes periodic freezing and thawing.

It is widely known that seismic wave velocity depends on the temperature of the soil. It was pointed out that the most significant velocity variations happen when the

K. V. Kislov (✉) · V. V. Gravirov
Institute of Earthquake Prediction Theory and Mathematical Geophysics of the Russian Academy of Science (IEPT RAS), Profsoyuznaya Str. 84/32, Moscow 117997, Russia
e-mail: kvkislov@yandex.ru

V. V. Gravirov
The Schmidt Institute of Physics of the Earth of the Russian Academy of Sciences (IPE RAS), B. Gruzinskaya Str. 10, Moscow 123242, Russia

T. B. Yanovskaya et al. (eds.), *Problems of Geocosmos—2018*, Springer Proceedings in Earth and Environmental Sciences, https://doi.org/10.1007/978-3-030-21788-4_22

Earth ground freezes or thaws. These phenomena happen due to the water-ice-water phase transitions within rocks of different dispersity and lithology. But there is a lack of understanding of seismic waves attenuation which depends on the depths of the frozen layer. The thickness of the frozen layer affects the accuracy of seismo-acoustic data in mining, seismo-acoustic logging, and geological surveys used to study the structure of geological formations, coal-seam fractures, assessment of the geological environmental stress state and shallow tomography. The electrical properties of frozen soils have to be further investigated. The calculated results of the average depth of frost penetration or the measured maximum depth of the frozen layer for a given area are often inaccurate.

The monitoring system was created to obtain geophysical data related to the thickness of the frozen top layer of soil. The general principles of solving the problem are described below. Promising scientific experiments are a detailed study of the propagation of seismic and acoustic waves in the near-surface layer of the Earth; electric properties of frozen soil of different thickness etc. It is very important and highly relevant issue for seismic exploration and shallow geophysics in general, for improving the resolution of seismic stations, studying seismic noise and for seismic microzoning of areas with frozen soils [1]. Monitoring of the thickness of the frozen soil could be used for the data processing in the structural seismology for corrections (including dynamic correction) that take into account the heterogeneity of the upper part of the geological environment [2]. Methodical studies of seismic wave propagation in shallow seismology remain relevant as the correct interpretation of the data depends on the accuracy of observations. The influence of subsurface on seismic signal can be so severe that static corrections are inadequate [3–5].

2 Monitoring of the Frozen Layer

It is necessary to build a measuring system that will be capable to collect accurate data vertically, at required time intervals, remotely and completely automatically. Such system will provide averaged data for large areas and, if necessary, highlight local features, and verify the correctness of the theoretical values of the frozen layer thickness. The frozen layer can be well below the ground surface when the top layer is melted due to positive daytime temperature, such as in spring and autumn. In this case the thickness of the frozen layer is the main factor. The distance from the surface of the Earth is a secondary factor. It should be noted that measurements would not make sense if they are recorded in a single prepared place. Measurements have to be taken with care covering a large area with sensors arrangements to cater for different conditions. The measurement time interval of a few hours is clearly too long. We must have a measuring system consisting of several groups of sensors (thermistor chains), located at a considerable distance from each other and under different conditions (different types of ground, self-ventilation of the site, vegetation, shading). The maximum depth of soil freezing for the Moscow region is known. The size, maximum for 10 years of measurements, was measured by the Danilin freezemeter [6] at an open, snow-cleaned site. It is 140 cm. If a depth of snow is 50 cm, the thickness of the frozen layer is reduced 3–5 times. Therefore to simplify the implementation of the freezemeter model, we have chosen the monitoring depth of no more than 30 cm at all. Other technical requirements for monitoring system were formalized as follows:

1. Measurements have to be carried continuously; the sampling time intervals should be set up from 10 min to 12 h.
2. The depths resolution should be at least 2 cm.
3. The sensitivity of the temperature sensors should be accurate to at least 0.5 °C.
4. The number of buried probes in the thermistor chains should be 4 or greater; the distance between probes should be at least 2 m.
5. A hardware or software averaging of the thickness of the frozen layer should be performed for each thermistor chain and, if necessary, for all chains.
6. The measurements should be carried out remotely, without removing the sensors from boreholes, that is, without disturbing the thermal regime, snow and vegetation.
7. The buried probes should have a lower thermal conductivity than the surrounding soil.
8. Reduction of the number of commuting wires should be welcomed.

3 Freezemeter

The electronic freezemeter has been developed and its production prototype has been manufactured according to the present requests [7]. An original technology was used, which allows to avoid the difficult and expensive digital systems by applying a translation of the measured analog input signals from thermosensors to a digital form without any analog-to-digital converters. The block diagram of the electronic freezemeter is shown in Fig. 1.

Working with almost any new system naturally creates methodical difficulties. For example, if seven sensors of the thermistor chain are placed in a forest, which also forms a thick layer of snow, and the eighth sensor is placed in the open area, then how will this reflect the fact that 7/8 of the study area is covered with the forest? How to locate the sensors and how to average the thickness of the frozen layer are the main questions. Linking the measurement results to the synthetic maps can help. The main source of heat entering the Earth ground is the radiant energy of the Sun, which is absorbed by the soil surface. The heat is translated to the underlying layers, and some thermal to heat the air and evaporate water. If you know the amount of incoming heat and the basic characteristics of the Earth ground, then it is quite simple to calculate the temperature at different depths. Such algorithms take into account the type of Earth soil, its humidity and density, the thickness of the snow cover, the latitude, as well as variations of the air temperature. The theoretical calculations are verified by experimental measurements at one or two points using pull-out thermometers [8].

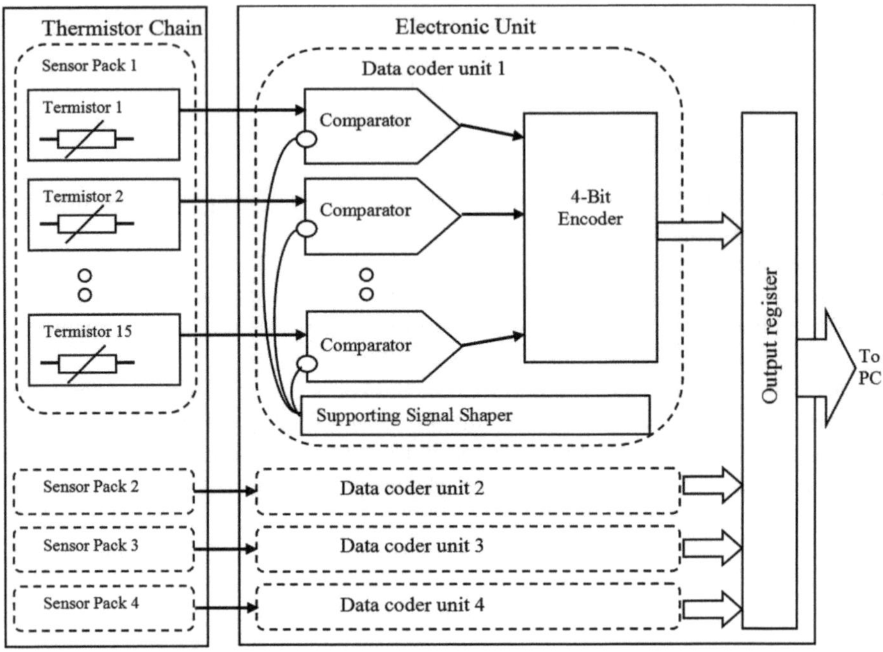

Fig. 1 The block diagram of the electronic freezemeter ELFREEZMETER-2

There is also the possibility of interactive calculations of the ground temperature distribution [9]. These calculations could be used for drawing the synthetic maps of up to 1 km × 1 km grid. Such calculations allow to obtain an assessment of the soil current state, but and also to predict its state according to average climatic changes in temperature and humidity. This data is useful, for example, for estimating the heat balance between the Earth ground and the atmosphere. It is clearly necessary to have more detailed and accurate experimental data of monitoring of the thickness of the frozen layer, to achieve the above-mentioned scientific and applied purposes.

4 Technique of Experiment

Thus, for collecting the informative data on the thickness of the frozen layer, it is necessary first of all to determine the required range of frequencies of seismic waves. From the wavelength range we can find the necessary spatial resolution. Based on the dynamic range of the seismic phenomena under study, we can determine the meaningful variation of the frozen layer thickness and the necessary resolution of the freezemeter in the vertical direction. Further planning of the experiment is impossible without prior knowledge of the territory where the experiment will be undertaken. In order for the freezemeter data to be representative, the thermistor chains and their buried sensor blocks have to be arranged according to local conditions. These include: terrain, maximum seasonal freezing of the ground, soil types, their moisture and density, vegetation cover. The conditions must be known with the above spatial resolution.

Researchers may need different types of data. In a simple case the designing of a seismic experiment requires the knowledge of average thickness of the frozen layer over the entire territory. In this case you will need to choose a location for the installation of a single thermistor chain, which has to be placed according to the "average" concept. Several buried packages of the sensors will prevent the monitoring of random errors and smooth out local conditions such as vegetation type, and thickness of the snow layer, insolation and ventilation of the site.

It may be necessary to monitor the frozen layer between the reception and excitation points of seismic signal and/or examine an object. In this case, several thermistor chains may be required, each of which is responsible for a certain average area. For a long-term monitoring, relationships between the thicknesses of the frozen soil layer at different sites can be found, and hence the amount of thermistor chains can be reduced.

5 Conclusion

First tests of the mockup have confirmed the operability of the proposed instrument and method. As expected, the thickness of the frozen soil layer and the dynamics of the freezing depth are very much dependent on the conditions of the sensor installation site (ground type, its moisture and density, thickness of a snow, vegetation cover, insolation and ventilation). One of the buried sensor packs was installed next to the Danilin freezemeter, under the same recommended conditions, i.e. on the site cleared during the whole period of measurements from vegetation and snow at a distance equal to twice the standard depth of seasonal soil freezing. In the initial period of registration, the instruments' responses were exact coincidence. In the later period the freezing point in this place fell below the measuring range of the mockup. At the same time, the buried sensor pack, which is protected by vegetation and snow, showed much less freezing. The maximum noted difference has made 20% of thickness of a frozen layer. Results on the temperature processes dynamics are also obtained. For example, when there was a sharp increase in depth of frost penetration (about 10 cm per two days) at the open place (see Fig. 2), the sensor, located under the protection of trees and snow of more than a meter thick, did not notice any dynamics.

This reaffirms the urgent need for detailed planning and preparation to monitoring and conducting of experiments.

The above considerations can be used in experiments in any region where the thickness of the frozen layer may affect the data for any type of seismic waves and other characteristics of geophysical fields.

Fig. 2 The buried sensor pack. We had to use the air lines, because under the snow the wires were attacked by mice

In our case, we considered seismic signals, excited by passing heavy trucks (the road was at a distance of 300 m from the registration point) and by freight weighing 500 kg falling from height of 4 m (at a distance of 350 m from the registration point). The frequency range of the seismic signal of interest was from 20 to 100 Hz. It would be wrong to expect that in this range, with such small thickness of the frozen layer (this year the depth of freezing between the points of signal excitation and recording did not exceed 26 cm) there would be any noticeable influence of the presence of the frozen layer on the experimental results.

References

1. Yushkin, V.I.: Experimental studies of engineering seismological characteristics of frozen soils of Eastern Siberia. Ph.D. thesis, Geological and Mineralogical Sciences, Irkutsk (2004). (in Russian)
2. Melnikov, V.P., Skvortsov, A.G., Malkova, G.V., Drozdov, D.S., Ponomareva, O.E., Sadurtdinov, M.R., Tsarev, A.M., Dubrovin, V.A.: Seismic studies of the frozen ground in Arctic areas. Russ. Geol. Geophys. **51**(1), 136–142 (2010)
3. Yadari, N.E., Ernst, F., Mulder, W.: Near-surface attenuation estimation using wave-propagation modeling. Geophysics **73**(6), U27–U37 (2008)
4. Mayor, J., Margerin, L., Calvet, M.: Sensitivity of coda waves to spatial variations of absorption and scattering: radiative transfer theory and 2-D examples. Geophys. J. Int. **197**, 1117–1137 (2014)
5. Pilz, M., Parolai, S.: Statistical properties of the seismic noise field: influence of soil heterogeneities. Geophys. J. Int. **199**, 430–440 (2014)
6. Vtyurina, U.A.: Recommendations for studying the processes of seasonal freezing and thawing of soils. Stroyizdat (1986). (in Russian)

7. Kislov, K.V., Gravirov, V.V., Likhodeev, D.V., Kotov, A.N.: Instrumental complex for measuring the depth of soil freezing. In: II-nd All-Russian Scientific and Practical Conference "Scientific Instrumentation—Current State and Prospects of Development", Proceedings, Kazan, 98–100 (2018). (in Russian)
8. Raifer, A.B., Alekseyenko, M.I., Burtsev, P.N., Zastenker, A.I., Belogurov, Y.A., Nepomnyashchy, S.I.: Handbook on Hydrometeorological Instruments and Devices, p. 432. Gidrometeoizdat, Leningrad (1976). (in Russian)
9. Climate guide of St. Petersburg. The temperature of the soil and its distribution in depth in St. Petersburg and the Leningrad region. http://www.atlas-yakutia.ru/weather/spravochnik/temp_grunt/climate_sprav-temp_grunt_260630687.php. Accessed 2018/11/03. (in Russian)

PEMSDAS—Portable Expandable Modular Seismic Data Acquisition System

V. V. Gravirov, K. V. Kislov, A. L. Sobisevich and L. E. Sobisevich

Abstract A common input problem in seismic data acquisition is to facilitate the wide dynamic range. In order to address the issue we have developed the new low-cost standalone portable data acquisition system based on a 24-bit analog-to-digital converter featuring low-power consumption. The new device is built on a modular-block principle allowing one to flexibly change the number of completely independent input data channels and the type/model of a controller so to meet the specific requirements of tasks being solved. A working prototype of the data acquisition system as an extension board for the Raspberry Pi 3 microcomputer is presented. The system has been initially designed for operation in the mobile seismic stations.

Keywords ADC · Data acquisition system · Seismic monitoring · Raspberry Pi

1 Introduction

Seismologists normally acquire the new knowledge by means of analysis of experimental data (nowadays—big data) so to unravel the mystery of the Earth's interior via seismic waves. However, the process of seismic data acquisition is far from easy while one needs the really wide dynamic range so to register the whole picture of seismic phenomena: from tidal variations to microseisms.

It is essential to emphasize that seismic phenomena are characterized by the wide frequency range too. For now, there are no seismic instruments capable of precise registering both tidal oscillations with a semidiurnal period as well as a high-frequency seismic noise.

V. V. Gravirov (✉) · A. L. Sobisevich · L. E. Sobisevich
The Schmidt Institute of Physics of the Earth of the Russian Academy of Sciences (IPE RAS), B. Gruzinskaya Str. 10, Moscow 123242, Russia
e-mail: gravirov@mail.ru

V. V. Gravirov · K. V. Kislov
Institute of Earthquake Prediction Theory and Mathematical Geophysics of the Russian Academy of Science (IEPT RAS), Profsoyuznaya Str. 84/32, Moscow 117997, Russia

T. B. Yanovskaya et al. (eds.), *Problems of Geocosmos—2018*, Springer Proceedings in Earth and Environmental Sciences, https://doi.org/10.1007/978-3-030-21788-4_23

So the number of channels for registration and their parameters may vary from a single vertical high-frequency channel in one case to multichannel registration for a number of three-component seismic sensors covering different frequency ranges [1]. One may expect upcoming and multi-parameter sensors combined in one station demanding larger variety of data streams to be captured.

Digital registration is often associated with wired connection to a personal computer (PC) and this could be inconvenient, for example, in case of spatially distributed mobile seismic stations or a temporary seismic network.

Commercial offers contain a significant variety of data acquisition systems aimed for seismological applications. However, there are only a few manufacturers of high-quality data acquisition systems on the global scale and so in Russia. And the core of each system is for sure the analog-to-digital converter (ADC). It is a common practice to use the Effective Number of Bits (ENOB) as the important parameter to determine the accuracy of ADC. To estimate the ENOB there exist a number of different methods yielding results that may vary. The concept of ADC accuracy encompassed following aspects: Offset and Gain Error, SINAD (Signal to Noise and Distortion—i.e. the signal-to-noise ratio that takes into account distortion), and an external or internal source is used (corresponding deviations from reference voltage value, temperature drift of the reference source, etc.) [2]. At the same time the number of commercially available versions of a 24-bit data acquisition systems is actually insignificant, and, as a default rule, their frequency range is limited to tens or, in extreme cases, by hundreds of Hertz [3, 4]. Furthermore, the limited availability implies a higher price of the final product.

In fundamental geophysical studies, many practical tasks turn out to be "piece goods" and for each one it is often necessary to have its own digital data acquisition system with some specific characteristics [5, 6]. That is why researchers are still pushed to develop new versions of data acquisition systems almost from scratch.

Obviously it is possible to obtain the high-quality domestic external ADC modules manufactured by the two Russian companies: L-Card and Zetlab. Unfortunately, for now the production of the L-Card external 24-bit module (E-24) [7] has been suspended for good, although many scientific organizations still use data acquisition systems based on it. Luckily the Zetlab Company continues to supply the ZET-220 external 24-bit ADC [8]. Its apparent advantages are the possibility of simultaneous use of up to 16 analog inputs, as well as the possibility to autonomous operation without connecting to a PC, allowing saving the acquired data in the internal buffer (flash disk) with the opportunity of its further transmission via the USB or Ethernet interfaces.

The obvious drawbacks of the described module include a significant level of channel-to-channel crosstalk and an increased level of conversion noise at sampling frequencies of 100 Hz and above.

2 System Development

The above situation provides the basis for the development of the original low-cost and high-precision modular seismic data acquisition system [9]. The low-cost option is achieved through a single-channel data acquisition system design, with the possibility of attachment of additional independent channels, and by implementation of the ADC chip developed by widely known manufacturers.

We had chosen the ADC with charge balancing, so called Sigma-Delta ADC. Their main advantage is the very high accuracy as well as the resolution of the conversion due to an extremely low level of intrinsic noise. The clock rate (which is also the sample rate here) is many times higher than the frequency of the input signal. Conventional converters require the sample rate to be more than twice the highest input frequency, while Sigma-Delta converters require much more. Obviously, the more pulses are produced the better is the approximation of the input signal. One of the principles embodied in such converters, which makes it possible to reduce the error introduced by noise, and consequently to increase the resolution—is the averaging of the measurement results over a large time interval [10]. Through its structure, the Sigma-Delta modulator is a low-pass filter for the input analog signal and a high-pass filter for the sampling noise, both of which have the same cutoff frequency. As a result, the noise is concentrated in the high-frequency range of the spectrum and is easily removed by the digital low pass filter (LPF), which is also part of the ADC.

Following the analysis of the commercially available ADC chips the decision to use products of Analog Devices Company has been made [11].

Thus, the main advantages of the Sigma-Delta ADC [12] are:

- Low cost,
- High resolution,
- Excellent differential nonlinearity,
- Low power consumption,
- Excessive oversampling,
- Effective filtering of quantization noise,
- Ideal for signal processing devices of a variety of sensors [13, 14].

We have used the external precision reference-voltage source and a buffer output cascade, that allows us to exclude the ADC's influence on the reference-voltage source at the transformation moments. The conversion board is connected to the control microcomputer via the 40-pin GPIO (General Purpose Input Output) connector of the Raspberry Pi 3 microcomputer. The size of the ADC board fits the dimensions of the Raspberry Pi main board (see Fig. 1). Good thing is that the ADC board implies compatibility with various models of the host microcomputer.

Along last year we have developed and tested the prototypes of the PEMSDAS data acquisition system and the main characteristics of its multi-channel version are provided in Table 1.

Considering an arbitrary moment of time the microcomputer can control the only one ADC modules among simultaneously connected. By setting address with

Fig. 1 View of the PEMSDAS board attached over the Raspberry Pi 3 microcomputer

Table 1 The basic features of multi-channel version of the PEMSDAS

Parameter	Value
Conversion resolution	24-bit
Input voltage (differential input)	0–5 V (10 V)
Number of input diff. channels	1–16
Gain of input signal	1
Cutoff frequency of the input LPF	8 Hz
Sample frequency (fixed values)	16 or 100 Hz
Programmable sample frequency	0.01–25 Hz
Power consumption of the ADC module (1 channel)	<0.316 W
Power consumption of all system (1 channel)	<1.6 W
Power supply voltage of the analog part	5 V
Power supply voltage of the digital part	3.3 V
Cross-channel signal penetration	None

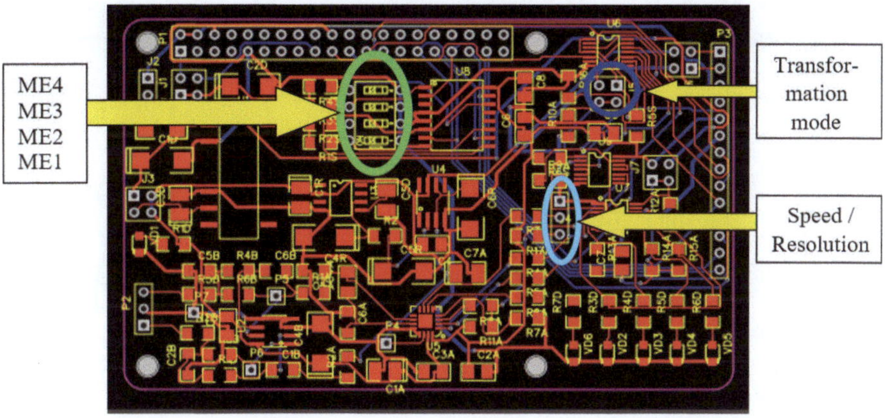

Fig. 2 Locations of micro switches to set the address and the operating mode jumpers

ME1–ME4 buses one can set the active state of the module. On the ADC module, the internal number (address) is set by the built-in micro switches (see Fig. 2).

ADC modules can operate in two different modes: independent (sequential) conversion on each module and simultaneous start of conversion on all connected ADC modules with sequential reading of data from each module. The user can also choose one of two conversion modes: with maximum Speed (sampling frequency 100 Hz) or with maximum Resolution (sampling frequency 16 Hz). Jumpers on the ADC board select all operating modes. An example of the ADC module output in the Speed mode with input grounded is shown in Fig. 3. Figure 4 demonstrates the Noise histogram for this mode.

It is strongly recommended to start the self-calibration mode of the ADC after starting operation and at each power-up. In order to reduce the effects of constant bias and take into account the gain errors of the input circuits of the ADC the self-calibration is also highly suggested. The duration of the self-calibration is around 187–313 ms. During self-calibration, the input signals are forcibly disconnected from the ADC inputs and have no effect on the calibration results. However, when the self-calibration is completed, the ADC output already has correct conversion data that can be used.

Also, the ADC module allows to measure upon request the temperature of the ADC chip so to assess the correctness of its operation. In this case, the ADC input is disconnected from the input circuits of the module and connected to the internal temperature measurement circuit. Thus, at a temperature of 25 °C, the input signal of the ADC corresponds to 106 mV. The temperature coefficient of the temperature measurement circuit is 360 μV/°C, so one may evaluate the temperature of the microchip as follows:

$$T = 25 + \frac{V_{in} - 0.106}{3.6 \times 10^{-4}}$$

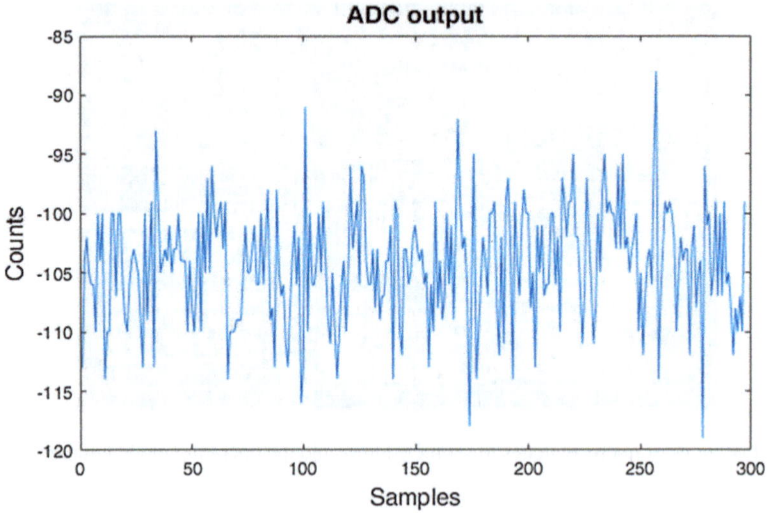

Fig. 3 Measurements of constant voltage, ADC module in speed mode

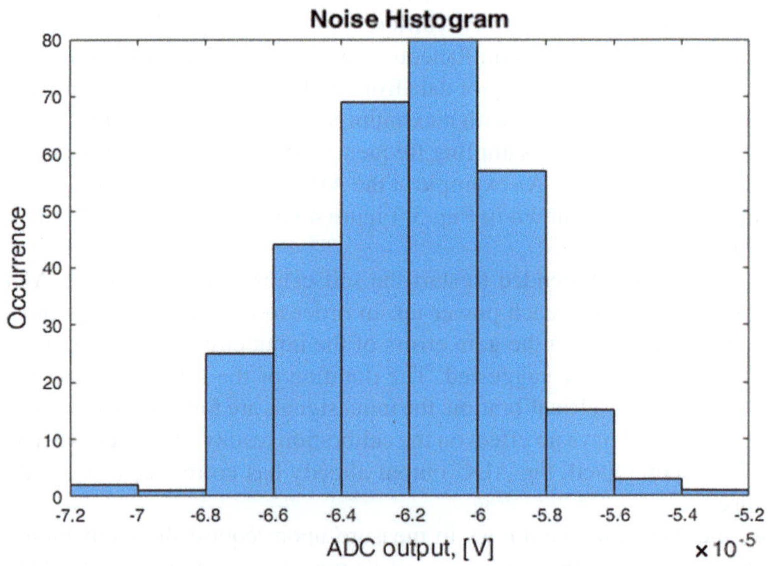

Fig. 4 Noise histogram of ADC module in speed mode with a grounded input

3 Conclusion

Future plans incorporate but are not limited to the development of the controller board so in order to significantly reduce the overall energy consumption of the system and make it even more mobile. Another important and upcoming activity is the full-scale testing of the developed device and possible hardware debugging to improve the overall system parameters. Also the extension in the number of characteristics of the developed system to be compared with its analogs is expected. In a general comparison between the two different ADCs, it is necessary to take into account not only the effective number of bits (considering how it has been evaluated), but also additional parameters such as signal-to-noise ratio, noise level and nonlinear distortion.

Thus the Portable Expandable Modular Seismic Data Acquisition System (PEMSDAS) may represent potential interest for registration of multi-parameter geophysical data. Forthcoming activities incorporate the implementation of the PEMSDAS-equipped geophysical instruments at the Geophysical Observatory in Northern Caucasus [6] operating under general supervision of the Schmidt Institute of Physics of the Earth of the Russian Academy of Sciences.

References

1. Kislov, K.V., Gravirov, V.V.: On the metrological support of the long-period seismology. In: Proceedings of the 10th International Conference "Problems of Geocosmos", St. Petersburg State University, Petrodvorets, pp. 183–189 (2014)
2. Downs, R.: ElectronicDesign page: ENOB isn't always the best figure of merit. http://electronicdesign.com/article/analog-and-mixed-signal/ENOB-Isn-t-Always-The-Best-Figure-Of-Merit/2.aspx. Accessed 2018/10/31
3. Kolyasev, V.A., Molin, S.M.: Features of the application of sigma-delta ADC. In: Instrument Making in the XXI Century. Integration of Science, Education and Production. Collection of Materials of the VII All-Russian Scientific and Technical Conference with International Participation, Dedicated to the 50th Anniversary of the Instrument-Making Faculty (2011). (in Russian)
4. Zhmud, V.A.: Application of DAC and ADC in control systems of higher accuracy. Automatics & Software Enginery, no. 4(6), pp. 68–79 (2013). (in Russian)
5. Likhodeev, D.V., Gravirov, V.V., Kislov, K.V.: Precision differential thermometers for studying thermal processes on the basis of the North Caucasus geophysical observatory. Sci. Technol. Dev. **97**(1), 15–24 (2018). https://doi.org/10.21455/std2018.1-2. (in Russian)
6. Sobissevitch, A.L., Gridnev, D.G., Sobissevitch, L.E., Kanonidi, K.K.: Instrumental equipment of geophysical observatory at North Caucasus. Seismic Instrum. **44**, 12–25 (2008). https://doi.org/10.3103/S0747923908010027
7. Butkevich, V., Nevzorov, V.: L-Card products: domestic ADC/DAC boards with signal processor. Electron. Sci. Technol. Bus. (3), 32–33 (1999)
8. Zetlab product page: ADC DAC ZET 220. https://zetlab.com/shop/izmeritelnoe-oborudovanie/moduli-atsp-tsap/atsp-tsap-zet-220. Accessed 2018/04/20

9. Gravirov, V.V., Kislov, K.V., Likhodeev, D.V., Numalov, A.S.: Precision autonomous modular 24-bit geophysical data acquisition system. In: Proceedings of II All-Russian Scientific and Practical Conference "Scientific Instrument Making—Current State and Prospects development", Kazan, pp. 101–103 (2018). (in Russian)
10. Shakhnovich, I.: Sigma-delta ADC: architecture, principles, components. Electron. Sci. Technol. Bus. (4), 18–22 (2006)
11. Analog devices product selection page: Analog-to-Digital converters from analog devices. http://www.analog.com/ru/products/analog-to-digital-converters.html. Accessed 2018/04/20
12. Grishin, E.I.: Application of ADC for measuring different physical quantities. In: Prospects for the Development of Information Technology. Proceedings of the XXXVI International Scientific and Practical Conference, pp. 68–72 (2017)
13. Gravirov, V.V.: Application of optimal filters for seismic signals detection on long-period channel of gyrotiltmeter. Seismicheskie Pribori **45**(1), 23–33 (2009). (in Russian)
14. Gravirov, V.V., Kislov, K.V.: Development of a hybrid earthquake early warning system based on single sensor technique. NED Univ. J. Res. Thematic Issue on Earthquakes, 1–20. Karachi, Pakistan (2012)

Part IV
Solar-Terrestrial Physics

Influence of Different Ionospheric Disturbances on the GPS Scintillations at High Latitudes

V. B. Belakhovsky, Y. Jin and W. J. Miloch

Abstract In this work we compare the influence of auroral particle precipitation and polar cap patches (PCP) on scintillations of the GPS signals in the polar ionosphere. We use the GPS scintillation receivers at Ny-Ålesund, operated by the University of Oslo. The presence of the auroral particle precipitation and polar cap patches was determined by using data from the EISCAT 42 m radar on Svalbard. We analyzed more than 100 events for years 2010–2017, when simultaneous EISCAT 42 m and GPS data were available. For some of the events, the optical aurora observations on Svalbard were also used. We consider the following types of the auroral precipitation: (i) the dayside and morning precipitation, (ii) precipitation on the nightside during substorms, (iii) precipitation associated with the arrival of the interplanetary shock wave. All considered types of ionospheric disturbances lead to enhanced GPS phase scintillations. For the polar cap patches, the morning and daytime precipitation (i), and precipitation related to the shock wave (iii), the phase scintillations index reaches values less than 1 rad. We observe that auroral precipitation during substorms leads to the greatest enhancement of the phase scintillation index (up to 3 rad). Thus, the substorm precipitation has the strongest impact on the scintillation of GPS radio signals in the polar ionosphere.

Keywords Ionosphere · Aurora · GPS receivers · Substorm

1 Introduction

The Global Navigation Satellite Systems (GNSS) play an important role for the modern society. However, the ionosphere as a medium for the radio waves propagation can have a negative influence on the quality of received signal. Irregularities in plasma density distribution can lead to the fast fluctuations of amplitude and phase

V. B. Belakhovsky (✉)
Polar Geophysical Institute, Apatity, Russia
e-mail: belakhov@mail.ru

Y. Jin · W. J. Miloch
Department of Physics, University of Oslo, Oslo, Norway

T. B. Yanovskaya et al. (eds.), *Problems of Geocosmos—2018*, Springer Proceedings in Earth and Environmental Sciences, https://doi.org/10.1007/978-3-030-21788-4_24

of the signal which is referred to as ionosphere scintillations [1]. The strong scintillations reduce the quality of the signal and even lead to the signal loss. Thus, the investigation of GPS scintillations is an important aspect of space weather. The level of scintillations is characterized by the phase ($\sigma\Phi$) and amplitude (S4) scintillation indexes. Amplitude scintillations are caused by the plasma irregularities with spatial scale ranging from tens to hundreds of meters, while the phase scintillations are caused by the irregularities with the sizes from hundreds of meters to several kilometeres.

The most powerful disturbances in the polar ionosphere are particle precipitation and polar cap patches (PCP). The PCP is identified as density increases above approximately 200 km. It is well known that the appearance of these structures is accompanied by the increase of the airglow intensity in 630.0 nm spectrum lines [2]. Its origin is caused by the reconnection on the dayside of the magnetosphere and penetration of plasma through the polar cap into the ionosphere [3].

It is shown in the paper [4] that PCP can produce GPS scintillations quite comparable with scintillations during the particle precipitation with appearance of strong green aurora. Thus, in the present work we address the following question: what disturbances in the polar ionosphere (particle precipitation or polar cap patches) have stronger impact on the GPS signals scintillations?

2 Data

We focused on the geophysical observations on Svalbard. The Ny Alesund (NYA) GPS scintillation receiver of the University of Oslo was the main instrument used in our study. Upon availability of data, the Skibotn (Norway, mainland) GPS receiver was also used. For describing the ionospheric plasma parameters (density, velocity, ion and electron temperature) we used the Svalbard EISCAT 42 m radar. The beam of this radar is directed along the geomagnetic field. For some convenient cases the optical observations of the aurora on Svalbard was used. IMAGE magnetometer data was used for the geomagnetic field observations. OMNI database was used for the evaluating the solar wind and interplanetary magnetic field parameters.

3 Data Analyses

We have identified more than 100 cases for years 2010–2017 when the data from the EISCAT 42 m radar was available. This report presents only some typical examples.

In this study, we considered the influence of four geophysical phenomena on the GPS scintillations: morning-daytime precipitation, nighttime substorm precipitation, precipitation associated with the interplanetary shock arrival and polar cap patches.

We focused mainly on the phase scintillation index because amplitude scintillation index (S4) practically has no large variations at high latitudes. The presence of the

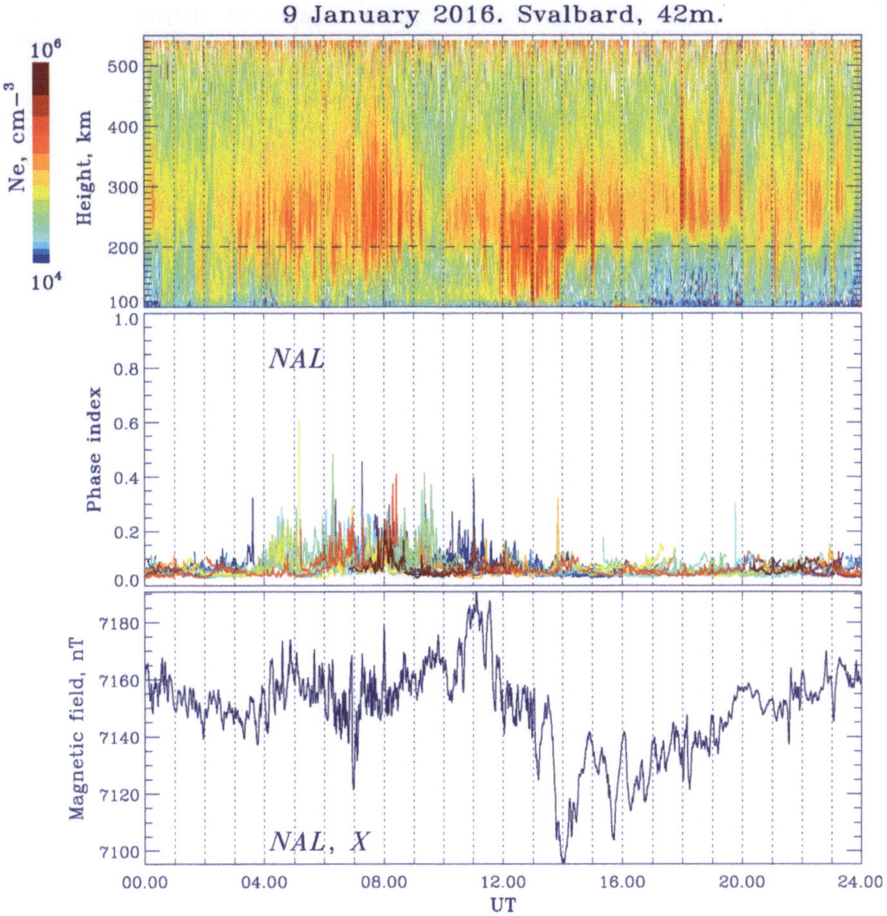

Fig. 1 Ionosphere density determined by the EISCAT 42 m radar in LYR, phase scintillation index calculated form NYA GPS receiver, X-component of the geomagnetic field at NYA station for the 9 January 2016

particle precipitation into the ionosphere associated with the appearance of the aurora was determined as the density increase between 100 and 200 km altitudes according to the EISCAT radar data. The presence of the polar cap patches was determined as a strong density increase above 200 km altitude.

Figure 1 shows an example of the morning-daytime precipitation for 9 January 2016. From 04 to 14 UT we see the ionosphere density increase at altitudes 100–200 km associated with the precipitation. The phase scintillation index has the value 0.4–0.5.

The example of the substorm precipitation and the GPS scintillations response to it is shown in Fig. 2 (11 December 2015). It can be noticed that the amplitude of the substorm at X-component of the geomagnetic field reach the value more than

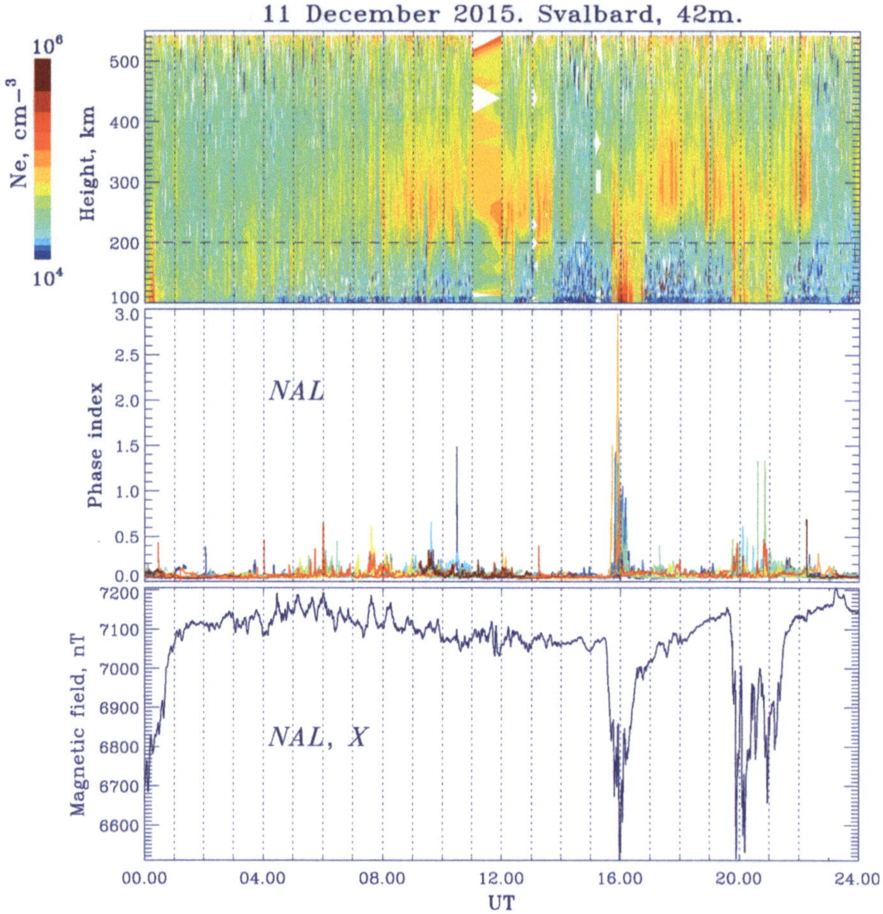

Fig. 2 Ionosphere density determined by the EISCAT 42 m radar in LYR, phase scintillation index calculated form NYA GPS receiver, X-component of the geomagnetic field at NYA station for the 11 December 2015

1000 nT at Hornsund (HOR) station, at NAL station the amplitude of the substorm was 600 nT. The substorm was accompanied by the strong increase of aurora intensity in different spectrum lines (not shown). The phase index reaches the value more than 3 approximately at 16 UT.

The example of the interplanetary shock influence on scintillations of GPS signals is shown in Fig. 3 for the event of 14 December 2015. The interplanetary shock is accompanied by an abrupt increase of the solar wind velocity, density, temperature, module of the interplanetary magnetic field (IMF) according to the OMNI database, abrupt growth of the SYM-H index. The NYA stations at moment of interplanetary shock arrival located on the dayside (16 MLT). We do not have the EISCAT data available for most of the shock events, but it is well known that the interplanetary

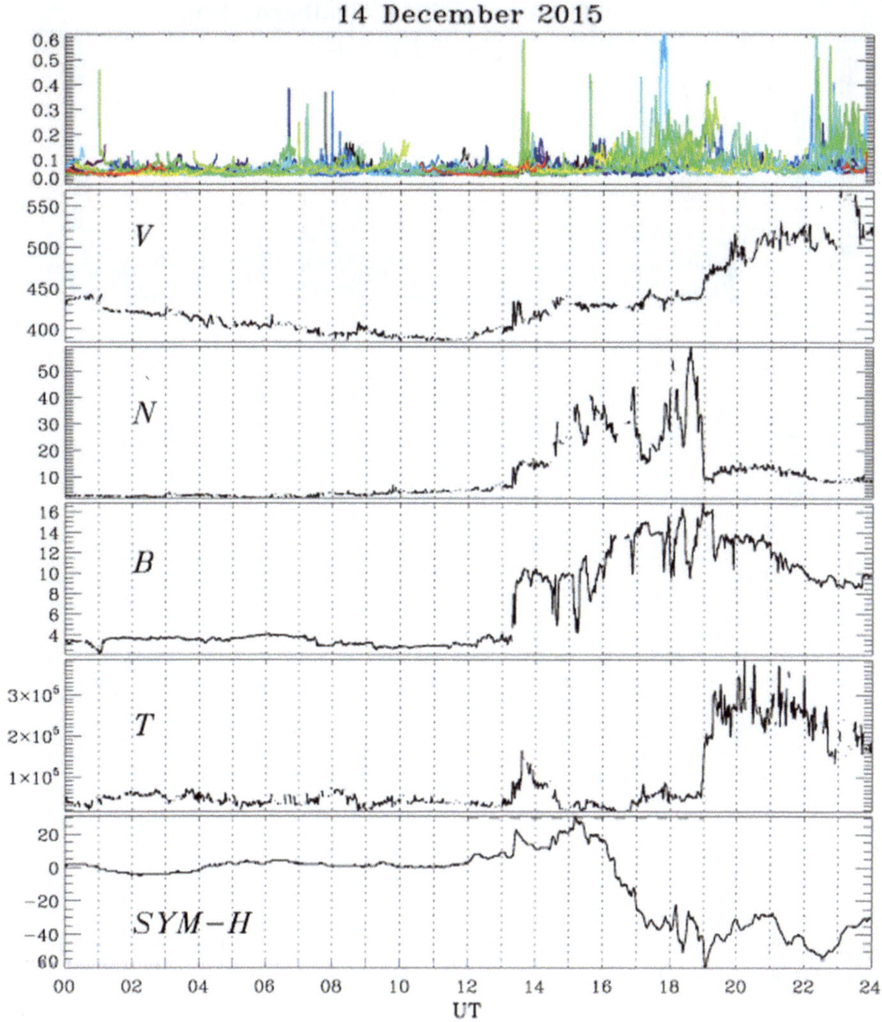

Fig. 3 The phase scintillation index determined from the GPS receiver at NYA station; solar wind parameters according to the OMNI database: velocity V [km/s], density N [cm^3], module of the interplanetary magnetic field B [nT], temperature [K]; SYM-H index

shock interaction with the Earth's magnetosphere leads to strong particle precipitation into the ionosphere [5]. Here the phase index reaches the value near 0.6. For the considered interplanetary shock cases the phase index reaches similar values.

The example of the polar cap patch (PCP) is shown on a Fig. 4 for the 20 January 2014. The PCP is observed at 08–10 UT as a density increase above 200 km according to the EISCAT data. The PCP is also identified in aurora intensity variations as forms propagating from the polar to low latitudes in 630.0 nm (red line) emission (not

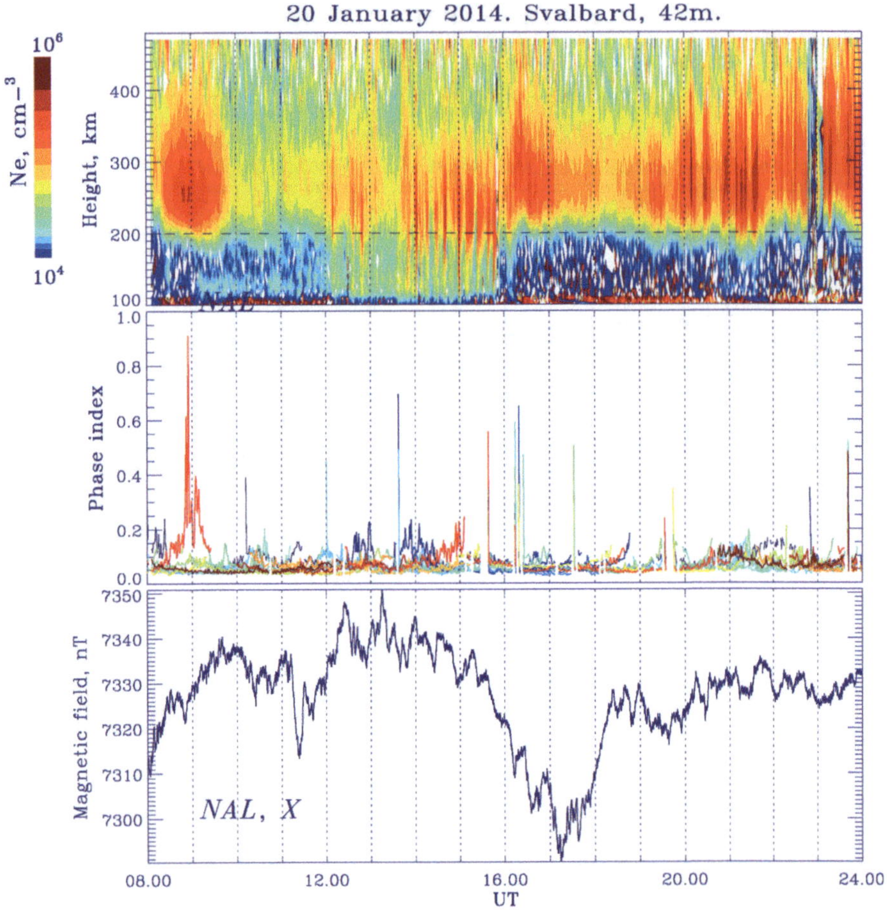

Fig. 4 Ionosphere density determined by the EISCAT 42 m radar in LYR, the phase scintillation index calculated form NYA GPS receiver, X-component of the geomagnetic field at NYA station for the 20 January 2014

shown). The Bz-component of IMF was negative during the appearance of PCP. The phase scintillation index reaches the value 0.9. For the all consider PCP cases phase index has the value less than 1.

4 Conclusion

We find that all considered geophysical phenomena (morning-dayside precipitation, nighttime substorm precipitation, shock induced precipitation, polar cap patch) give rise to increased scintillation levels. But the particle precipitation during substorm lead to the strongest scintillations (the phase scintillation index reaches values even close to 3) of the GPS signals in the polar ionosphere. Thus, the substorm precipitation has the strongest impact on the scintillation of GPS radio signals in the polar ionosphere.

Acknowledgements The authors thank the Norwegian Polar Research Institute at Ny-Ålesund for assisting us with the GPS receiver in Ny-Ålesund, Bjørn Lybekk and Espen Trondsen for the instrument operations. The IMF data are provided by the NASA OMNIWeb service (http://omniwegsfc.nasa.gov).
The authors wish to thank IMAGE (http://www.ava.fmi.fi/image/), EISCAT groups for the available data. EISCAT is an international association supported by research organizations in China (CRIRP), Finland (SA), Japan (NIPR and STEL), Norway (NFR), Sweden (VR), and the United Kingdom (NERC). Data from EISCAT can be obtained from the Madrigal database http://www.eiscat.se/madrigal.
This study is supported by the RSF grant № 18-77-10018.

References

1. Basu, S., Groves, K.M., Basu, S., Sultan, P.J.: Specification and forecasting of scintillations in communication/navigation links: current status and future plans. J. Atmos. Solar- Terr. Phys. **64**(16), 1745–1754 (2002)
2. Hosokawa, K., Shiokawa, K., Otsuka, Y., Nakajima, A., Ogawa, T., Kelly, J.D.: Estimating drift velocity of polar cap patches with all-sky airglow imager at Resolute Bay, Canada. Geophys. Res. Lett. **33**, L15111 (2006). https://doi.org/10.1029/2006GL026916
3. Lorentzen, D.A., Moen, J., Oksavik, K., Sigernes, F., Saito, Y., Johnsen, M.G.: In situ measurement of a newly created polar cap patch. J. Geophys. Res. **115**, A12323 (2010). https://doi.org/10.1029/2010JA015710
4. Jin, Y., Moen, J., Miloch, W.: GPS scintillation effects associated with polar cap patches and substorm auroral activity: direct comparison. J. Space Weather Space Climate **4**, A23 (2014)
5. Zhou, X.-Y., Strangeway, R.J., Anderson, P.C., Sibeck, D.G., Tsurutani, B.T., Haerendel, G., Frey, H.U., Arballo, J.K.: Shock aurora: FAST and DMSP observation. J. Geophys. Res. **108**, 8019 (2003). https://doi.org/10.1029/2002JA009701

The Study of Auroral Oval Position Changes in Terms of Moving of the Earth's Magnetic Pole

Yuri A. Kopytenko, Sergey A. Chernouss, Alevtina A. Petrova, Mikhail V. Filatov and Maxim S. Petrishchev

Abstract The auroral oval is actually a natural coordinate system to which theoretically and practically the events of space geophysics in the ionosphere and magnetosphere are attached. The oval was built according to the International Geophysical Year more than 50 years ago and probably is related to the magnetic pole of the Earth, which has shifted over this time by more than 1500 km. It should be expected that the configuration and position of the auroral oval could also change. The purpose of this paper is to study the possible displacement of the auroral oval for various geomagnetic activity under current conditions of the position of the Earth's magnetic pole. The article shows the aurora position relative to the magnetic pole at the present time by analyzing direct optical measurements at the stations of the Kola Peninsula (Lovozero) and Northern Scandinavia (Sodankyla, Kiruna).We have used the model of predicting auroral oval NORUSKA (Svalbard 2) for the comparison of experimental and calculated results. Numerical optical observations from the high apogee satellites during modern time also were used for definition of the oval position. To achieve the goal, precision data and a three-dimensional component model of the Earth's magnetic field of SPbF IZMIRAN were used, taking into account the contribution of magnetic anomalies of the lithosphere in the height range from 80 to 400 km, obtained according to aeromagnetic, marine and aerospace imagery. The paper indicates that the displacement of the auroral oval position follows the displacement of the magnetic pole of the Earth and location of the magnetic anomalies of the lithosphere. These are the first results obtained on a limited number of

Y. A. Kopytenko · A. A. Petrova · M. S. Petrishchev (✉)
Pushkov Institute of Terrestrial Magnetism, Ionosphere and Radio Wave Propagation of the Russian Academy of Sciences, St.-Petersburg Branch (SPbF IZMIRAN), Universitetskaya nab., 5B, St. Petersburg 199034, Russia
e-mail: ms_petr@mail.ru

Y. A. Kopytenko
e-mail: office@izmiran.spb.ru

S. A. Chernouss · M. V. Filatov
Federal State Budgetary Institution, Polar Geophysical Institute, Khalturina str., 15, Murmansk 183010, Russia
e-mail: chernouss@pgia.ru

T. B. Yanovskaya et al. (eds.), *Problems of Geocosmos—2018*, Springer Proceedings in Earth and Environmental Sciences, https://doi.org/10.1007/978-3-030-21788-4_25

measurements, and the work should be continued using more significant statistical material on a wide network of optical observation stations.

Keywords Displacement of auroral oval · Magnetic anomalies

1 Introduction

The auroral oval is actually a natural coordinate system to which theoretically and practically the events of space geophysics in the ionosphere and magnetosphere are attached. The oval was built according to the International Geophysical Year more than 50 years ago [1–3] in geomagnetic coordinates at a time when the difference in the position of the magnetic and geomagnetic poles was small, and in many calculations it could be neglected. In this paper, the following terms are used. Geomagnetic poles are the points where the axis of the magnetic dipole intersects the surface of the Earth. Since the magnetic dipole (Fig. 1, blue points) is only an approximate model of the Earth's magnetic field, geomagnetic poles differ in their location from the true magnetic poles (Fig. 1, red points), in which the magnetic inclination is 90°. Probably, the oval is related to the magnetic pole of the Earth [4, 5], which shifted during this time to more than 1500 km (Fig. 1). It should be expected that the configuration and position of the auroral oval could also change [6].

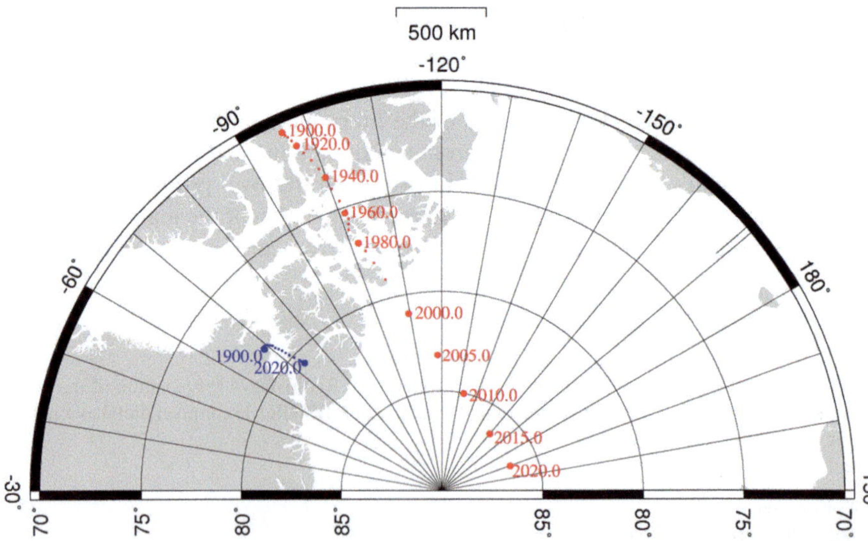

Fig. 1 Positions of the north geomagnetic (blue points) and magnetic (red points—shortest way) poles for 1900–2020 estimated from the 12th Generation IGRF. Image credit the British Geological Survey

The purpose of this paper is to analyze and to measure directly the position of the auroral oval during the day for various perturbations of the magnetic field under current position of the Earth's magnetic pole.

To achieve the goal, experimental data and a component model of the Earth's magnetic field were used, taking into account the contribution of magnetic anomalies of the lithosphere [7, 8], and direct measurements of the spatial and temporal distribution of auroras by the all-sky camera from observations of individual stations and optical observations of the auroral oval apogee satellites [9]. In addition, we used the aurora forecast model "NORUSCA" based on the data obtained in IZMIRAN and PGI [9–11].

The average position of the auroral oval, obtained mainly according to the International Geophysical Year (1957–1958), was compared with direct measurements of the position of auroras in corrected geomagnetic coordinates during the modern era by analyzing direct optical measurements made by the all-sky cameras at the Observatory of the PGI "Lovozero" on the Kola Peninsula (67.98N, 35.08E) and in the Nordic Observatories Kiruna (Sweden) (67.84N, 20.41E) and Sodankyla (Finland) (67.37N, 26.63E). The equatorial boundary of the auroral oval lies here for Kp values 2–5 which gives the most reliable model of the oval position appropriate for our study. That is why these observatories have been chosen for the analysis.

The position of the auroral oval boundaries at Kiruna Station according to the NORUSCA model [10, 11] based on the Starkov-Feldstein data is shown in Fig. 2 (continuous lines) in Cartesian coordinates: latitude - time together with the position of the equatorial boundary of the oval (redpoints) and the polar boundary of the oval (blue points) according to the all-sky camera [12]. For four days out of seven considered, a situation was observed similar to that shown in Fig. 2. In three other cases, displacements of the oval boundaries also take place, but they are insignificant. Thus, on the basis of comparison of model data and direct measurements, one can assume that the position of the equatorial boundary of the oval during dark nights in the present epoch sometimes is shifted to the southeast by several hundred kilometers in relation to the position of the equatorial boundary calculated using the NORUSCA model based on the International Geophysical Year data.

Since all-sky cameras record the aurora only at night, we see only the night part of the oval. And it would be desirable to know the changes in the position of the oval as a whole. This can be done by modern methods of aurora observation by cameras [9] installed on high-apogee satellites operating in the far ultraviolet region of the spectrum and allowing for round-the-clock photography of auroras. A comparison of the auroral ovals obtained on the basis of the network of all-sky cameras during the International Geophysical Year period and obtained using high-apogee satellites is shown in Figs. 3 and 4 [9–11, 13, 14].

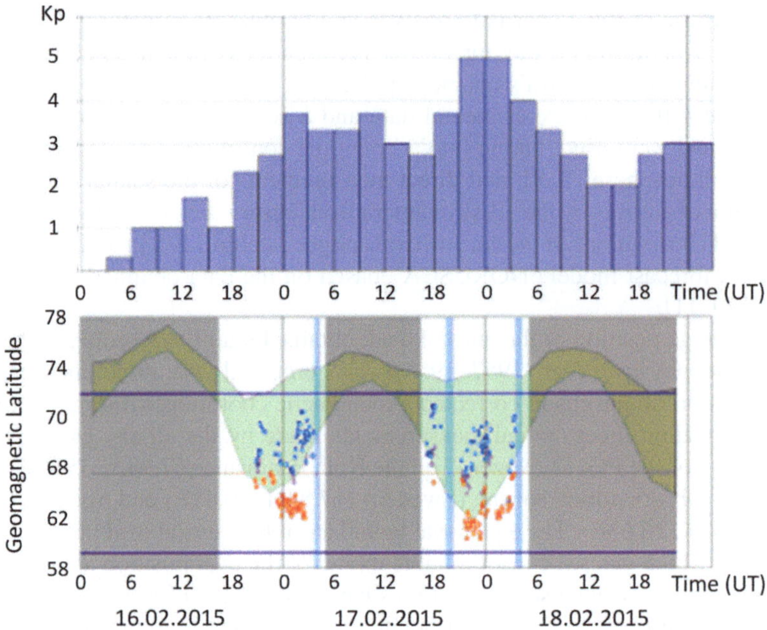

Fig. 2 The time series of Kp indices (upper panel) and the position of the auroral oval according to the NORUSCA model [10, 11] based on the Starkov-Feldstein data and direct measurements of the oval borders at the Kiruna Observatory (lower panel) on February 16–18, 2018. Model oval borders are designated by lines, direct measurements are marked by blue (Northern border) and red (South border) points. Horizontal straight lines indicate the boundaries of the all-sky camera's field of view at the height of the lower boundary of the discrete auroras

2 Features of Changes in the Components of the Induction Vector of the Earth's Magnetic Field at the Heights of the Ionospheric Layers

In order to study the configuration of the auroral oval under the Earth's magnetic pole displacement, the features of changes in the induction vector of the Earth's magnetic field (EMF) from 1957 to 2017 were studied.

For this purpose, the values of the vertical, horizontal, northern, and eastern components of EMF, declinations and inclinations obtained using SPbF IZMIRAN component model [7, 8] were calculated within the Polar Cap in the altitude range from 80 to 400 km. The component model is based on the observed and obtained values of the components calculated from the measurements of the modulus near the surface of the Earth. For the most ancient tectonic structures, the influence of the structure of magnetic anomalies of the lithosphere at the heights of the ionospheric layers manifests in the full values of EMF components.

Analysis of changes in the configuration of the auroral oval for the period from 1957 to 2017 was carried out on the basis of a comparison of the position of the oval

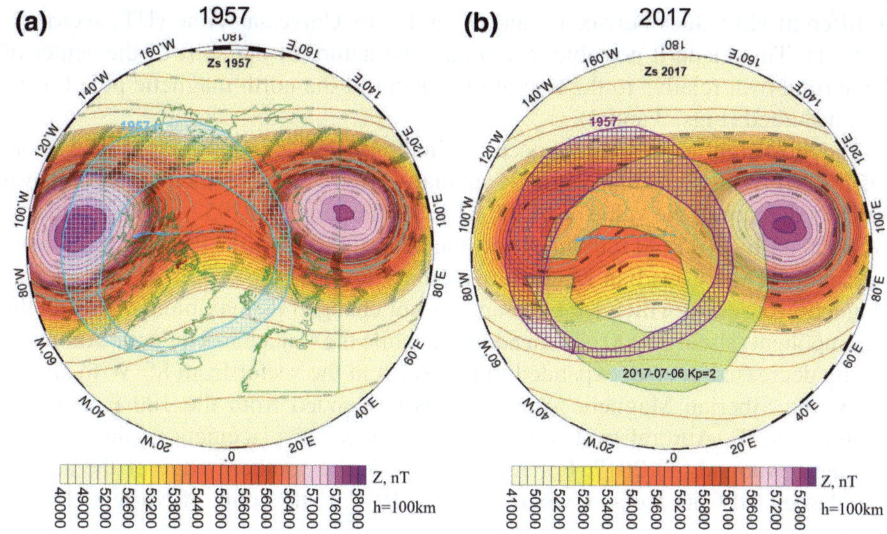

Fig. 3 Position of the auroral oval and distribution of the Z-component isolines of the Earth's magnetic field at 100 km in 1957 (**a**) and in 2017 (**b**). For comparison, Figure (b) shows the position of the auroral oval in 1957

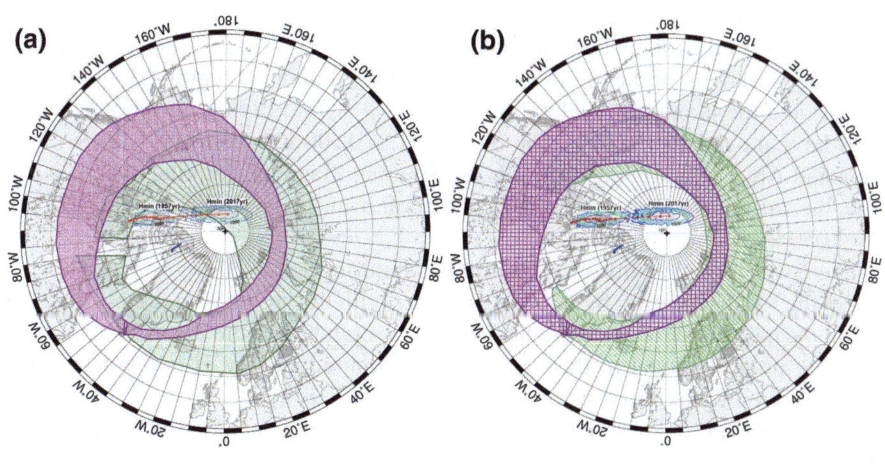

Fig. 4 Offset of the auroral oval against the displacement of the magnetic pole at 100 km for different values of Kp in 2017: Kp = 2 (**a**); Kp = 5 (**b**). The figures depict the ovals of the aurora in 1957 (violet grid) and 2017 (green hatching) and the movement of the H-component minima (red line) over this period

at different Kp values between 2 and 9 for 16–18 Universal Time (UT) sector [1, 2, 9–11]. This made it possible to estimate the nature of the shift of the center of the auroral oval relative to the line of movement of the north magnetic pole for the period studied (Figs. 3 and 4).

The most significant changes occurred in the vertical Z-component (Fig. 3) from 1957 to 2017. The Canadian magnetic anomaly, to the center of which the auroral oval was timed, reached the value of 61200 nT in Z-component in 1957, which decreased to 58000 nT by 2017. At the same time, the auroral oval in 2017 [9–11] moved from the center of the Canadian Magnetic Anomaly towards the center of the Siberian Magnetic Anomaly, which for 60 years has not changed its maximum values in the Z-component (about 61000 nT). In addition, while the Canadian Magnetic Anomaly, having decreased in size, expanded to the North in the sector from 80°W to almost 85°W, the Siberian Magnetic Anomaly has descended from the 100°E meridian to almost 82°E. Auroral oval followed the collinear displacement of the Siberian Magnetic Anomaly in 2017. For Kp indices 2-5, the southern edge of the auroral oval descended from Iceland, where it was in 1957, to the south of Norway (Figs. 3 and 4).

From 1957 to 2017 there was a shift of the $Z = 58000$ nT isoline of the Siberian Magnetic Anomaly to the south by $L = 600$ km. At the same time, the auroral oval shift to the south for 1957–2017 amounted the same value ($L = 600$ km), for analyzed Kp values from 2 to 5 (Fig. 4) taken from [9, 10].

Migration of the magnetic pole for the period from 1957 to 2017 may be depicted by the displacement of the minimum horizontal H-component region ($H \leq 1000$ nT) (Fig. 4) and the maximum inclination region ($I \geq 89°$). For 60 years, both of these areas have moved eastward by the distance $L \sim 1600$ km from the islands of the Canadian Arctic Archipelago towards the Siberian Magnetic Anomaly. At the same time, the shift of the eastern edge of the auroral oval for 1957–2017 to the east was $L = 1050$ km.

There was a spatial shift of the auroral oval during 1957–2017:

- in the south direction by 600 km, together with the shift of the Z isoline;
- in the east direction by more than 1000 km, collinear to the displacement of the magnetic pole, identical to the region of the minimum of the H-component ($H \leq 1000$ nT) and of maximum inclination ($I \geq$ °).

Based on experimental data from 2013 to 2017, we concluded that there was a spatial shift of the auroral oval during 1957–2017.

As a result, for the analyzed period, the auroral oval shifted in the south-east direction from the center of the Canadian magnetic anomaly towards the Siberian magnetic anomaly.

3 Discussion and Conclusions

Based on experimental data, the present work indicates that the displacement of the position of the auroral oval occurred during the displacement of the magnetic pole and changes in the magnitude and configuration of the Z-component in the centers of the Canadian and Siberian magnetic anomalies.

These are the first results obtained on a limited number of measurements, and the work should be continued using more significant statistical material on a wide network of optical observation stations.

Probably the cause of the detected shift of the auroral oval is the significant decrease in the power of the source supporting the Canadian magnetic anomaly over the past 50 years.

The practical value of this study lies in the fact that in some cases, when calculating and forming models of the Earth's magnetic field, it is necessary to use not only the data of high-altitude satellite measurements and the analytical description by Gauss series, which allow to calculate the components of the main geomagnetic field only. It is necessary to apply more realistic values of the surface magnetic field measurements taking into account the contribution of magnetic anomalies of the lithosphere, which is considerable up to a height of 400 km.

It is also obvious that charged particles precipitate to the lower layers of the ionosphere along the real magnetic field lines. Thus, the uselessness of computational models taking into account the characteristics of the oval for magnetospheric particles precipitation studies [3] may be true. But this is not right for modern ionospheric studies. For example, consideration of the modern configuration of the auroral oval and the forecast of its further displacement will give us a necessary link for the operation of space-based navigation systems [13, 14]. And it gives a new life for the auroral oval.

Acknowledgements We thank the Editor Nadezhda Zolotova and the anonymous reviewers for their careful reading of our manuscript and their many insightful comments and suggestions which helped us to improve the manuscript.

References

1. Khorosheva, O.V.: Spatio-temporal distribution of auroras, 82 p. Nauka Publishing, Moscow (1967) (in Russian)
2. Feldstein, Y.I., Starkov, G.V.: Dynamics of auroral belt and polar geomagnetic disturbances. Planet. Space Sci. **15**(2), 209–230 (1967)
3. Lazutin, L.L.: The auroral oval is a beautiful but outdated paradigm. Solar-Terr. phys. **1**(1), 23–35 (2015) (in Russian with English abstract)
4. Zvereva, T.I.: Motion of the Earth's magnetic poles in the last decade. Geomag. Aeron. **52**(2), 261–269 (2012)
5. Thébault E., et al.: International geomagnetic reference field: the 12th generation. Earth, Planets Space **67**, 79 (2015)

6. Lazutin, L.L.: The impact of magnetic storms on the technosphere and the effect of the north magnetic pole shift, Troitsky variant, 17 July 2012. TrV–№ 108, p. 10 (2012)
7. Petrova, A.A.: Digital maps of the components of the magnetic field induction vector. In: Proc. IZMIRAN, Moscow, pp. 412–416 (2015) (in Russian)
8. Kopytenko, Y. A., Petrova, A. A.: The development and use of a component model of the Earth's magnetic field for magnetic cartography and geophysics. Fundamentalnaya i Prikladnaya Gidrofizika **9**(2), 88–96 (2016) (in Russian with English abstract)
9. Aurora—30 minute forecast/ OVATION-Prime Model// Current space weather conditions on NOAA Scales. https://www.swpc.noaa.gov/products/aurora-30-minute-forecast#
10. Sigernes, F., Dyrland, M., Brekke, P., Gjengedal, E.K., Chernouss, S., Lorentzen, D.A., Oksavik, K., Deehr, C.S.: Oval real-time prediction—SvalTrackII. OpticaPurayAplicada (OPA) **44**, 599–603 (2011)
11. Sigernes, F., Dyrland, M., Brekke, P., Chernouss, S., Lorentzen, D.A., Oksavik, K., Deehr, C.S.: Two methods for predicting auroral manifestations. J. Space Weather Space Climate (SWSC). **1**(1), A03 (2011). https://doi.org/10.1051/swsc/2011003
12. The Kiruna All-Sky camera. http://www2.irf.se/allsky/
13. Chernouss, S.A., Shagimuratov, I.I., Alpatov, V.V., Filatov, M.V., Budnikov, P.A.: Using of the auroral oval model in modern time. In: Proceedings of VI international conference atmosphere, ionosphere, safety, part 2, Kaliningrad, pp. 58–62 (2018)
14. Chernouss S.A., Shagimuratov I.I., Ievenko I.B., Filatov M.V., Efishov I.I., Shvets M.V., Kalitenkov N.V.: Auroral perturbations as an indicator of ionosphere impact on navigation signals. Russ. J. Phys. Chem. B **12**(3), 562–567 (2018)

Geomagnetic and Telluric Field Variability as a Driver of Geomagnetically Induced Currents

O. Kozyreva, V. Pilipenko, E. Sokolova, Ya. Sakharov and D. Epishkin

Abstract We analyse temporal and spatial variability of geomagnetic and telluric electric fields in the Eastern Fennoscandia. These variations are compared with available geomagnetically induced current (GICs) measurements in electric power transmission lines at Kola Peninsula and Karelia. The information on impedance tensors is provided by the electromagnetic sounding array BEAR. We synthesize the telluric E-field from geomagnetic field variations, using the complex impedance tensor. We analyze geomagnetic and E-field variations at several sites for different space weather events: magnetic storm and Pi3 pulsations. We compare the spectral content of geomagnetic, telluric, and GIC variations. This comparison shows that the commonly used as a measure of GIC disturbances time derivative of geomagnetic field dB/dt does not totally control them. The suppression of high-frequency component in the GIC spectrum as compared with dB/dt variations is determined by the frequency-dependent character of the crustal geoelectric response.

Keywords GIC · Magnetic storm · Telluric fields

1 Introduction

As advanced technologies are implemented more widely, they have become increasingly subject to failures and interruptions due to the influence of space weather. This natural tendency cannot be avoided, and these vulnerabilities may increase as

O. Kozyreva (✉) · E. Sokolova
Institute of Physics of the Earth, Moscow, Russia
e-mail: kozyreva@ifz.ru

V. Pilipenko
Geophysical Center, Moscow, Russia

Ya. Sakharov
Polar Geophysical Institute, Apatity, Russia

D. Epishkin
Nord-West Limited, Moscow, Russia

T. B. Yanovskaya et al. (eds.), *Problems of Geocosmos—2018*, Springer Proceedings in Earth and Environmental Sciences, https://doi.org/10.1007/978-3-030-21788-4_26

these systems continue to evolve. Magnetic storms and substorms have potential to cause serious failures of space and ground technological systems. One of the most significant factors of space weather for terrestrial technological systems is electric geomagnetically induced current (GIC) in technological conductor systems related to abrupt changes of the geomagnetic field dB/dt [4, 6]. GICs associated with great magnetic disturbances were found to be dangerous for technological systems, causing malfunctioning of railway equipment [2, 10], disruption of ground and transatlantic communication cables, and reduction of the lifetime of pipelines.

Besides that, there are numerous examples of the noticeable consequences of space weather for extended high-voltage power grids. GICs caused saturation, growth of harmonics, overheating and even damage of high-voltage transformers. The most intense currents (over hundreds of Amperes) have been measured in the neutral leads of transformers at auroral latitudes during magnetic storms and substorms [7]. There is no general rule of how strong GIC should be to become harmful, since there are many types of transformers with different sensitivity to GIC. For some power transformers only a few Amperes are needed to shift the transformer operation from a linear regime [11].

Ongoing expansion of high-voltage power networks, growth of linkage between them, increase of load, and transition to low-resistive transmission lines with higher voltage lead to an increased probability of accidents during space weather events. Moreover, catastrophic failures are not necessarily required in order to have a detectable economic impact because of the way that wholesale electricity markets operate. The most likely consequence of a major geomagnetic storm, if not mitigated, is a widespread system voltage collapse [8]. Therefore, even if power infrastructure hardware is not lost during severe space weather events, GICs in regional power grids can still have broad flow-on effects throughout the global economy [3].

The highest risk of GIC may be related not directly to large-scale auroral currents, but too much weaker, but fast, small-scale processes. Though the power of such processes is many orders of magnitude lower than the power of magnetospheric storms and substorms, their rapidly varying electromagnetic fields can induce a significant GIC [1, 12]. So, information about GIC may be important from a fundamental scientific viewpoint, revealing a fine structure of fast geomagnetic variations during storms and substorms.

The difference of the electric field potential in the surface layers of the crust is responsible for GICs in grounded electric power systems. However, direct information on geoelectric fields is not easily available. While the geomagnetic field variations are monitored by world-wide array of magnetometers (>300), regular long-term observations of telluric electric fields are not so common yet. Moreover, information on GICs is property of commercial companies and is not available for the world scientific community for in-depth analysis.

Commonly, it was assumed that intensity of GIC $\boldsymbol{J}$ is proportional to time derivative of the geomagnetic field, $\boldsymbol{J} \sim d\boldsymbol{B}/dt$ [12]. However, this relationship is valid for a closed circuit with a conductive resistance in a free space only. In realistic situations, a contour where GICs flow, is formed by transmission lines, ground contacts, terminating transformers, and ground. The electric parameters of those elements—resistivity and induction, as well as their dependence on frequency, are known very approximately. The actual relationship between the spectral contents of magnetic variations $\Delta \boldsymbol{B}$, telluric electric field $\boldsymbol{E}$, and GIC $\boldsymbol{J}$ is a question which our paper tries to answer.

2 Geomagnetic, Magnetotelluric and GIC Data

The system to monitor the impact of GIC on electric power transmission lines has been deployed at Kola Peninsula and Karelia by the Polar Geophysical Institute (Apatity) and Center for Physical and Technical Problems of North's Energetic (Apatity) [9]. The system consists of 4 stations at a 330 kV power line and a station at a 110 kV power line (see map in Fig. 1). Each station records with a 1 min sampling rate a quasi-DC current in dead-grounded neutral of the autotransformer in the power line.

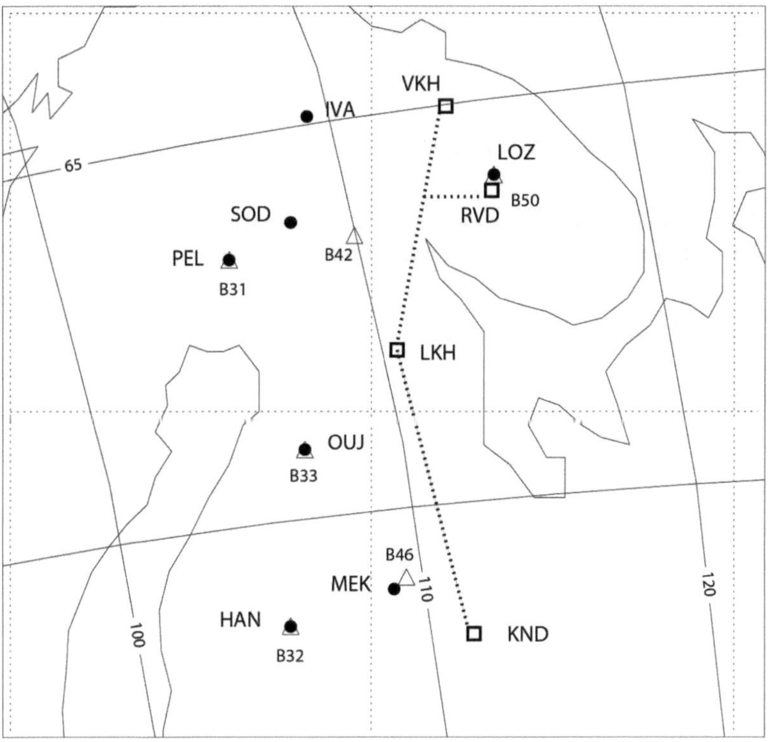

Fig. 1 Map showing magnetic IMAGE stations (black dots), selected BEAR deep sounding sites (empty triangles), and GIC recording stations (empty squares). Dotted lines denote power transmission lines. Thin solid lines denote the geomagnetic coordinates

To characterize the general geomagnetic disturbance level in the region under study we use the regional electrojet index EI (http://space.fmi.fi/image/www/). Local geomagnetic variations are monitored using 10 s data from the IMAGE magnetometers located in the vicinity of GIC recording stations (Fig. 1).

The telluric electric fields were calculated, using the available impedance data. The spatial distribution of crustal conductivity in the Eastern Fennoscandian Shield is very inhomogeneous. The comparison between impedances of several characteristic sites is given in Fig. 2. At period $T = 90.5$ s the largest |Z| is near PEL (B31), ~10 mV/km nT. The ~4 times lower |Z| is near OUJ (B33), ~2.4 mV/km nT. Rather high |Z| ~ 7.2 mV/km nT is near LOZ (B50). Thus, for the same magnetic disturbance, the telluric E-field contrast between different points may be as high as several times. Therefore, for precise estimates even a small deviation between the location of geomagnetic observatory, MTS sites, and GIC stations might be important. Essential magnitudes of the off-diagonal elements of the tensor $\boldsymbol{Z}$ (Fig. 2) indicate that the geoelectric properties of the crust are strongly anisotropic.

Constrained by data availability we used the following groups of the IMAGE, BEAR, and GIC stations: closely located LOZ—B50—RVD, and the magnetic and GIC stations at the same geomagnetic latitudes IVA—VKH.

2.1 Telluric Field Calculations E

The telluric E-field has been synthesized according to the following scheme. The algorithm of telluric $\boldsymbol{E}$(t) field synthesis from geomagnetic field variations $\Delta\boldsymbol{B}(t) =$

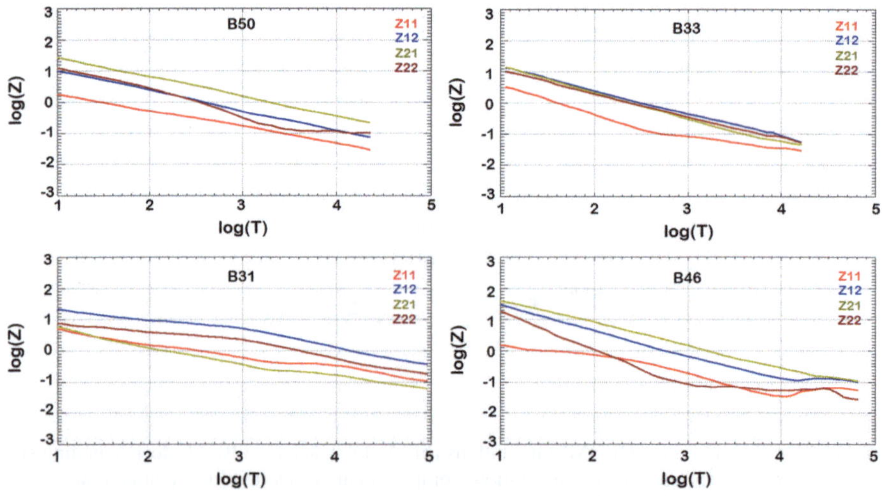

Fig. 2 The dependence of amplitudes of impedance tensor elements $Z_{i,j}$ (where x is Northward, y is Eastward) on period T at several sites of BEAR array

$\{X(t), Y(t)\}$ uses the standard frequency domain relationship between electric $\boldsymbol{E}(f)$ and magnetic $\mathbf{B}(f)$ fields via the complex impedance tsor $\boldsymbol{Z}(f)$, as follows

$$\begin{pmatrix} E_x(f) \\ E_y(f) \end{pmatrix} = \mu_o^{-1} \begin{pmatrix} Z_{xx} & Z_{xy} \\ Z_{yx} & Z_{yy} \end{pmatrix} \begin{pmatrix} X(f) \\ Y(f) \end{pmatrix}. \quad (1)$$

The elements of the impedance tensor in the range of periods T from 8 s to 55,000 s are known from the results of the BEAR experiment in 1998 [5]. The Fourier transform is applied to detrended magnetic records in a running time window to produce a set of spectral estimates of $\Delta\boldsymbol{B}(f)$ and from (1) the corresponding spectral estimates of telluric field $\boldsymbol{E}(f)$. The inverse Fourier transform performed for each running window. Thus, for a specific moment of time we have several electric field estimates, which are averaged to get final synthetic electric field time series E(t). The program was successfully tested on the synthetic magnetic and electric fields of COMDAT project. Further, we use the synthesized telluric fields together with geomagnetic data for several space weather events to search for factor responsible for the GIC excitation.

3 Magnetic Storm 2012, Nov. 12–14

This strong magnetic storm (|SYM-H| > 100 nT) was caused by fast solar wind streams. During the period Nov. 14, 00–04 UT a series of short-term auroral intensifications occurred as evident from the local electrojet intensity (EI) index (up to ~1500 nT) (Fig. 3). Two irregular substorm-related intensifications reaching maximum at IVA-LOZ occurred at ~01 UT and ~0220 UT. At a lower latitude (OUJ) maximum of magnetic disturbance was observed at ~0150 UT. The spike of EI-index at ~0330 UT is due to a more distant activation.

During this period intense fluctuations of GIC were recorded in all elements of the power transmission line. The most intense GICs, with peak-to-peak amplitudes up to ~60 A, were recorded at VKH. We compare GIC intensity J at VKH, magnetic field variations X and time derivative dX/dt at IVA, and telluric fields Ex, Ey at IVA/B50 (synthesized via impedance at B50 site) (Fig. 4). The bursts of J occur in coherent way with intensifications of dX/dt (up to ~100 nT/s) and E-field (up to several V/km) throughout all stations.

Even visual comparison shows that fluctuations of dX/dt are more high-frequency as compared with E and J. The spectral analysis confirms this conjecture (Fig. 5). To compare in detail the spectral features of geomagnetic, telluric, and GIC variations we have normalized their spectra. The spectral components at frequencies ~4 and ~6 MHz are highlighted in the spectrum of $J(f)$ and X'(f) ($X' =$ dX/dt). These components are due to the contribution from the fast fluctuations in the Pi3 range embedded in the auroral electrojet intensification. But, in general, the spectrum of X′(f) strongly deviates from the spectra of $J(f)$ at higher frequencies (>5 MHz) and

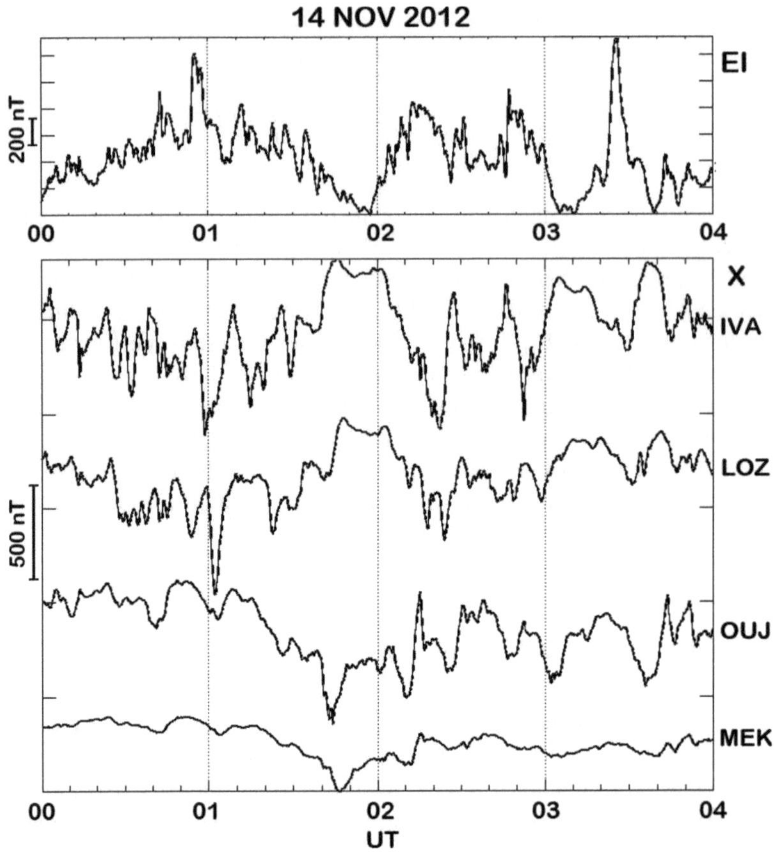

Fig. 3 Variations of the geomagnetic field during magnetic storm at 2012, Nov. 14, 00–04 UT: regional EI-index, *X*-component of magnetic disturbance (North) at the IMAGE stations IVA, LOZ, OUL, and MEK

is much closer to $Ex(f)$. Thus, the ground geoelectric properties play a role of a low-pass filter for geomagnetic variability d*B*/d*t*.

4 Pi3 Pulsations During the Magnetic Storm on 2013, June 29

The magnetic storm on June 27–29 started with the interplanetary shock arrival on June 27, ~15 UT. The interplanetary magnetic field (IMF) *Bz* since June 28, 08 UT gradually turned southward ($Bz < 0$) and kept steady at the level about −10 nT till June 29, 12 UT. Such IMF orientation produced driving of the magnetosphere into

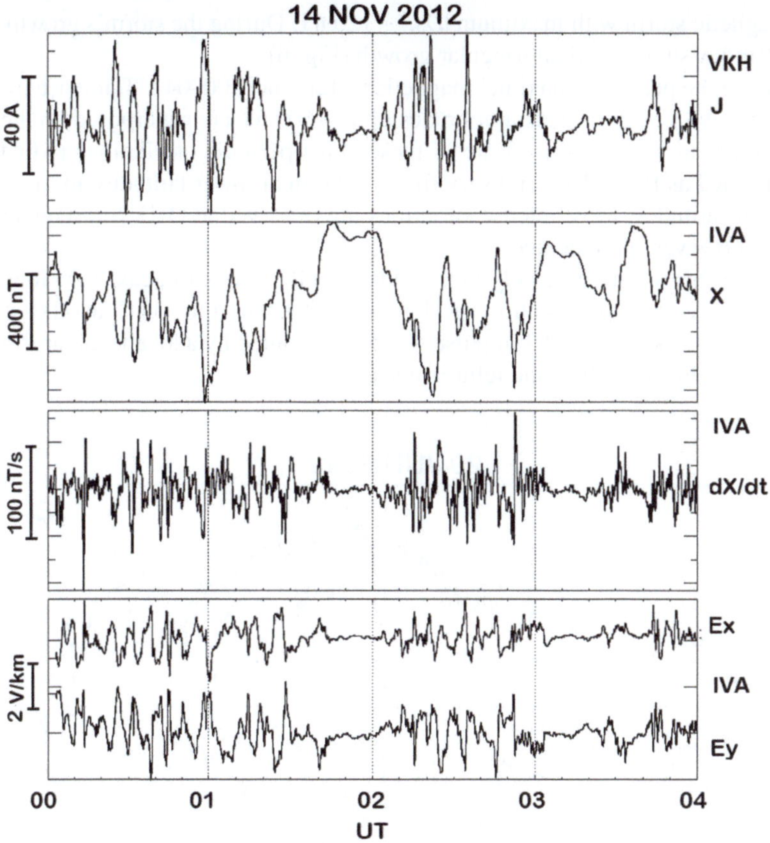

Fig. 4 GIC intensity *J* (A) at VKH, magnetic field variations *X* (nT) and time derivative *dX/dt* (nT/s) at IVA, and telluric fields *Ex* (V/km) and *Ey* (V/km) from IVA/B50

Fig. 5 Normalized spectra for the period 2012, Nov. 14, 01–03 UT: *J*(f) at VKH, *X*(*f*), X′(*f*), and *E*x(*f*) at IVA

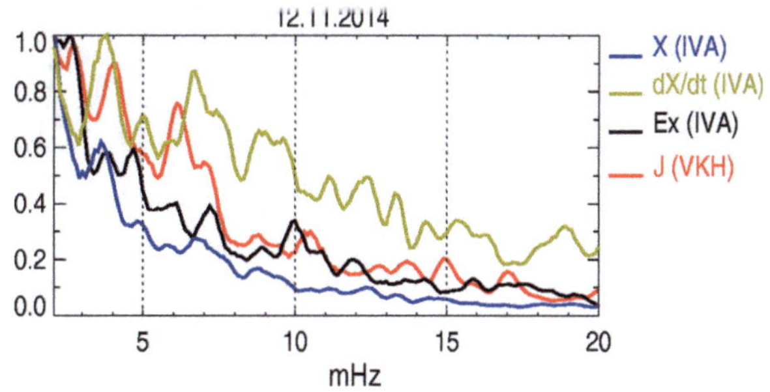

the magnetic storm with maximum $|Dst| \sim 120$ nT. During the storm's growth phase the EI-index shows gradual irregular growth (Fig. 6).

During the period of maximal magnetic disturbance, 00–04 UT, intense irregular variations with peak-to-peak amplitudes up to ~600 nT are superposed on the magnetic bay (Fig. 6). The time scale of these quasi-periodic oscillations (which may be classified as Pi3 pulsations) vary from ~20 min at lower latitudes up to ~10 min at higher latitudes. Most intense variations are observed at OUJ. Subsequent paragraphs, however, are indented.

During this storm extremely high values of GIC (peak-to-peak variations up to ~200 A) were recorded at station VKH on June 29, from 01 to 03 UT, at other stations GICs were weaker. Each Pi3 impulse was accompanied by a localized burst of GIC, magnetic field variability, and telluric current.

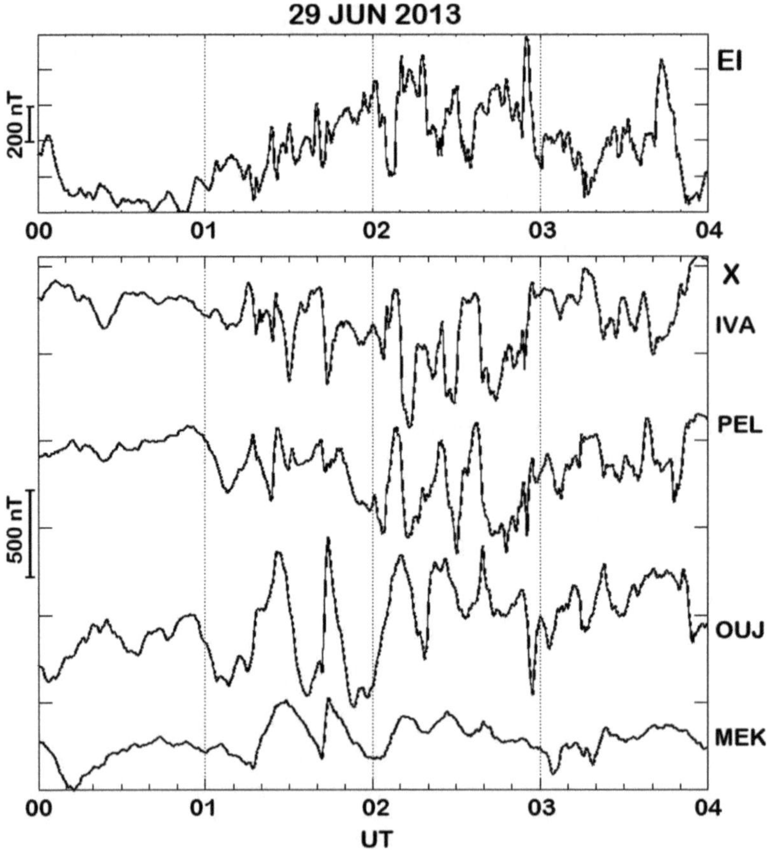

Fig. 6 Geomagnetic activity on 2013, June 29, 00–04 UT: EI-index, *X*-component of magnetic field variations at IMAGE stations LOZ, PEL, OUL, MEK

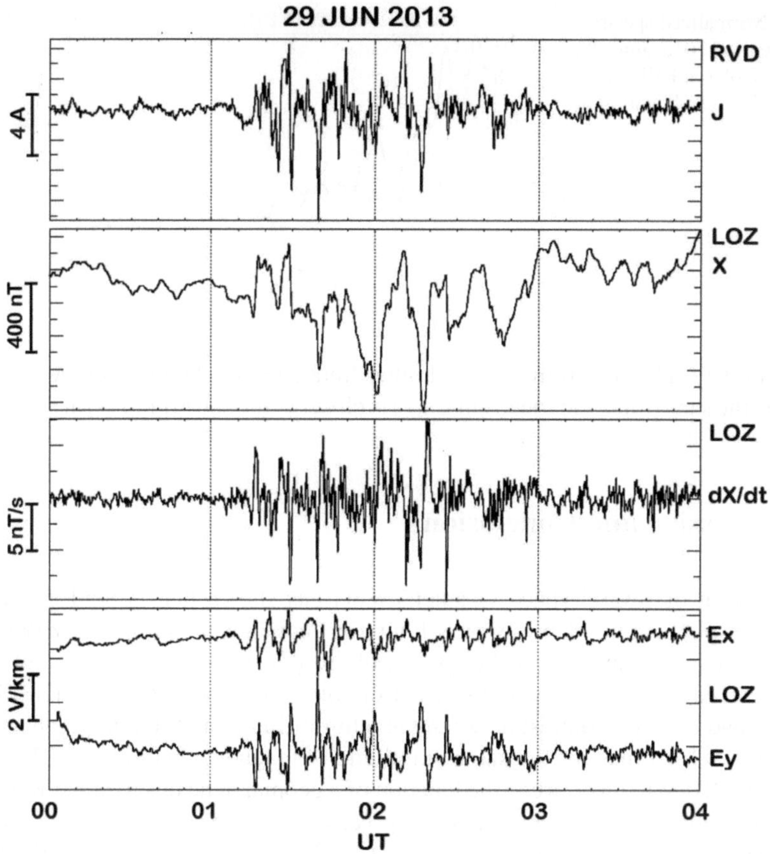

Fig. 7 Comparison of *J* at RVD with magnetic disturbances *X* and time derivatives of magnetic field d*X*/dt at near-by station LOZ, and *E*-field at LOZ/B50 during the Pi3 event on June 29, 20

The telluric field demonstrates irregular variations with different amplitudes at different locations (not shown). For example, the most intense E-field variations (>15 V/km) are observed at PEL, though at OUJ and PEL amplitudes of magnetic variations are nearly the same. This asymmetry is most probably caused by the ~2.5 times higher magnitude of crustal impedance at PEL as compared with that at OUJ.

For a detailed comparison of spectral features of GIC and terrestrial electromagnetic field we use observational data from two neighboring stations RVD and LOZ. Visual examination of *J* at RVD together with *X* and d*X*/dt at LOZ, and *E*-field at LOZ/B50 (Fig. 7) shows that for the interval under analysis the time derivative d*X*/dt correlates with variations of GIC not much better than just magnetic variation *X* does. This fact is confirmed by the spectral analysis (Fig. 8). Normalized overlaid spectra of all timeseries have a decaying character, with some enhancements at ~4 and ~7 MHz, provided by contributions from transient Pi3 impulses embedded in the

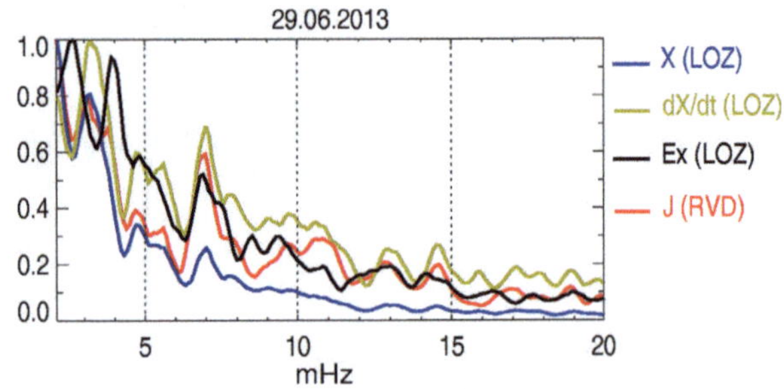

Fig. 8 Normalized spectra for the period 2013, June 29, 01-03 UT of J at RVD, X and dX/dt at LOZ, and Ex at LOZ

magnetic bay. The spectrum $X'(f)$ at higher frequencies is larger than the spectrum $J(f)$. At the same time, the spectrum Ex(f) is closer to the spectrum $J(f)$ at $f > 7$ MHz.

5 Discussion and Conclusion

Examination of several space weather events has shown that ground geoelectric properties play a role of a low-pass filter for geomagnetic variability *dB/dt*. The high frequency component of the spectra of "response"—telluric fields and GICs, is suppressed as compared with the spectrum of a driving *dX/dt* fluctuations. In the Fennoscandia region contrast between the telluric E-field in different points may be as high as ~3–4 times for the same amplitude of magnetic disturbance. Therefore, even a small deviation between the location of a geomagnetic observatory, MTS site, and GIC station might be important for precise studies. Spectral analysis revealed enhancements at typical Pi3 frequencies (~2–3 and ~ 5–7 MHz) of GIC due to contribution of sporadic impulses embedded in the magnetic storm evolution.

Acknowledgements This study is supported by the grant 16-17-00121 from the Russian Science Foundation. We acknowledge the data from the deep sounding project BEAR and IMAGE array (http://space.fmi.fi/image). The maintenance of the GIC recording system by the Polar Geophysical Institute and Center of Physical and Technical Problems of the Northern Energetics is supported by the grant 17-48-510199 from the Russian Fund for Basic Research and Murmansk Region Government.

References

1. Belakhovsky, V.B., Pilipenko, V.A., Sakharov, Ya.A., Selivanov V. N.: Characteristics of the variability of a geomagnetic field for studying the impact of the magnetic storms and substorms on electrical energy systems. Izv. Phys. Solid Earth, **54**(1), 52–65 (2018)
2. Eroshenko, E.A., Belov, A.V., Boteler, D., Gaidash, S.P., Lobkov, S.L., Pirjola, R., Trichtchenko, L.: Effects of strong geomagnetic storms on Northern railways in Russia. Adv. Space Res. **46**, 1102–1110 (2010)

3. Forbes, K.F., St. Cyr, O.C.: Space weather and the electricity market: An initial assessment. Space Weather **2**, S10003 (2004). https://doi.org/10.1029/2003SW000005
4. Knipp, D.J.: Synthesis of geomagnetically induced currents: commentary and research. Space Weather **13**, 727–729 (2015)
5. Korja, T., et al.: Crustal conductivity in Fennoscandia—a compilation of a database on crustal conductance in the Fennoscandian Shield. Earth Planets Space **54**, 535–558 (2002)
6. Lanzerotti, L.J.: Space weather effects on technologies. Space Weather Geophys. Monogr. Ser. AGU. **125**, 11–22 (2001)
7. Pirjola, R. Kauristie, K., Lappalainen, H., Viljanen, A., Pulkkinen, A.: Space weather risk. Space Weather **3**, S02A02 (2005). https://doi.org/10.1029/2004SW000112
8. Pulkkinen, A., et al.: Geomagnetically induced currents: science, engineering and applications readiness. Space Weather **15**, 828–856 (2017)
9. Sakharov, Ya.A., et al.: Registration of GIC in power systems of the Kola Peninsula. In: Proceedings of 7-th International symposium on Electromagnetic Compatibility and Electromagnetic Ecology, St-Petersburg, pp. 291–293 (2007)
10. Sakharov, Ya.A., et al.: Geomagnetic disturbances and railway automatic failures. In: Proceedings of 8th International Symposium on Electromagnetic Compatibility, St-Petersburg, pp. 235–236 (2009)
11. Vakhnina, V.V., Kuvshinov, A.A., Shapovalov, V.A., Chernenko, A.N., Kretov, D.A.: The development of models for assessment of the geomagnetically induced currents impact on electric power grids during geomagnetic storms. Adv. Electr. Comput. Eng. **15**, 49–54 (2015)
12. Viljanen, A., Nevanlinna H., Pajunpää, K., Pulkkinen A.: Time derivative of the geomagnetic field as an activity indicator. Ann. Geophys. **19**, 1107–1118 (2001)

[illegible]

Pitch-Angle Diffusion of Energetic Protons upon Their Interaction with EMIC Waves: Comparison of Calculation Results with THEMIS and NOAA/POES Data

Tatyana Popova, Andris Lyubchich, Andrei Demekhov and Alexander Yahnin

Abstract We analyze several cases of THEMIS spacecraft observations of EMIC waves in proton and He+ frequency bands in various local time sectors and L shells. We selected the events during which the energetic proton precipitation was observed by low-orbiting POES spacecraft in the conjugate region. By using the data on the spectral intensity of EMIC waves, the geomagnetic field, and the plasma density we calculated the pitch-angle diffusion coefficients of energetic protons for various energies at low pitch-angles. The calculations were performed under the assumption that the ambient plasma consisted of three ion species, i.e., protons, He+, and O+ ions. The results of calculations were compared with the energy of precipitated protons. On the whole, the energies of precipitated energetic protons are in agreement with those which can be expected on the basis of the calculated pitch-angle diffusion coefficients.

Keywords EMIC wave · The energetic proton precipitation · The pitch-angle diffusion

1 Introduction

The inner magnetosphere, including the plasmasphere, radiation belts and ring current, is of great interest to researchers. This is confirmed by launching new satellite missions and a variety of theories and assumptions about processes taking place in this region.

One of such processes is the pitch-angle diffusion of energetic charged particles under their interaction with electromagnetic waves. The resonant wave-particle interactions are described by the quasi-linear theory [1–3] in the case of noise-like wave

T. Popova (✉) · A. Lyubchich · A. Demekhov · A. Yahnin
Polar Geophysical Institute, Apatity, Russia
e-mail: tarkada@yandex.ru

A. Demekhov
Institute of Applied Physics, Russian Academy of Sciences, Nizhny Novgorod, Russia

T. B. Yanovskaya et al. (eds.), *Problems of Geocosmos—2018*, Springer Proceedings in Earth and Environmental Sciences, https://doi.org/10.1007/978-3-030-21788-4_27

spectrum. If the pitch-angle becomes small enough, the particle gets into the loss cone, in which case it moves along the magnetic field line to low altitudes, where it interacts with the Earth atmosphere particles. Such particle doesn't come back to the equatorial plane. Thus, the pitch-angle diffusion is an important mechanism of the energetic charged particle loss. To estimate it quantitatively, one has to calculate of the pitch-angle diffusion coefficient.

In our paper we consider the interaction of energetic (ring current) protons with electromagnetic ion-cyclotron (EMIC) waves. We use THEMIS spacecraft data in the magnetosphere and calculate the pitch-angle diffusion coefficient by using the measured EMIC wave spectra. For comparison of the energies of resonant interaction yielded by these calculations with the energies of precipitating protons we use the data from low-orbiting satellites NOAA POES.

2 Satellite Data

The THEMIS mission spacecraft (three probes THEMIS-A, -D, -E) move successively in an elongated elliptical orbit and cross the distances of up to 14 Re near the equatorial plane of the magnetosphere. All satellites have an identical set of instruments to measure electric and magnetic fields and parameters of cold and hot plasma. The background magnetic field and EMIC wave parameters were obtained with the instrument FGM (Flux Gate Magnetometer) [4] that measures low-frequency fluctuations of the magnetic field (up to 64 Hz. To determine the cold plasma density and fluxes of ions, the instruments i(e)ESA (ion and electron electrostatic analyzers) [5] and SST (Solid State Telescope) were used. These instruments measure particles in the energy range from eV to 30 keV for electrons and to 25 keV for ions (i(e)ESA) and particles in the energy range from 25 keV to 6 meV (SST).

The NOAA Polar Orbiting Environmental Satellites (POES) series spacecraft have polar circular orbits at an altitude of about 800 km. Satellite detectors allow to measure fluxes of protons and electrons. Precipitating particles with energies from 50 to 20 keV are observed by the Total Energy Detector (TED) instrument; trapped and precipitating particles with energies from 30 to more than 200 meV are measured by the Medium Energy Proton and Electron Detector (MEPED) instrument [6].

In the region geomagnetically conjugate with the magnetospheric spacecraft measuring EMIC waves, the low-altitude satellites often observe the precipitation of energetic protons which precipitate into the loss cone as the result of interaction with these waves [7].

3 The Formula for the Pitch-Angle Diffusion Coefficient

We calculate the pitch-angle diffusion coefficient by using the formula derived in [8] under several simplifying assumptions. First, the EMIC waves were assumed to propagate in a plasma with the wave vector parallel to the geomagnetic field. Effective wave-particle interaction takes place under the resonance condition:

$$\omega - k \cdot v_{||} = \Omega_p \quad \text{or} \quad v_{||} = V_{\text{res}} = \left(\omega - \Omega_p\right)/k,$$

where Ω_{p} is the gyrofrequency of charged particle (proton), ω and k are frequency and the field-aligned component of the wave vector. The energy of a resonant particle is then determined as

$$W_{\text{res}} = (M \cdot V_{\text{res}}^2)/2,$$

where M is the proton mass.

The second simplification is that the spectrum of EMIC waves is described by a Gaussian function (or by a sum of Gaussians):

$$W(\omega) = A \cdot \exp\left\{-[(\omega - \omega_m)/\delta\omega]^2\right\},$$

where ω_m is a central frequency, and $\delta\omega$ is the halfwidth of the spectral maximum. We fitted be observed wave spectrum by the equation above. Using the obtained free parameters, the frequency range is determined from the condition:

$$\sigma\delta\omega = (\omega_2 - \omega_1)/2,$$

where σ is an arbitrary parameter, i.e. $\omega_1 = \omega_m - \sigma\delta\omega$ and $\omega_2 = \omega_m + \sigma\delta\omega$.

The energy density of the wave magnetic field can be found from the expression:

$$W_{\text{tot}} = (\Delta B)^2/8\pi,$$

where

$$(\Delta B)^2 = 8\pi \int_{\omega_1}^{\omega_2} W(\omega)d\omega.$$

Under these assumptions, the pitch-angle diffusion coefficient of energetic protons is calculated as:

$$D_{\alpha\alpha} = \frac{\pi}{2} \cdot \frac{1}{v} \cdot \Omega_p \cdot \sum_{j=1}^{N} \left(1 - \frac{\text{x}}{\text{y}} \cdot \frac{\cos\alpha}{\beta}\right)^2 \cdot \frac{(\Delta B)^2}{\delta\text{x} \cdot B_0^2} \cdot \frac{|\text{dx}/\text{dy}|}{|\beta\cos\alpha - \text{dx}/\text{dy}|} \cdot e^{-\left(\frac{x - x_m}{\delta\text{x}}\right)^2}, \tag{1}$$

where summation is performed over all roots of dispersion relation satisfying the resonance condition. Since we consider only EMIC waves we will have $N = 1$.

In (1) α is a pitch-angle of a particle, $\nu = \sqrt{\pi} \cdot \mathrm{erf}(\sigma)$ is a normalization constant, $\beta = v/c$ is a normalized proton velocity, c is the speed of light, B_0 is the background magnetic field; $\mathrm{x} = \omega_j/\Omega_p$, $\delta\mathrm{x} = \delta\omega/\Omega_\mathrm{p}$ and $\mathrm{y} = \mathrm{ck_j}/\Omega_\mathrm{p}$.

In the case when the cold plasma consists of electrons, protons (p), He+ and O+ ions, the dispersion relation is [9]:

$$\mathrm{y}^2 = \mathrm{x}^2 - \frac{Y}{\alpha^* \varepsilon}$$

where $\mathrm{Y} = \mathrm{x} \cdot \left\{\frac{1}{1+\varepsilon \cdot \mathrm{x}} + \frac{\eta_\mathrm{p}}{\mathrm{x}-1} + \frac{\eta_\mathrm{He}}{4\mathrm{x}-1} + \frac{\eta_\mathrm{O}}{16\mathrm{x}-1}\right\}$, $\varepsilon = \mathrm{m_e}/\mathrm{M}$, and the group velocity normalized to the speed of light is:

$$\frac{\mathrm{dx}}{\mathrm{dy}} = \frac{1}{c} \cdot \frac{\mathrm{d}}{\omega}\mathrm{dk} = \frac{2\alpha^*\varepsilon \cdot y}{2\alpha^*\varepsilon \cdot \mathrm{x} - Z},$$

where $Z = \frac{1}{(1+\varepsilon \cdot \mathrm{x})^2} - \frac{\eta_\mathrm{p}}{(\mathrm{x}-1)^2} - \frac{\eta_\mathrm{He}}{(4\mathrm{x}-1)^2} - \frac{\eta_\mathrm{O}}{(16\mathrm{x}-1)^2}$.

The phase velocity $\mathrm{V_{ph}/c} \equiv \mathrm{u} = \omega/\mathrm{kc}$ and the group velocity of the wave depend on several parameters characterizing the conditions of the propagation of this wave: $\alpha^* = \Omega_\mathrm{e}^2/\omega_\mathrm{pe}^2$, $\eta_p = \mathrm{N}_p/\mathrm{N}_e$, $\eta_\mathrm{He} = \mathrm{N_{He}}/\mathrm{N}_e$, and $\eta_O = \mathrm{N}_O/\mathrm{N}_e$, where Ω_e is the electron gyrofrequency, ω_pe is the electron plasma frequency, m_e is the mass of electron, N_p, N_He, N_O are densities of proton, He+ and O+ ion, respectively.

According to statistical studies [10], the percentage of ions in plasma can vary in a wide range. The fraction of helium ions can reach 50%; the fraction of oxygen ions can be a few percent. In each case we selected specific ion composition values so that the frequencies of the detected EMIC waves satisfied the condition for the wave propagation.

4 The Event of 05 August 2010 (Case 1)

We calculated the pitch-angle diffusion coefficient for the events when one of the THEMIS satellites detected the EMIC waves. For identification of these waves we used the following signatures. First, these magnetic field fluctuations must be in the frequency range below the proton gyrofrequency; and second, the perpendicular component of fluctuations must be more intense than the parallel one. Of the events satisfying the above criteria we selected those for which a conjugate (within 1 h of MLT) flight of one of NOAA POES satellites was available. In total, we analyzed 7 cases.

One of the selected events occurred on 05 August 2010 (1310–1405 UT) when the THEMIS A spacecraft was in the afternoon sector of the magnetosphere (12.5–13.9 h MLT) at distances 4.2–5.8 $\mathrm{R_E}$ and the EMIC waves were detected above the He+ gyrofrequency (Fig. 1). The noise-like bursts below the He+ gyrofrequency that occurred after 14 UT are not considered in this paper. The cold plasma density

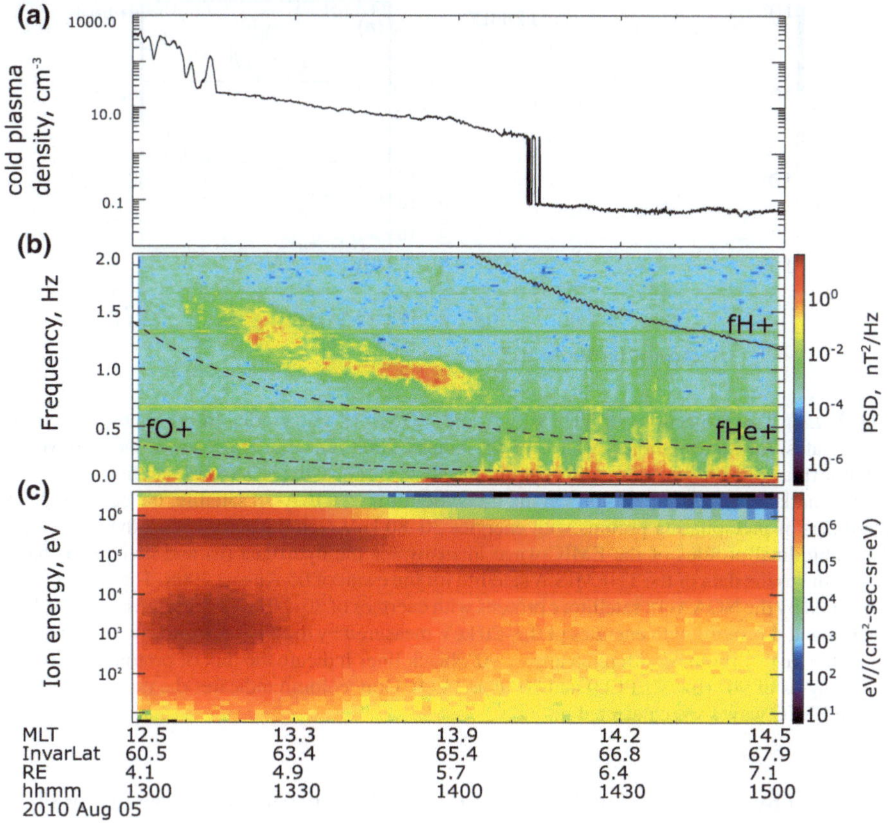

Fig. 1 The THEMIS A data for the event of 13–15 UT on 05 August 2010: **a** the cold plasma density, **b** the spectrogram of transverse fluctuations of the magnetic field (solid, dotted and dash-dotted lines show the gyrofrequencies of H+, He+, and O+ ions, respectively), **c** The ion energy spectrogram. Grey area marks the time interval when the conjugate flight of a low-altitude satellite took place (see Fig. 2)

varied from 18.6 to 5.8 cm^{-3} at this time, and these density values correspond to the region outside the plasmasphere. The ESA and SST instruments allow us to measure the fluxes of ions at different energies. In Fig. 1c it is seen that at time of the wave detection the three ion populations exist: the low-energy (500 eV to 3 keV), high-energy (10^5–10^6 eV) and the intermediate energy (~8 keV).

During this event in the region conjugate with the EMIC wave observations, a flight of low-altitude NOAA-18 satellite took place. The conjugation is confirmed by using the combination of the Earth's main field model (IGRF) and the T89 model [11]. In Fig. 2 (left) the projections of THEMIS-A and NOAA-18 trajectories mapped to the altitude of ionosphere (110 km) are shown, and it is seen that the conjunction in MLT is good enough (about 0.5 h). Precipitating proton flux is marked by color on the trace of NOAA-18. It corresponds to the proton flux with the pitch-angle about 0°

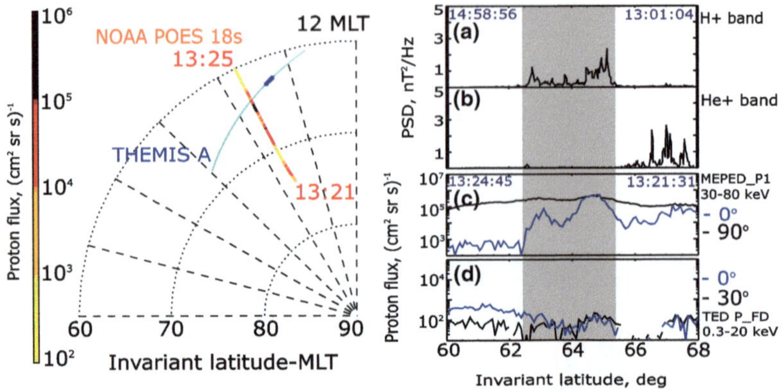

Fig. 2 *Left* The projections of THEMIS-A and NOAA-18 trajectories to the ionosphere in the coordinates "invariant latitude-MLT" for 05 August 2008. On the trajectory of the THEMIS-A satellite (blue line) the position of this satellite during the flight of the NOAA-18 satellite (1321–1325 UT) is marked by dark blue color. On the trajectory of the NOAA-18 satellite, the colors indicate the fluxes of precipitating protons with energies 30–80 keV (see the color bar on the left). *Right:* The latitudinal profiles of the EMIC wave intensity for the hydrogen (**a**) and helium (**b**) bands according to the data of the THEMIS A satellite for the event of 05 August 2008. The proton fluxes measured by the NOAA-18 satellite: **c** protons with energies of 30–80 keV measured by the MEPED instrument, **d** protons with energies of 0.3–20 keV measured by the TED instrument. The blue lines show the flux of protons with pitch angles ~0°, black lines indicate the flux of protons values with pitch angles of 90° (for MEPED) and 30° (for TED). Gray column indicates the range of latitudes where EMIC waves were detected

measured by channel 1 (30–80 keV) of the MEPED instrument and shown in Fig. 2 (right, panel (c)). Fluxes of lower-energy protons measured by the TED instrument are shown in Fig. 2 (right, panel (d)). It is clearly seen, that at latitudes of EMIC wave observations (62.5–65.5°, marked by a grey column) the precipitating proton flux has pronounced enhancements (especially in higher-energy range). Besides, there is a noticeable similarity of spatial variations of the precipitating flux and the EMIC wave intensity shown in Fig. 2 (right, panel (a)).

We calculated the pitch-angle diffusion coefficient for 11 instants within the interval of the EMIC wave observations. For each instant the background magnetic field, the cold plasma density and the spectral characteristics of waves were determined. We assumed that the cold plasma ion composition was: 5% O+, 5% He+ and 90% H+. The results of calculations are presented in Fig. 3.

The maximum values of the pitch-angle diffusion coefficient shown on the left panel of Fig. 3 vary within a rather wide range (from 0.005 to 0.075 s^{-1}). The diffusion regime is parameterized by [2]

$$z = 2 \cdot \sqrt{\mu_c/(D_{\alpha\alpha} \cdot T_b)},$$

where μ_c is the loss cone boundary, T_b is the bounce period. The strong (weak) diffusion corresponds to $z < 1$ ($z > 1$). In our case, $z = 0.054$ to 0.196. Therefore,

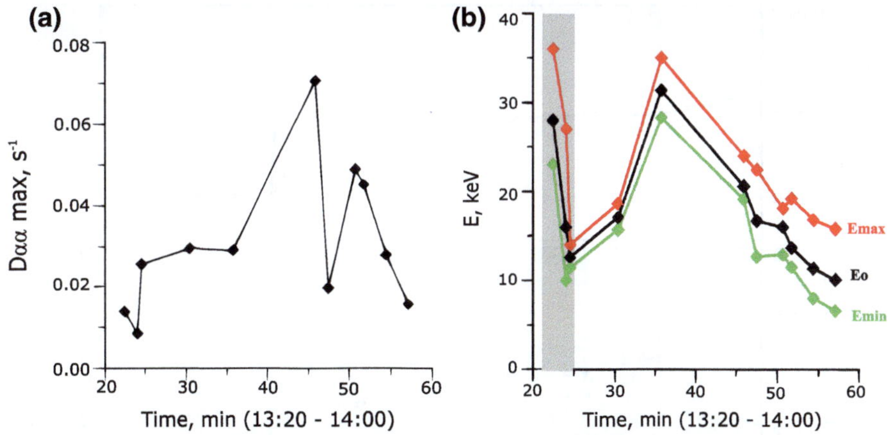

Fig. 3 The maximum values of the diffusion coefficient at each of 11 selected points of time (*left*) and the corresponding resonant energies (*right*). On the right panel: the black line indicates the energy at which the pitch-angle diffusion coefficient is maximum, and the green and red lines show the energies corresponding to the diffusion coefficient which is *e* times smaller than the maximum value

even the lower value of the calculated diffusion coefficient corresponds to the strong diffusion regime, i.e., to an almost filled loss cone. Although the maximum diffusion time $T_D = D_{\alpha\alpha}^{-1} \sim 200$ s is greater than the proton bounce period ($T_b \sim 60$–100 s), the strong diffusion is reached due to the smallness of the loss cone. This estimate agrees with the observation results shown in Figs. 2 and 5 (see Sect. 5): the trapped and precipitated fluxes are equal to each other in Fig. 2 and are almost equal in Fig. 5.

The relatively wide range of $D_{\alpha\alpha}$ is related to the time variations of the wave spectrum and cold plasma density. For example, the greatest value was obtained when the wave spectrum was very narrow and the waves had large amplitude. On the right panel of the same Figure the range of values of the resonant energy associated with the calculated diffusion coefficients is shown. The first three values correspond to the time when the low-altitude satellite NOAA-18 flew in the conjugate area. From the comparison of Figs. 2 (right) and 3 one may conclude that the energy range of precipitating protons, which is in the range of both TED and MEPED instruments (from a few to tens keV) agrees with the range obtained from calculations of the resonant energy (10–37 keV).

5 The Event of 13 October 2010 (Case 2)

Let us consider another event (13 October 2010), when THEMIS A was situated at MLT = 9–10 h at distances 4.8–5.5 R_E and observed EMIC waves outside the plasmasphere (the cold plasma density varied from 5.2 to 2.2 cm^{-3}, see Fig. 4).

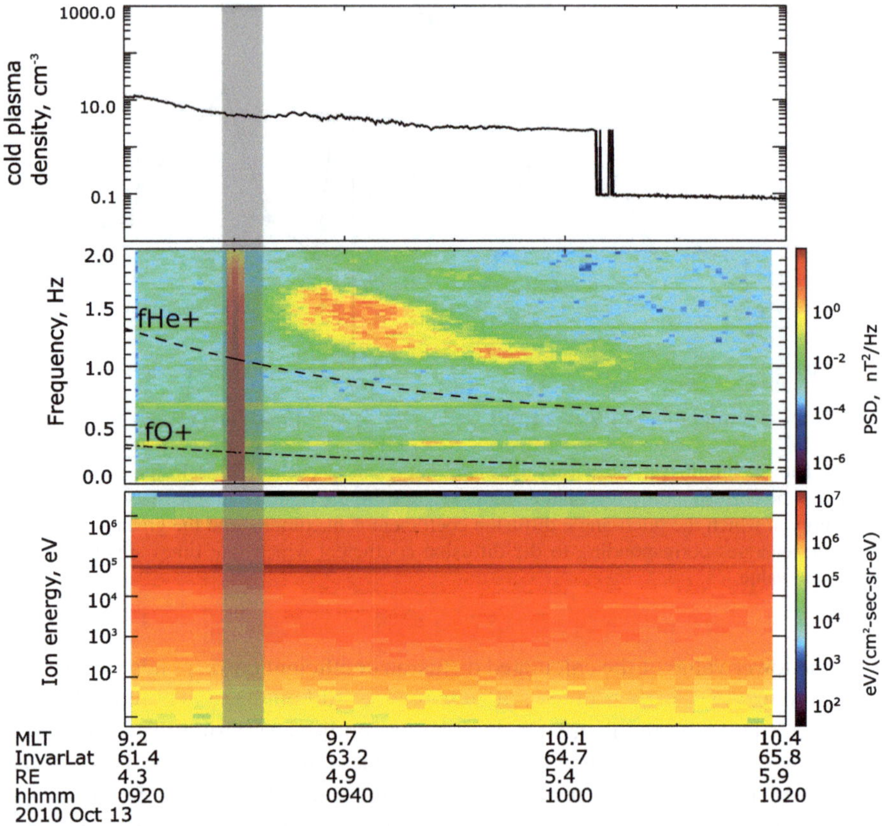

Fig. 4 THEMIS A data on the cold plasma density, EMIC waves, and proton energy spectra during the event of 0920-1020 UT on 13 October 2010. Grey area marks the time interval when the conjugate flight of a low-altitude satellite took place (see Fig. 5 (*left*))

The energy spectra of protons measured by the ESA and SST instruments show two populations with maximum fluxes at about 7 and 70 keV.

In this case the low-orbiting satellite NOAA-16 crossed the conjugate region a few minutes before THEMIS A entered the region of the EMIC wave observations. Traces of the two spacecraft are shown in Fig. 5 (left). The NOAA-16 satellite observed precipitating protons in two channels of the MEPED instrument: P1 (30–80 keV) and P2 (80–250 keV) (Fig. 5 (right)). No detectable fluxes were observed with the TED instrument. From Fig. 5 (right) it is clearly seen that the latitudes of the precipitation region and the EMIC wave observations nearly coincide.

The diffusion coefficients for this case were calculated for 15 time moments; the cold plasma composition was chosen the same as in case 1. Figure 6 demonstrates the results of calculations. As one can see, the resonant energy obtained from the

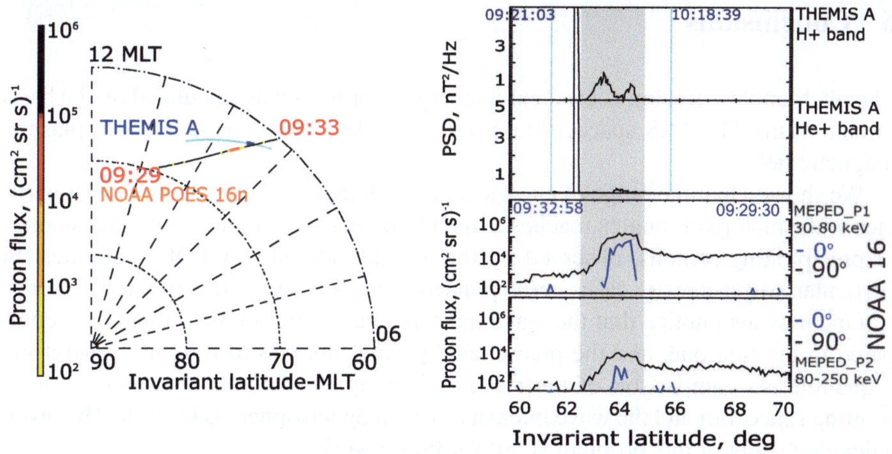

Fig. 5 *Left* Traces of THEMIS A and NOAA-16 satellites in the MLT-Invariant latitude coordinates. *Right* Latitudinal profiles of the EMIC wave intensity in H+ and He+ bands measured by THEMIS A as well as precipitating and trapped proton fluxes observed in channels P1 and P2 of the MEPED instrument onboard the NOAA-18 satellite

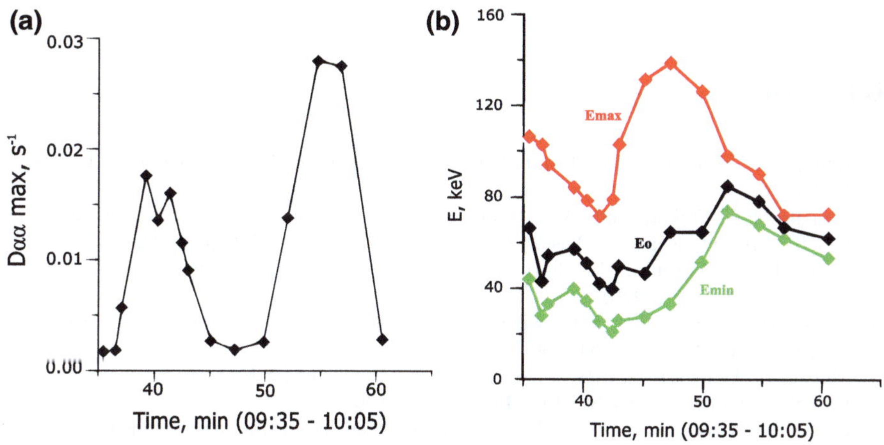

Fig. 6 The same as Fig. 3, but for the event of 0935-1035 UT on 13 Oct 2010

calculations varies from 20–140 keV. This is in agreement with the energy range of the precipitating protons detected by the low-orbiting satellite.

Similar correspondence of measured and calculated energies were obtained on the basis of analysis of other considered events.

6 Conclusions

The pitch-angle diffusion coefficient of energetic protons was calculated on the basis of data of the THEMIS spacecraft (spectra of EMIC waves, parameters of plasma, magnetic field).

We showed that the range of energies, in which the diffusion coefficient for particles with small pitch-angles reaches a maximum, is in accordance with the energies of precipitating protons, observed by the low-altitude NOAA POES satellites. In particular, lower-density cases corresponded to higher-energy precipitated protons.

One may also notice that the wave intensity was higher in the second case compared to the first one, and the precipitated proton flux was also higher. Therefore, a quantitative comparison of the values of precipitated fluxes measured by low-orbiting spacecraft and the wave intensity in the magnetosphere is feasible. However, a detailed study of this problem is left for future work.

Acknowledgements We thank the THEMIS Project Team (project manager V. Angelopoulos) for providing free access to project data (http://themis.ssl.berkeley.edu), and NOAA for providing POES satellite data (https://www.ngdc.noaa.gov/stp/satellite/poes/dataaccess.html). This work was supported by the Russian Science Foundation, project no. 15-12-20005.

References

1. Kennel, C.F., Petschek, H.E.: Limit of stably trapped particle fluxes. J. Geophys. Res. **71**(1), 1–28 (1966)
2. Bespalov, P.A., Trakhtengerts, V.Y.: Cyclotron instability in the Earth radiation belts. In: Leontovich, M.A. (ed.) Report of Plasma Physics, vol. 10, pp. 155–192. Plenum, New York (1986)
3. Cornwall, J.M., Coroniti, F.V., Thorne, R.M.: Turbulent loss of ring current protons. J. Geophys. Res. **75**, 4699–4709 (1970)
4. Auster, H.U., Glassmeier, K.H., Magnes, W., et al.: The THEMIS fluxgate magnetometer. Space Sci. Rev. **141**, 235–264 (2008)
5. McFadden, J.P., Carlson, C.W., Larson, D., et al.: The THEMIS ESA plasma instrument and in-flight calibration. Space Sci. Rev. **141**, 277–302 (2008)
6. Evans, D.S., Greer, M.S.: Polar Orbiting Environmental Satellite Space Experiment Monitor-2: Instrument Descriptions and Archive Data Documentation. NOAA Technical Memorandum version 1.3. NOAA Space Environment Center. Boulder, Colo. (2004)
7. Popova, T.A., Yahnin, A.G., Demekhov, A.G., Chernyaeva, S.A.: Generation of EMIC waves in the magnetosphere and precipitation of energetic protons: comparison of the data from THEMIS high Earth orbiting satellites and POES low Earth orbiting satellites. Geomag. Aeron. **58**(4), 469–482 (2018)
8. Summers, D.: Quasi-linear diffusion coefficients for field-aligned electromagnetic waves with applications to the magnetosphere. J. Geophys. Res. **110**, A08213 (2005). https://doi.org/10.1029/2005JA011159
9. Summers, D., Thorne, R.M.: Relativistic electron pitch-angle scattering by electromagnetic ion cyclotron waves during geomagnetic storms. J. Geophys. Res. **108**(A4), 1143 (2003). https://doi.org/10.1029/2002JA009489
10. Young, D.T., Geiss, J., Balsiger, H., et al.: Discovery of He^{2+} and O^{2+} ions of terrestrial origin in the outer magnetosphere. Geophys. Res. Lett. **4**(12), 561–564 (1977)
11. Tsyganenko, N.A.: A magnetospheric magnetic field model with a warped tail current sheet. Planet. Space Sci. **37**, 5–20 (1989)

Sunset Variations of the Bottom Edge of the Ionosphere During the Proton Precipitations on and After 29 September, 1989

G. F. Remenets, M. I. Suhovey and V. A. Shishaev

Abstract One of the most powerful solar proton precipitations (SPP), which began on 29 September 1989, was at the same time the longest one. This peculiarity gave to us the possibility to study on the base of VLF data the precipitation effect during a sunset on the dynamics of the electric properties of the lowest part of the ionosphere in the terms of complex reflection coefficient for a sequence of 8 days. The input data for the analysis were the relative variations of amplitudes and variations of phases for 3 VLF monochromatic, ground based signals with radio path North Norway—Kola Peninsula having a moderate length, 885 km. A self-consistent method for solution of an inverse VLF problem was used. The sunset dynamics of electric conductivity of the temporally modified bottom ionosphere was stated for all days of the SPP. The solution of such inverse VLF problem was possible due to the multi frequency measurements for the radio path and the synchronous registration of amplitudes and phases.

Keywords Proton precipitation on 29 September 1989 · Sunset effect · Inverse VLF problem · Self-consistent method

1 Introduction

Our paper is devoted to the terrestrial electromagnetic sounding of the lowest part of ionosphere during the natural disturbances. Such investigation may be named as a solution of inverse radio physics problem. Thirty years ago a self-consistent (S-C) method for solution of an inverse VLF problem (SIvlfP) in non-stationary radio path conditions was developed, and it was used for an analysis of 3 different VLF disturbances on 29 September 1989: an ultra-energetic relativistic electron precipitation (04:00–10:00 UT), a sudden ionospheric disturbance due to the Sun

G. F. Remenets (✉) · M. I. Suhovey
Saint-Petersberg State University, Saint-Petersburg, Russia
e-mail: g.remenets@spbu.ru

V. A. Shishaev
Polar Geoghysical Institute, Apatity, Murmansk Reg., Russia

T. B. Yanovskaya et al. (eds.), *Problems of Geocosmos—2018*, Springer Proceedings in Earth and Environmental Sciences, https://doi.org/10.1007/978-3-030-21788-4_28

X-ray flare at 11 UT and a solar proton precipitation (SPP)—one of the most powerful event of this kind for 70 years [1], which began at 12 UT. Its analysis was stopped at 17 UT [2], at the maximum of precipitation. The SIvlfP gave the dynamics of the complex refection coefficient with its modulus *R* and phase recalculated into the effective height (altitude) *h*. The radio path was purely auroral with moderate length (885 km, North Norway—Kola Peninsular, id est. Aldra station—Apatity town, Murmansk region). The radio signals were monochromatic with frequencies 10.2, 12.1, 13.6 kHz. The effective height *h* for a real waveguide at fixed time is a height of an effective model empty waveguide with the abrupt bottom and top boundaries. This waveguide is effective in the sense that the phase pass of a given propagating mode (a ray in our case) is the same as in the real waveguide.

The peculiarity of the SIvlfP is its independence on the nature of a geophysical disturbance, and this solution does not demand a priori knowledge of the ionosphere parameters before a disturbance. A "ray" (hop) version of theory of the wave propagation in the near ground (terrestrial) waveguide [3, 4] was used in the solutions pointed above: the diffraction wave above the spherical model of the Earth, 1st and 2nd rays reflecting from a top virtual boundary one and two times correspondingly. We worked in the conditions for which the used radio frequencies are much less the frequency of effective electron collisions with the atmosphere neutrals, therefore, the bottom edge of ionosphere is an inhomogeneous electric conductor depending on the altitude, and the radio waves react on the time changes of electric conductivity.

For the same radio path pointed above the same S-C method [2, 5, 6] was used for analysis of the SPP on 16 February 1984 [7] and for 18 sunset variations in 8 August–8 September, 1974, half of which referred to the quiet geophysical conditions [8]. In the case of SPP which continued according to the VLF data from 9 to 13 UT the effective height $h(t)$ fell down from 63–67 to 44–47 km at 10 UT. The modulus of the first sky wave (1st ray) slightly changed about 0.6 during the event. Figure 1 represents these statements, and in Fig. 2 the comparison of the experimental phase variations with the calculated ones are given.

The sunset variations are represented in Fig. 3, [8].

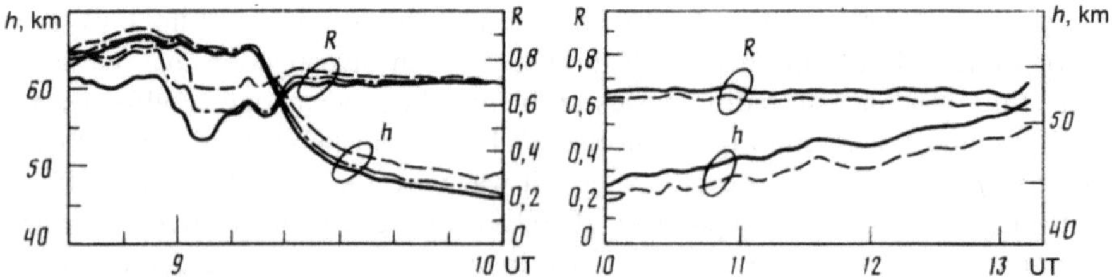

Fig. 1 The dynamics of the modulus of the reflection coefficient of the 1st sky wave (1st hop) from above $R(t, \psi(h))$ and the effective altitude $h(t)$ of reflection for the beginning stage (left panel) and for the restoration stage (right panel) [7]. The angle of the wave incidence $\psi(h)$ depends on $h(t)$ according to the geometrical relations for spherical waveguide model and is greater than 1.4 rad, Three curves for each magnitude estimate the accuracy of the SIvlfP method applied to the 16 February, 1984 data

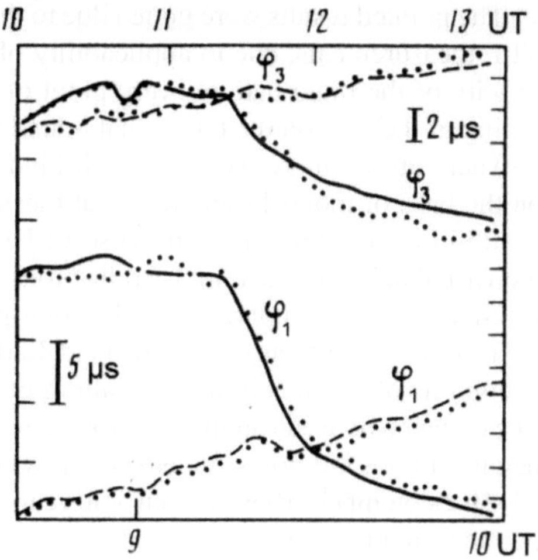

Fig. 2 The comparison of the calculated curves (the continuous and dashed ones) with the experimental curves (the dotted ones) for the phase variations while the initial and restoration stages of the SPP on 16 February, 1984, [7]. The calculated results were gotten according to the *R*(*t*, *h*) and *h*(*t*) represented in Fig. 1

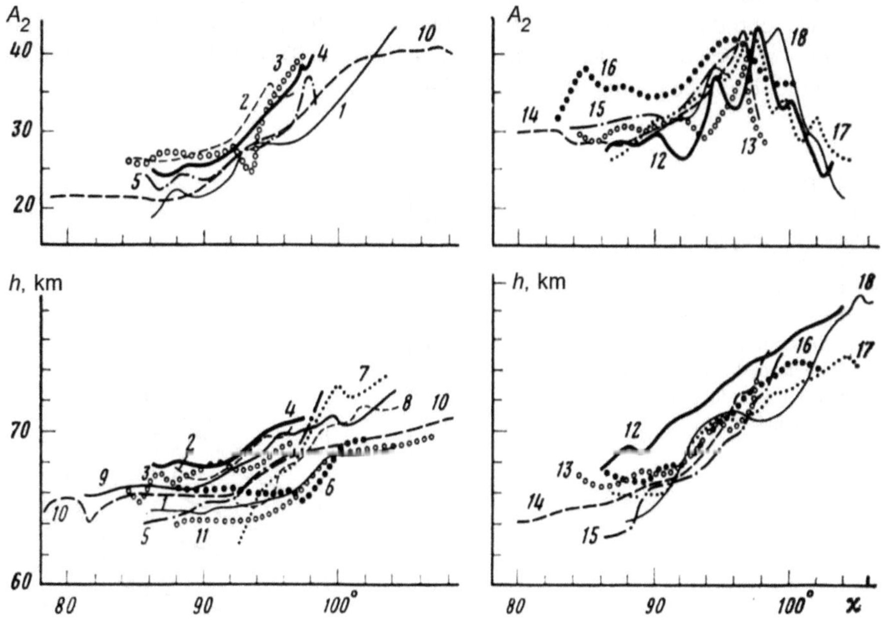

Fig. 3 The variations of amplitude (12.1 kHz), top panel, in relative units, and of the corresponding effective heights h(t) km, bottom panel, at sunset for the polar radio path North Norway, (Aldra)–Kola peninsula, (Apatity), [8]. The numbers of the curves correspond to different dates of August and September. On the left of the panel the moderately disturbed dates are collected, and on the right—the quiet days are represented

The pointed results were gotten due to the several frequency amplitude and phase VLF measurements, due to applicability of a supposition about the prolong homogeneity of the radio path at every point of the disturbance time and due to sliding falling of the rays on the top virtual boundary with the altitude *h*.

There are no more works of such kind with the inverse VLF problem solution on the base of monochromatic signal measurements. But there are two unique and beautiful works in which the impulse VLF sounding was used [9, 10]. In the first one the virtual height of the impulse reflections from above at the Berd station (Antarctic Continent) were compared with the solar proton fluxes (E > 10 MeV) which were measured on the board of hand maid Earth's satellite "Explorer-41" [10]. In the conditions of practically normal falling of an impulse on the disturbed ionosphere (the distance between an impulse source and a receiver was equal 20 km) the virtual height did not fall down under 60 km, with the maximum values 77–80 km while 25–29 September, 1969. About ten years after the inverse VLF problem for the pointed impulse sounding was solved [10] and the corresponding $N_e(z)$ for 6 days of polar night from 27 September to 2 October, 1978 were represented on Fig. 4, which we reproduced, the review [11]. The $N_e(z)$ profiles (the continuous curves) of this Fig. 4 are practically identical at first two days: on 27 and 28 September. On the following 4 days the profile lifted at 2–3 km per day.

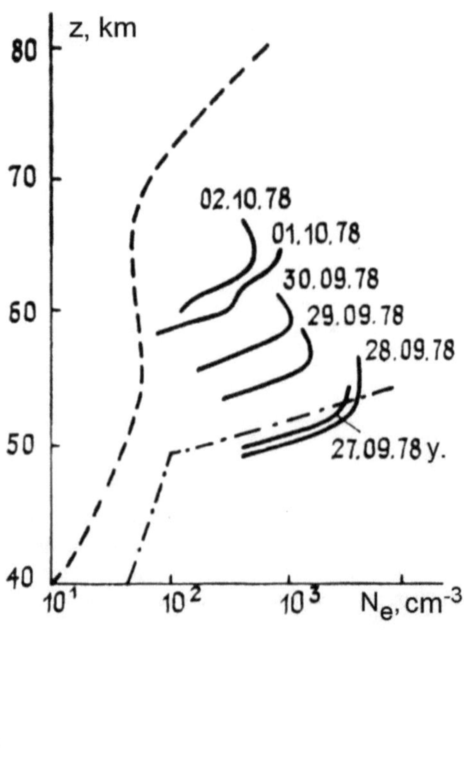

Fig. 4 The sounding of polar ionosphere by the VLF impulses in Antarctic Continent (the Berd station) and the corresponding $N_e(z)$ for the days of polar night (the continuous curves) [10]. The dashed curves [11], which are the model ones, are not discussed here

2 A Problem and Its Solution

We have reported about all science sources devoted to the VLF sounding of the polar ionosphere disturbed by the SPPs and the sunset effect independently. Now we pass to our problem about the same VLF sounding but in the conditions when two causes of ionization are working simultaneously: the SPP and the sunset for the radio path. The SPP on 29 September, 1989 continued not less than 10 days, Fig. 5, [12]. In the present report we characterize the electric conductivity of the disturbed low polar ionosphere by 2 parameters: by an effective height (effective altitude of reflection) $h(t)$ and by a value of reflection coefficient modulus $R(t, \psi\ (h))$ for the first ionosphere ray in a model of effective waveguide with $h(t)$. The argument of complex $R(t, \psi\ (h))$ is equal to π for sliding angles ψ of ray incidence ($\psi > 1.4$ rad.) and for a middle working frequency 12.1 kHz. Our problem is to get the time functions $R(t, \psi(h))$ and $h(t)$ according to 6 input experimental VLF functions: 3 amplitudes A_i(t) an 3 phase variations ϕ_i(t) caused by the geophysical disturbances, $i = 1, 2, 3$; the radio path and the working frequencies are pointed in the Introduction. Our analysis is limited by the sunset hours (15:30–18:30 UT). For this time the most "bad" approximation is done: at every time moment a model waveguide is supposed to be homogeneous along the radio path. Our estimation has shown that the corresponding systematic error is not greater than the apparatus phase error of the experimental data used (1 μs).

We shall compare quantitatively the daily variations of solar proton fluxes, which began on September 29, 1989, 12 UT, with the sunset changes of our output parameters $R(t)$ and $h(t)$ for a completely auroral radio path Aldra–Apatity which we pointed above. For realization of this purpose we have used the satellite data [12], the experimental VLF data of the Polar Geophysical Institute—RAS, Apatity, Murmansk

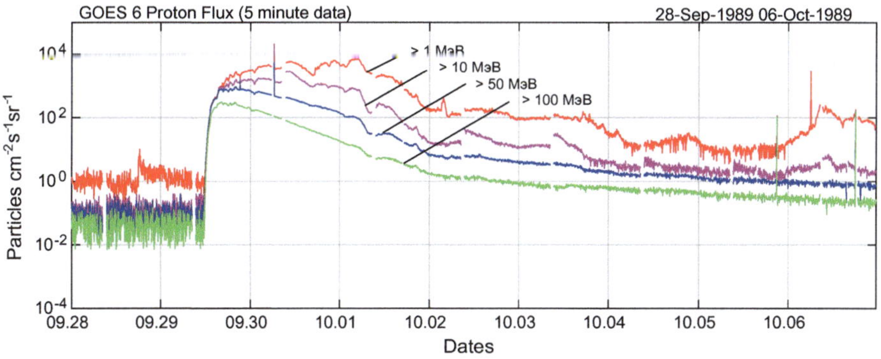

Fig. 5 Satellite data on the proton fluxes with different energies (>1 MeV, >10 MeV, >50 MeV and >100 MeV) for the period from 28 September (09.28) to 6 October (10.06), 1989, [12]. The event began on 29 Sept. (09.29), and to 5 Oct (10.05). the proton fluxes were greatly weakened. On 6th Oct. the fluxes of electrons with energy less than 50 MeV increased again significantly. This figure was plotted by us according to the data [13]

region and the S-C method of SIvlfP [2]. This investigation is a natural continuation of the analysis which was begun in this pointed work.

The VLF amplitude and phase variations corresponding to the proton precipitations of Fig. 5 are represented by the data in the Fig. 6. We used this data in our analysis only partly, id est. we analyzed the sunset hours 15:00–19:00 for every date. In Fig. 7 such data on 28 and 29 September are represented.

We solved the inverse problem and established the following: (i) The presented graphs in Figs. 8 and 9 give an error estimate of the method by comparing the analysis of sunset VLF variations for the positive direction of time (solid lines) with the analysis for the negative direction of time (dashed lines). Each of these analyses, based on the solution of the inverse VLF problem by S-C method and which did not use any geophysical data, was fulfilled independently. (ii) With the change of proton flux density in the period from September 29 to October 2, 1989 the daily variation of the effective height $h(t)$ at sunset changed from 5 km on September to 10 km on October and the reflection coefficient of the first ionospheric ray at a sunset was practically constant. In this date period the value of effective altitude changed from 47 to 54 km at 15:00 UT, from 52 to 64 km at 19:00 UT. The value of reflection coefficient modulus has changed from 0.8 to 0.6, id est. to the value of 16 February 1984, Fig. 1. (iii) From October 3 to October 5, the proton flux density did not yet come back to its undisturbed value but weakened significantly.

In a real wave guide there is none ray mode as for high frequencies, because of a wave length value being comparable with the altitude scale of inhomogeneity of electric conductivity. But it was shown numerically several decades ago that an effective height $h(t)$ was placed inside the altitude layer of electric conductivity which was significant for the radio wave reflection. Roughly speaking the h is an analog of

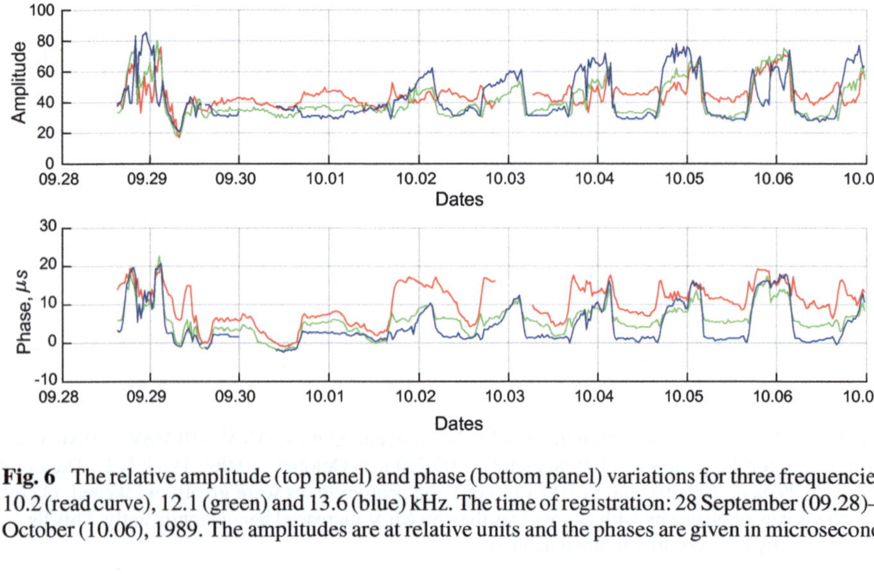

Fig. 6 The relative amplitude (top panel) and phase (bottom panel) variations for three frequencies: 10.2 (read curve), 12.1 (green) and 13.6 (blue) kHz. The time of registration: 28 September (09.28)–6 October (10.06), 1989. The amplitudes are at relative units and the phases are given in microseconds

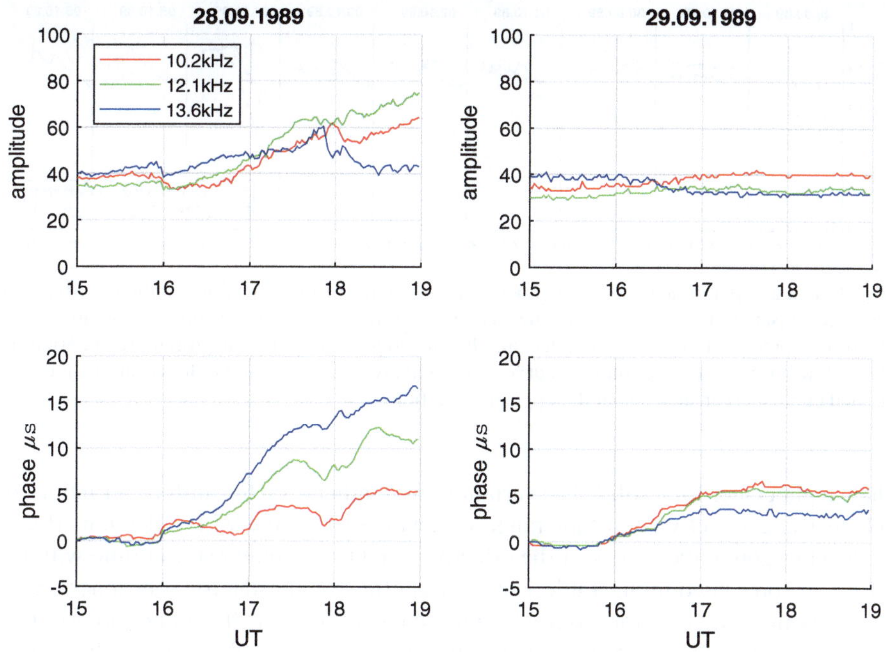

Fig. 7 The time fragments of the VLF variations before (the left half) and on (the right half) the solar proton precipitation 29.09.1989. These fragments correspond to the sunset hour for the radio path

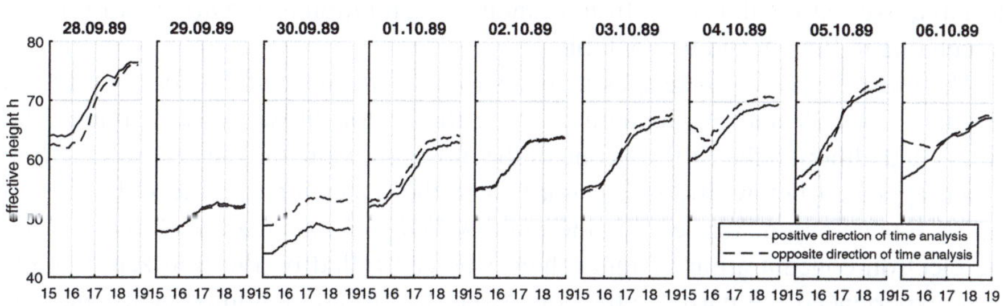

Fig. 8 The computation results: Dynamics of the effective height $h(t)$ at sunset. The variations of effective height h at sunset (15:00–19:00 UT) for a week of measurements having begun with an absence of proton precipitations (28 September) and having continued with solar proton precipitation for the following days. The continuous curves—is an analysis in the positive direction of time. The dashed curves—is an analysis in the opposite direction

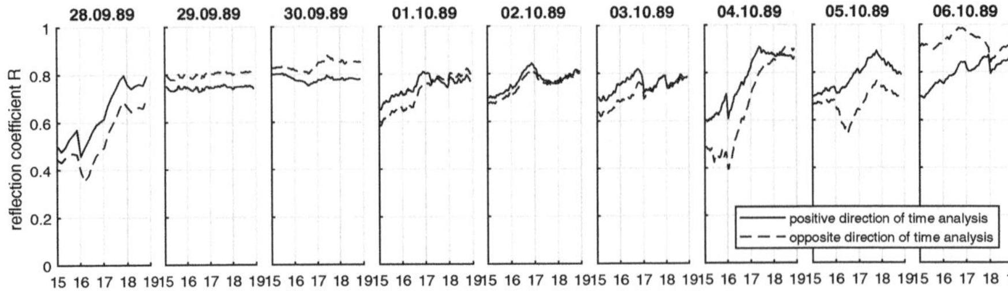

Fig. 9 The computation results: dynamics of the modulus of the reflection coefficient $R(t)$ from above at sunset, 15:00–19:00 UT, for the week of measurements having begun with an absence of the proton precipitation (28 September) and having continued with solar proton precipitation for the following days. The continuous curves—is an analysis in the positive direction of time. The dashed curves—is an analysis in the opposite direction

the weight center for a solid. The variations of h indicate on the altitude variations of all electric conductivity altitude range which determines the VLF reflection effect.

Having gotten the values of the effective height at different sunset time with the SPP we represented in the Table 1 the sunset effect on the electric ionospheric conductivity in a term of h for different stages (days) of the SPP. It is important to stress on the point that the values before and after the sunset (the values of h in the second and third columns) were determined independently on any geophysical suppositions about the electron profile in the ionosphere. Deviation from the monotony decreasing of the proton flux with the proton energy less than 50 MeV on 6 October, Fig. 5, declared itself at the deviation from monotone restoration of the $h(t)$ and $R(t)$, as it is seen due to Figs. 8 and 9 and Table 1.

The most interesting result, as it seems to us, is a sunset effect for the effective height $h(t)$ on 29 and 30 September when the proton fluxes were extreme. That means that although the SPP was one of the most powerful, the electric conductivity which appeared below 50 km did not isolate the upper ionospheric D-layer from the VLF penetration. We see that the radio waves penetrated at the altitudes where the electron concentration of ionosphere was controlled by the Sun zenith angle, were experienced the partial reflections from the altitude inhomogeneity, may be, not monotonous and came to the receiver creating the effect pointed.

In Table 2 we represent the approximate correspondence between the values of h for different days of the EPP event and a minimal value of the proton energy $\boldsymbol{E}$ with which it, precipitating from above, may achieve the altitude h in the Earth's atmosphere according to the energy estimate [14].

The last point of our analysis is a question about the accuracy of our results. As the VLF magnitude variations are significantly larger than the apparatus errors (the signal/noise ratio is larger than 10 and the phase error about 0.5 μs) the main troubles of our procedure were the following: (i) while our deterministic analysis from one time point to another point several approximations of the theory [2, 15, 16] were used, (ii) practically there were none statistic data, (iii) the over fullness of the input experimental data (3 amplitude and 3 phase VLF input data against 4 unknown

Table 1 The values of effective attitude h for different dates at the case of fully lighted (day) and fully shadowed (night) radio path and the values of differences between the pointed values for the same date

Date	h, km at day-time	h, km at night-time	h day-night variations, km	The notes
28.09.1989	63	76	13	Without the proton precip.
29.09.1989	48	52	6	The first day of SPP
30.09.1989	47	53	6	The second day of SPP
01.10.1989	52	63	11	The third day of SPP
02.10.1989	55	64	9	The fourth day of SPP
03.10.1989	55	67	12	The fifth day of SPP
04.10.1989	60	71	11	The sixth day of SPP
05.10.1989	57	74	17	The seventh day of SPP
06.10.1989	57	68	11	The eighth day of SPP

Table 2 The correspondence for different dates between the effective attitude h values and the proton energy E values

Date	Sun lighted		Night	
	h, km	E, MeV	h, km	E, MeV
28.09.1989	63	10	76	3
29.09.1989	48	30	52	20
30.09.1989	41	50	53	20
01.10.1989	52	20	63	10
02.10.1989	55	20	64	10
03.10.1989	55	20	67	5
04.10.1989	60	10	71	5
05.10.1989	57	15	74	3
06.10.1989	57	15	68	5

Such energy is sufficient for a proton penetration from above at altitude h according to [14]

magnitudes—Δh, ΔR for every time step Δt and unknown initial values h_0, R_0) was used for the h_0 and R_0 determination. The minimization of a discrepancy function relatively these two parameters gave a solution of the inverse VLF problem which had the physics sense. The quality and accuracy of such solutions are represented in Fig. 10—the amplitude data, and in Fig. 11—the phase data. The dates and the frequencies used are pointed on the figures. This comparison of the experimental data with the calculated ones according to the SIvlfP, Figs. 8 and 9, is quite satisfactory. These results are the analogs of Fig. 2.

Our last remark relates to our induced supposition about homogeneity of the model waveguide at every time moment of a sunset. We may report that according

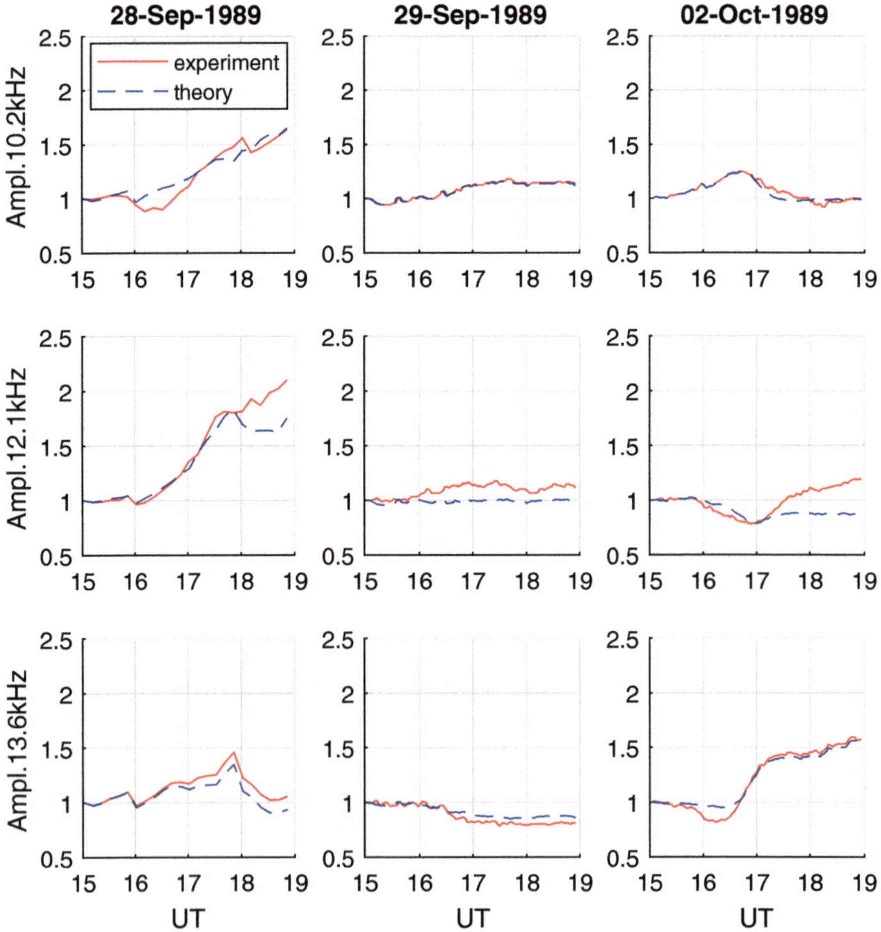

Fig. 10 Comparison of experimental and calculated amplitudes for 3 frequencies and 3 dates: before, on the first and the 4th day of the proton precipitation (28 September, 1989; 29 September, 1989 and 2 October, 1989). The analysis was done for positive direction of time

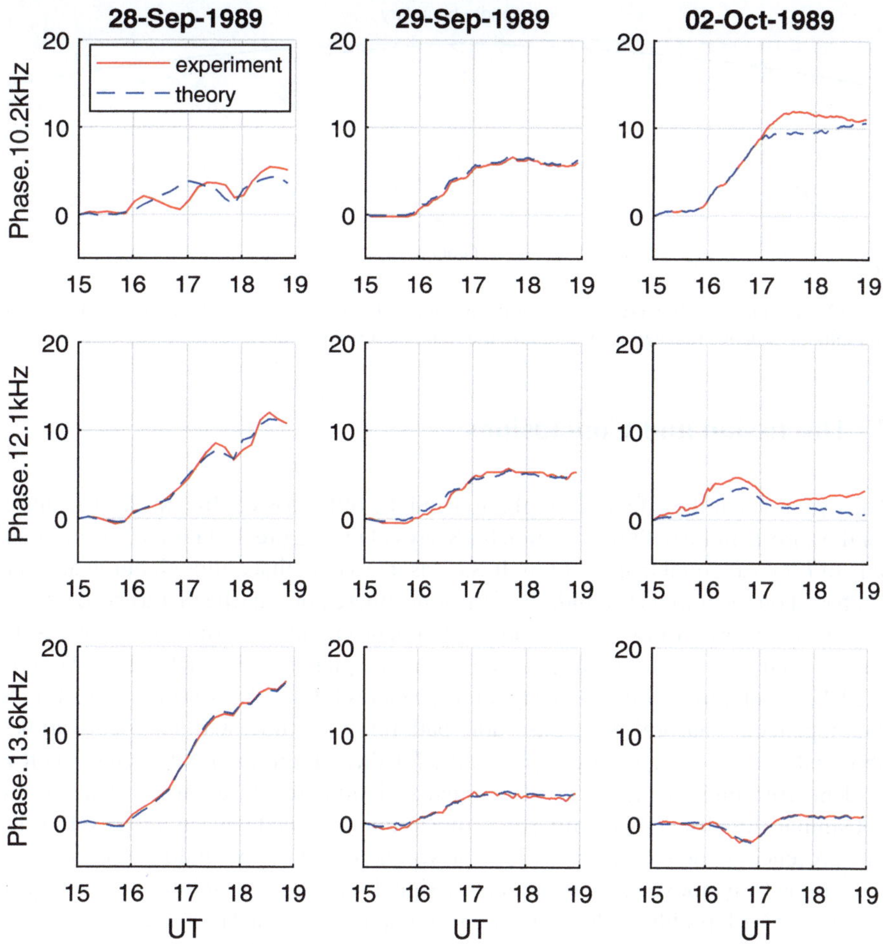

Fig. 11 Comparison of experimental and calculated phases for 3 frequencies and 3 dates: before, on the first day and the 4th day of the proton precipitation (28 September, 1989; 29 September, 1989 and 2 October, 1989). The analysis was done for positive direction of time

to our calculations, Fig. 12, the differences between optical lengths for red curves is equal to 0.2 μs and for blue curves is equal to 0.4 μs. For this figure the radiuses for two cylinders were 6370 km and (6370 + 70) km; the distance between a source and a receiver—900 km and in the model of inhomogeneous waveguide (the dashed curves) the difference between the boundary heights—20 km. As the apparatus phase errors were about 1 μs, then the pointed variations are not critical when the terms of ray summation are far enough from their anti phase state.

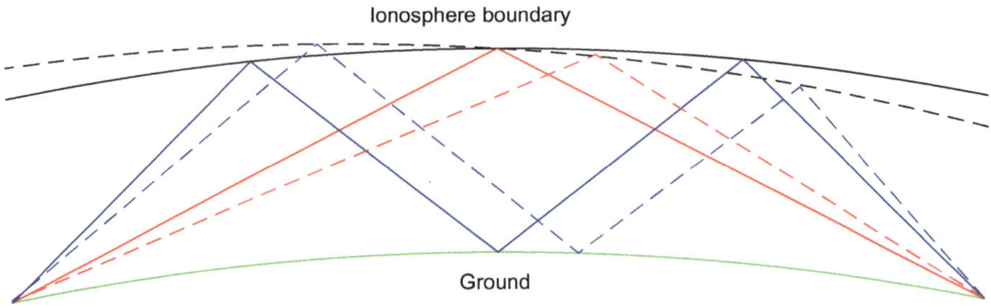

Fig. 12 A scheme of the 1st and 2nd hop trajectories for the homogeneous (continuous curves) and inhomogeneous (dashed curves) along the radio path waveguides

3 Discussion and Conclusions

The dynamics of the effective height *h*(*t*) and the reflection coefficient *R*(*t*) at sunset for the auroral radio path were gotten for 8 days of proton precipitations. The minimal vale of *h*(*t*) was equal to 45–47 km. It is pleasant to note that although our analysis is not based on the impulse sounding of the ionosphere bottom (the disturbed D-layer), as it is in the case of Fig. 4 [10], and is "monochromatic" sounding, our new results do not contradict to the old ones and they are supplemental to each other. We have used the "sounding" with 3 radio signal frequencies for the (10–14) kHz range for the "homogeneous" auroral West-East radio path for every time, with magnetic latitude about 64° and moderate length (885 km); for these radio path parameters, as it is evident from Figs. 6 and 7 the VLF signal variations for different frequencies, are not similar. An absence of similarity is due to interference between 3 terms, Fig. 12. The presence of the second sky wave (the second ray or "hop") with two radio wave reflections from above at every frequency signal increases the solution capability of the inverse VLF problem. Returning to the comparison of the Helm's electron profile results, Fig. 4, and our ones, Figs. 8 and 9 we see that they having the same physics nature but referring to the different dates (1978 and 1989 years) and to the radio paths sunlit differently. In one case the conditions were stationary, in the other case there was a sunset.

It was shown several year decades ago [17] that the effective height *h(t)* was inside the significant altitude range of the ionosphere relative to a concrete VLF magnitude. For example, if we demand that the accuracy of our calculations have to be not worthier than the accuracy of the experiment than the altitude width of the range is equal to 6, 4 and 5 km for the frequencies 10, 15 and 20 kHz correspondently (for a definite daytime profile $N_e(z)$) [18]. According to Fig. 4 the significant altitude range of the profiles *Ne*(*z*) represented is 5–7 km.

The sunset effect was seen at every day, although the proton precipitation on 29 September, 1989 was one of the strongest between the known ones. That means that (i) the bottom part of disturbed D-layer generated by the proton precipitation did not isolate the altitudes of regular D-layer for the VLF waves propagating at them from the ground and (ii) that the significant altitude range in the case was at several times greater than in the undisturbed conditions. The future analysis of the dawn time may be support these statements.

An increase of proton flux with the energies (1–50) MeV on 6 October significantly influenced the VLF data. The effective height $h(t)$ changed from 57–73 km on 5 October to 60–68 km on next day, Fig. 8. The $h(t)$ data on 3 and 6 October are practically identical.

References

1. Miroshnichenko, L., de Konong, C.A., Perez-Enriques, R.: Large solar event of September 29, 1989: ten years after. Space Sci. Rev. **91**(3), 615–715 (2000)
2. Remenets, G.F., Beloglazov, M.I.: Dynamics of an auroral ionospheric fringe at geophysical disturbances on 29 September 1989. Planet. Space Sci. **40**, 1101–1108 (1992)
3. Bremmer, H.: Terrestrial radio waves: Theory of propagation, 344 p. Elsevier, New York (1949)
4. Gyunninen, E.M., Zabavina, I.N.: Propagation of log waves and inhomogeneous ionosphere. In: Gyunninen, E.M. (ed.) Problems of Diffraction and Radio Wave Propagation, Issue 7, pp. 142–151. Leningrad State University, Leningrad (1968). (in Russian)
5. Remenets, G.F.: Unique ground VLF monitoring of the relativistic electron precipitations. In: Pudovkin, M. I., Besser, B.P., Riedler, W., Lyatskaya, A.M. (eds.) Problems of Geospace, Proceedings of an International Conference "Problems of Geocosmos" held in Petrodvorets, St. Petersburg, Russia, pp. 273–278, Austrian Academy of Science Press, Vienna (1997)
6. Remenets, G.F., Beloglazov, M.I.: About dynamics of the lower ionosphere during auroral disturbances by the VLF data. Geomagn. Aeron. **22**, 205–210 (1982). (in Russian)
7. Remenets, G.F., Beloglazov, M.I., Tamkun, L.G.: A modified method of VLF anomalies and a dynamics of a low ionosphere fringe at the time of the PCA event. Geomagn. Aeron. **29**, 45–50 (1989). (in Russian)
8. Remenets, G.F., Beloglazov, M.I.: An initial analysis of the dynamics of reflection properties of the low ionosphere at sunset for an auroral radio pass (according to the VLF data). Geomagn. Aeron. **25**, 69–72 (1985). (in Russian)
9. Helms, W.J., Swarm, H.M.: VLF step frequency sounding of the polar lower ionosphere. J. Geophys. Res. **74**(26), 6341–6351 (1961)
10. Field, E.C., Warren, R.E., Warber, C.R.: Calculation of ionospheric conductivity profiles by inverting VLF/LF reflection data: 1. Isotropic propagation. Radio. Sci. **18**, 452–460 (1983)
11. Remenets, G.F., Beloglazov, M.I.: Reflection properties of the lower polar ionosphere, and features of the excitation and propagation of superlong waves (SLW) at high latitudes (review). Radiophys. Quantum Electron. **28**(12), 1029–1040 (1985)
12. Coffey, H.E.: Solar-Geophysical Data Number 542, October 1989. Part 1 (prompt reports). Data for September, August 1989, and late data, p. 18. United States: N. p. (1989)
13. National Center for Environmental Information. https://satdat.ngdc.noaa.gov/sem/goes/data/avg/1989/09/goes06/csv/, Page g06_eps_5m_19890901_19890930.csv, https://satdat.ngdc.noaa.gov/sem/goes/data/avg/1989/10/goes06/csv/, Page g06_eps_5m_19891001_19891031.csv
14. Gordon, C.W., Canuto, V., Ian Axford, W.: Handbook of Astronomy, Astrophysics and Geophysics. The Earth: The Upper Atmosphere, Ionosphere and Magnet*osphere*, vol. 1, pp. 276–277. Gordon and Breanch Science Publishers

15. Remenets, G.F., Beloglazov, M.I.: Ultrarelativistic electrons in the near cosmos and X-ray aurora in the middle polar atmosphere. J. Geophys. Res. Space Phys. **118**(11), 6829–6838 (2013)
16. Remenets, G.F., Astafiev, A.M.: Solution uniquity of an inverse VLF problem: a case study of the ground-based, VLF radio signal disturbances caused the ultra-energetic relativistic electron precipitation and of their southern boundaries. Adv. Space Res. **56**(6), 878–889 (2016)
17. Galyuk, Y.P., Ivanov, V.I.: Determination of the propagation characteristics of VLF fields in the earth-ionosphere waveguide in the case of an anisotropic ionosphere inhomogeneous with height. Problemy Difraktsii i Rasprostraneniia Voln (16), 148–154 (1978). (in Russian)
18. Rybachek, S.T.: Effect of the essential region of the ionospheric layer on the characteristics of the propagation of ultra-long waves. Radiophys. Quantum Electron. **15**(9), 992–995 (1972)

Geomagnetic Disturbances with Sudden and Gradual Commencements in 1869–1954 According to Observatory Data and Catalogs of Storms

Sergey Sokolov, Vadim Soldatov and Vladimir Koshelevsky

Abstract Observations of variations in the geomagnetic field and the magnetic activity indices based on these observations are the only source of information about "geomagnetic weather" in the late 19th—first half of the 20th century. The aa indices, calculated from 1868, allow us to distinguish periods of magnetic storms of different magnitude and duration. Information on disturbances can be obtained directly from observational data at observatories, as well as from storm catalogs compiled from local observations on some of them. To understand how the geomagnetic activity changed during eight 11-year solar cycles (1869–1954), we use the aa-indices, the magnetograms and the catalogs of storms. On the basis of aa-indices we select periods during which the maximum value of aa-index (aa_{max}) exceeds 30 ($K_{pa\_max} > 3$) for more than three hours. These periods are considered disturbed, and it is taken into account that within themselves they could have quiet intervals with a duration of up to 18 h. Depending on the values of aa_{max}, each event is assigned a corresponding rank. Data on sudden commencement (SC) storms help to distinguish storms with SC among selected disturbances. The resulting sampling divides the disturbances by magnitude and by type of commencement—sudden or gradual (GC). A comparison of the obtained list with the catalogs data shows that the latter, as a rule, includes disturbances with $aa_{max} > 95$ ($K_{pa\_max} > 5$) only; smaller events are not included. Some storms in the directory are wrongly attributed to disturbances with GC, not SC. This is indicated by the data of the endorsed list of SCs (http://www.obsebre.es/en/rapid) and our review of the magnetograms of the observatories of St. Petersburg and Pavlovsk; they show almost all SCs from the above list. Our catalog includes 4619 storms: 1340 with SC and 3279 with GC. Using it, we have found out how the number of storms of different magnitude (with SC and GC) changed depending on the number of sunspots and at different phases of solar cycles. It is shown, in particular, that during the period considered there was a faster growth of storms with GC than with SC.

Keywords Geomagnetic disturbances · Sudden commencement · Gradual commencement

S. Sokolov (✉) · V. Soldatov · V. Koshelevsky
SPbF IZMIRAN, St. Petersburg, Russia
e-mail: sergnsokolov@yandex.ru

T. B. Yanovskaya et al. (eds.), *Problems of Geocosmos—2018*, Springer Proceedings in Earth and Environmental Sciences, https://doi.org/10.1007/978-3-030-21788-4_29

1 Introduction

The preserved observational data on geomagnetic field variations in the pre-satellite era provide indirect information on space weather and solar activity in the past. The aim of the work is to obtain a quantitative distribution over the years (1869–1954) of magnetic storms with a gradual (G storms) and sudden commencement (S storms) and to identify long-term trends in the frequency of their occurrence in 11–18 solar cycles. The reason for this work, which we started earlier [1], is the following. First, the information given in publications [2–9] about the storms that occurred during these years is often not consistent. Thus, the catalogs [2, 6] contain significantly fewer storms than those given in [3–5, 7–9]. This is due to the fact that the compilers of the catalogs used different criteria for the selection of disturbances, different verbal characteristics of storms and scales for assessing their intensity. Secondly, our review of the magnetograms of St. Petersburg and Pavlovsk [10] shows that some events in the catalogs [3–5] are mistakenly attributed to G storms, not to S storms. Almost all the moments of sudden commencements, given in the lists [11, 12], are visible on the magnetograms, but some of them have not been noticed before. Third, some disturbances, excluding large and very large storms (verbal characteristics adopted in the observatories of the USSR), were not included in the catalogs [3–5], although clearly visible on the magnetograms. In addition, as a rule, only storms corresponding [13] to the maximum values of the geomagnetic index $K_p \geq 5$ (and, accordingly, $K_{pa} \geq 5$) were usually recorded in all catalogs. Smaller amplitude (and most frequently occurring) perturbations were not included in them. We have compiled an adjusted storm catalog which is based on aa indices and includes disturbances with $K_{pa} > 3$ (aa > 30). Such a catalog has been compiled recently for extreme (aa $\geq$ 300) storms [9].

2 Data and Method

The paper uses: catalogs of storms in Russia (USSR) [3–5], and the UK [2, 6, 7]; aa index tables [12, 14] and "musical diagrams"; data on sudden commencements [11]; sunspots [15]; magnetograms (and their digital images) of observatories of Russia [10] and the UK [16]. For aa indices and their "musical diagrams", perturbed periods with a duration of more than three hours, which begin and end at aa > 30, are visually distinguished. It is taken into account that these periods may contain unperturbed intervals (aa $\leq$ 30) with a duration of up to 18 h [13]. Successive disturbances are considered as separate events if the interval between the end of the first and the beginning of the second was a day or more. Storms are considered separate also in cases when S storms (the beginnings of which are unambiguously identified in contrast to G storms) started on the background of the ongoing storms.

3 Results

Depending on the aa value of the index, each perturbation is assigned a rank (Table 1).

Figure 1 shows the changes in the annual number of storms in Pavlovsk in 1878–1940 (blue) according to the catalogs [3, 4], while red is the number of storms calculated by us using aa index for 1868–1941. Smaller number of disturbances in the catalogs is mainly due to the fact that they do not include storms with $K_{pa} \leq 5$.

Figure 2 shows the distribution by rank of all 1340 S storms (red) and 3279 G storms (blue) in 1869–1954. It is seen that disturbances of ranks 1 and 2 are about 60% (776) of all S storms and about 88% (2895) of all G storms. Our sampling includes all storms, without exception, from [2–7], in 1869–1954.

Table 1 The ranks of the storms, depending on the values of aa indexes and K_{pa}-indexes

Storm ranks used	aa_{max} (nT)	K_{pa_max}
1	31÷54	$3_{+}\div4_{0}$
2	55÷95	$4_{+}\div5_{0}$
3	96÷160	$5_{+}\div6_{0}$
4	161÷259	$6_{+}\div7_{0}$
5	260÷370	$7_{+}\div8_{0}$
6	≥371	$>8_{+}$

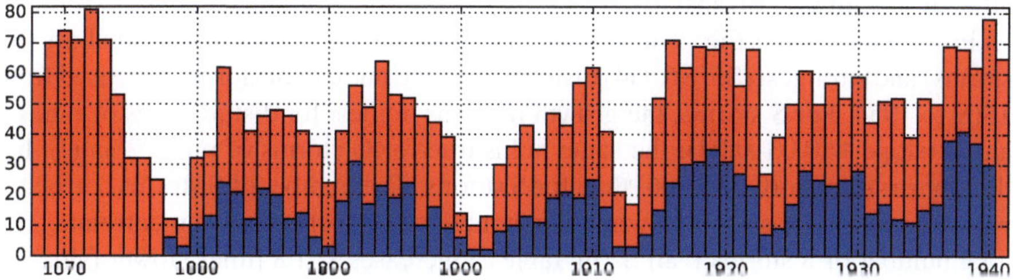

Fig. 1 The annual number of storms according to the catalogs [3, 4] (blue) and this work (red)

Fig. 2 Distribution of storms by rank

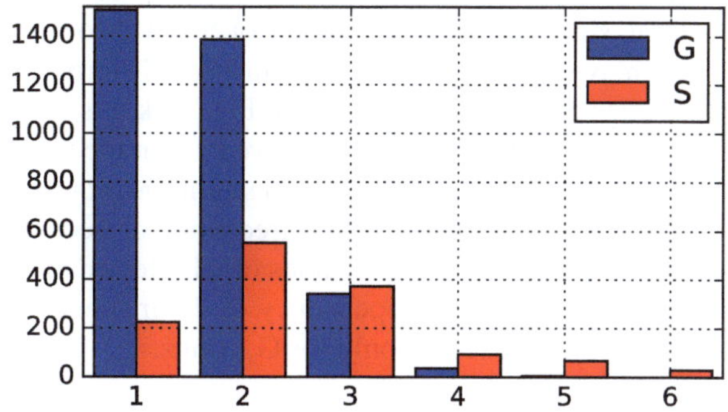

Table 2 Coefficients of linear approximation of the corresponding trends, and the correlation coefficients of the number of storms and sunspots

Storm rank	b	a	r
S1	2.712	−0.0022	0.408
S2	5.1021	0.0303	0.734
S3	2.7324	0.0366	0.735
S4	0.6679	0.0098	0.630
S5 + S6	1.0583	0.0019	0.553
All S	12.269	0.00764	0.861
G1	16.132	0.0322	0.468
G2	13.782	0.0539	0.125
G3	1.3688	0.0594	0.066
G4	0.4183	0.0003	0.190
All G	31.97	0.1413	0.384
All S + all G	41.728	0.1611	0.700

Figure 3 shows changes in the number of S storms in 1869–1954 and G storms of different ranks (gray) along with the change in the number of sunspots (red).

Table 2 shows the values of the coefficients a and b of the trend equations ($y = ax + b$) and the correlation coefficients r of the number of storms with the number of sunspots (Fig. 3). The bottom 12th line of the table gives these values for sampling all ("all S + all G") storms.

From Table 2 and Fig. 3 it follows that, with the exception of the weakest (S1 storms) and strongest (S5, S6, G4) disturbances, the number of storms increased from 1869 to 1954. For S storms, the growth rate was highest for disturbances of ranks 2 and 3 (slope $\approx$ 0.03–0.04); the same holds true for G storms of ranks 2 and 3, but the slope is steeper (0.05–0.06) than for S storms. The number of S storms correlates with the number of sunspots, while the number of G storms does not correlate. The total number of S storms ("all S" in Table 2) increases $\approx$ 1.8 times slower (slope $\approx$ 0.076) than the total number of G storms ("all G" in the Table 2, slope $\approx$ 0.141). For the entire S storms sample ("all S") the correlation with the number of sunspots is high ($r = 0.861$), and for the entire G storms sample ("all G") it is very low. This is natural, since the greatest number of S storms occurred in the years of the sunspot cycle maxima, and the largest number of G storms appeared in the third and fourth year after the highs (Fig. 3). From Fig. 3j, k it is seen that for 11–18 cycles of solar activity the number of S storms in years near the minimum number of sunspots does not change, while the number of G storms near these minima increases significantly.

Seasonal dependency of the frequency of occurrence of G storms (blue) and S storms (red) based on the samples of all selected events is shown in Fig. 4. It can be seen that the seasonal dependency of the frequency of storms with highs in spring and autumn is observed only for G storms.

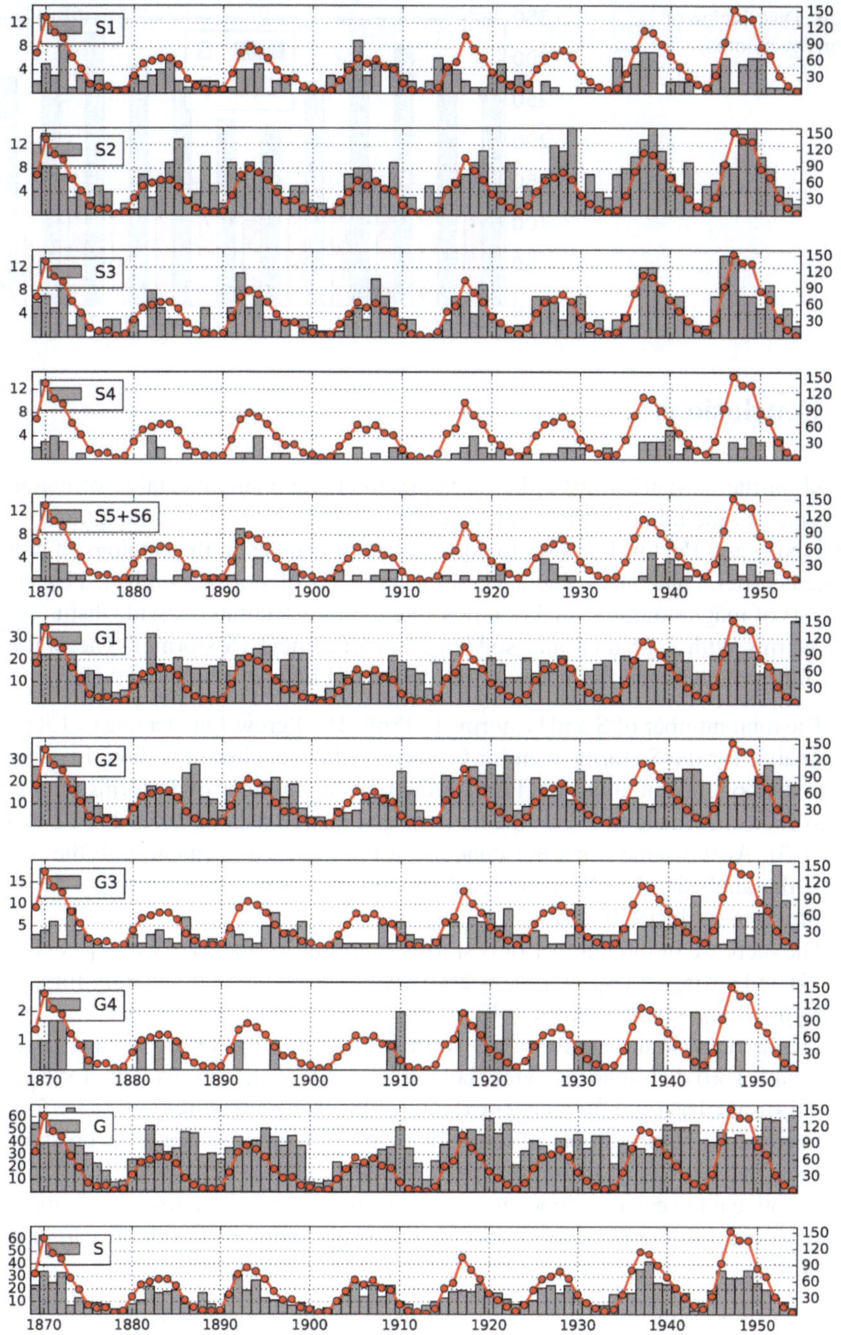

Fig. 3 The annual number of storms for each rank (S1–S4, S5 + S6, G1–G4); total number (S, G). Red line represents the number of sunspots

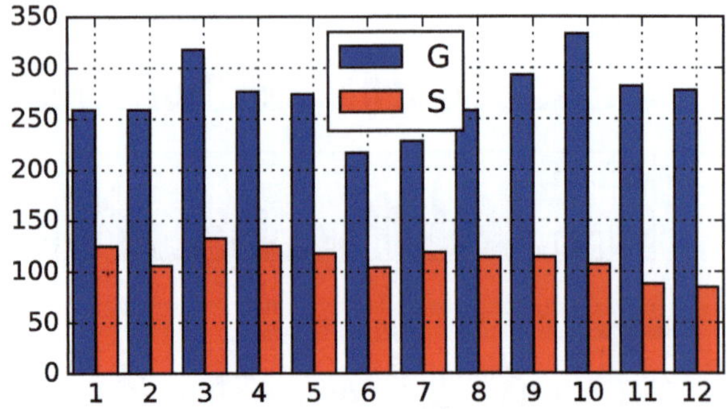

Fig. 4 Distribution of storms by months

4 Conclusions

Based on the aa (and/or ap) indices, we try to create a single catalog of magnetic disturbances including storms with aa > 30. Incorporation of magnetograms of St. Petersburg and Pavlovsk allows to correct some mistakes in the catalogs of storms created earlier. Data on the sudden commencements of storms, confirmed by visual analysis of magnetograms, makes it possible to more accurately divide disturbances into storms with a gradual and sudden commencement (S storms). The obtained long-term trends in the frequency of occurrence of both storms show that:

1. The total number of S and G storms in 1869–1954 grew, but at a rate $\approx$ 1.8 times smaller for the S storms than for G storms. This conclusion is directly opposite to the conclusion about the faster growth of the number of S storms compared to G storms made in [1] on the basis of unadjusted data of 1878–1954 catalogs [3–5]. At the same time, our conclusion is in good agreement with the results of the analysis of [17] of the local catalog of storms observed at the Moscow Observatory in 1950–2010.
2. The increase in the number of S storms is absent for the weakest ($K_{pa} < 4$, aa < 55) and strongest ($K_{pa} > 7$, aa > 260) disturbances; it occurs mainly for storms with $4 < K_{pa} < 6$ (55 < aa < 160).
3. The increase in the number of G storms is absent only for the strongest ($K_{pa} > 6$, aa > 160) disturbances, while being observed for the storms with $4 < K_{pa} < 6$ (55 < aa < 160) and, to a lesser extent, for the storms with $K_{pa} < 4$ (aa < 55).
4. Correlation of the number of storms with the number of sunspots occurs only for S storms, which is consistent with the results obtained in [9, 17].
5. A noticeable seasonal dependence of the frequency of occurrence of storms is observed only for G storms; this conclusion also coincides with the earlier ones [9, 17].

The results confirm the possibility of creating a single catalog of magnetic storms based on geomagnetic indices and local observations [13] and its use for studying space weather in the past.

References

1. Ptitsyna, N.G., Sokolov, S.N., Soldatov, V.A., Tyasto, M.I.: The upward trend in geomagnetic storm occurrences during the 20th century. In: Proceedings of the 11th International School and Conference "Problems of Geocosmos", pp. 278–284. St. Petersburg (2016)
2. Maunder, E.A.: Magnetic disturbances, 1882 to 1903, as recorded at the Royal Observatory, Greenwich, and their Association with Sun-spots. Mon. Notices Royal Astron Soc. **65**(1), 2–18 (1904)
3. Ben'kova, N.P., Kalinin, Ju.D.: Katalog magnitnykh bur' Slutskoj magnitnoj observatorii [Catalogue of magnetic storms at Slutsk magnetic observatory]. Kosmicheskiye dannye. Dekadnyj obzor [Space data. Decennial summary]. (125), 8; (126), 5; (127), 5, Gidrometeoizdat, Leningrad-Moscow (1941) (in Russian)
4. Afanas'yeva, V.I.: Magnitnaya aktivnost'. Spravochnik po peremennomu magnitnomu polyu SSSR. p. 166, Gidrometeorologicheskoye izdatel'stvo, Leningrad (1954) (in Russian)
5. Afanas'yeva, V.I., Bychkova A.K.: Svodnyy katalog magnitnykh bur' za 1949–1958 gody. pp. 7–93. IZMIRAN publ., Moscow (1979) (in Russian)
6. Sunspot and geomagnetic—storm data derived from Greenwich observations 1874–1954. Appendix 1. pp. 102–103. Her Majesty's Stationery Office, London (1955)
7. Moos, N.A.F.: Magnetic observations made at the Government Observatory for the period 1846 to 1905 and their discussion. Part II. The phenomenon and its discussions, pp. 452–453, 457–459, Government Central Press, Bombay (1910)
8. Cliverd, M.A., Clark, T.D.G., Clarke, E., Rishbeth, H.: Increased magnetic storm activity from 1868 to 1995. J. Atmos. Solar-Terr. Phys. **60**(10), 1047–1056 (1998)
9. Vennerstrom, S., Lefevre, L., Dumbović, M., et al.: Extreme geomagnetic storms—1868–2010. Solar. Phys. **291**(5), 1447–1481 (2016)
10. http://db.izmiran.nw.ru/magnetograms/. Last accessed 2018/04/01
11. http://www.obsebre.es/en/rapid. Last accessed 2018/04/01
12. http://isgi.unistra.fr. Last accessed 2018/04/01
13. Shevnin, A.D.: Analysis of joint catalogs of magnetic storms (1938–1958). Geomag. Aeron. **44**(5), 550–553 (2004)
14. ftp://ngdc.noaa.gov/STP/GEOMAGNETIC_DATA/AASTAR/aaindex. Last accessed 2018/04/01
15. https://www.ngdc.noaa.gov/stp/solar/solar-indices.html. Last accessed 2018/04/01
16. http://www.bgs.ac.uk/data/Magnetograms/home.html. Last accessed 2018/04/01
17. Obridko, V.N., Kanonidi, Kh.D., Mitrofanova, T.A., Shelting, B.D.: Solar activity and geomagnetic disturbances. Geomag. Aeron. **53**(2), 147–156 (2013)

Manifestation and Possible Reasons of ~60-Year Climatic Cycle in Correlation Links Between Solar Activity and Lower Atmosphere Circulation

Svetlana Veretenenko and Maxim Ogurtsov

Abstract Temporal variability of correlation links observed between characteristics of the lower atmosphere and solar activity phenomena at the multidecadal time scale remains one of unresolved problems of solar-terrestrial physics. In this work we continue studying possible reasons for this variability. Temporal behavior of correlation coefficients between troposphere pressure at extratropical latitudes and sunspot numbers was compared with long-term changes in the evolution of large-scale circulation, the intensity of the stratospheric polar vortex and global temperature anomalies. It was shown that a roughly 60-year periodicity revealed in solar activity influences on troposphere pressure (development of extratropical baric systems) is closely related to changes in the large-scale circulation regime which accompany transitions between strong and weak states of the stratospheric polar vortex. It was suggested that the detected correlation reversals are caused by changes of the troposphere-stratosphere coupling associated with changes of the vortex intensity. It was shown that the changes of the polar vortex state and the corresponding changes in the regime of large-scale circulation may be related to global temperature variations, with a possible reason for these variations being long-term changes of total solar irradiance.

Keywords Solar activity · Solar-atmospheric links · 60-year cycle

1 Introduction

Solar-climatic links are known to reveal a temporal variability which remains an unresolved problem so far and makes difficult forecasting future climate changes. In this work we continue studying possible reasons for this variability. According to

S. Veretenenko (✉) · M. Ogurtsov
Ioffe Institute, Russian Academy of Sciences, St. Petersburg 194021, Russia
e-mail: s.veretenenko@mail.ioffe.ru

M. Ogurtsov
Central (Pulkovo) Astronomical Observatory, Russian Academy of Sciences, St. Petersburg 196140, Russia

T. B. Yanovskaya et al. (eds.), *Problems of Geocosmos—2018*, Springer Proceedings in Earth and Environmental Sciences, https://doi.org/10.1007/978-3-030-21788-4_30

our previous results [1, 2], correlations links observed between pressure at extratropical latitudes and solar activity variations are characterized by a roughly 60-year periodicity, with a plausible agent being galactic cosmic rays. Climatic cycles with periods close to 60 years are observed in a large number of characteristics, such as global temperature anomalies [3], the Atlantic Multidecadal Oscillation [4], the Pacific Decadal Oscillation [5] etc. In particular, ~60-year variability is observed in sea-level pressure and temperature oscillations in the North Pacific [6], as well as in the alternation of cold and warm epochs in the Arctic [7] providing evidence for changes in the regime of large-scale circulation. So, the aim of this work is to consider possible reasons for a roughly 60-year variability observed in the Earth's climate and the dynamics of correlation links observed between troposphere pressure (development of extratropical baric systems) and solar activity characteristics.

2 Solar Activity Influences on Troposphere Pressure and Epochs of Large-Scale Circulation

According to Veretenenko and Ogurtsov [1], a roughly 60-year periodicity was detected in the amplitude and sign of correlation links between troposphere pressure at middle and high latitudes and sunspot numbers. Sunspot numbers do not represent a direct physical input to the atmosphere; however, the length of their time series allows us to study long-term variations of solar activity influences on atmospheric characteristics. Correlations with sunspot numbers may imply contributions of different agents, such as total and ultraviolet solar radiation, disturbances of solar wind and interplanetary magnetic fields etc. An important part in the mechanism of solar influence of the atmosphere seems to be played by variations of galactic cosmic rays (GCRs) (e.g., Pudovkin and Raspopov [8], Svensmark [9], Tinsley [10]). GSRs are strongly modulated by solar activity, with the correlation between yearly values of GCR fluxes and sunspot numbers reaching −0.8. Indeed, the data in [1, 2] confirmed a highly significant effect of GCR variations on the development of extratropical cyclones at Polar fronts of middle latitude at the decadal time scale. Thus, we shall use sunspot numbers as an indicator of solar variability taking into account that correlations with them may involve contributions of different agents, with GCR fluxes being one of most important factor for the evolution of extratropical baric systems.

It was found [1] that a ~60-year variation in correlation links of troposphere pressure with sunspot numbers is most pronounced in the Arctic and in the North Atlantic cyclogenetic zone near the eastern coasts of North America. The correlation coefficients in these regions vary in a clear antiphase which suggests a link between atmospheric processes developing in them as a response to solar variability. To understand possible causes of temporal variations (sign reversals) of solar activity effects, let us compare them with the evolution of large-scale circulation according to the Vangengeim-Girs classification [11]. This classification includes three main

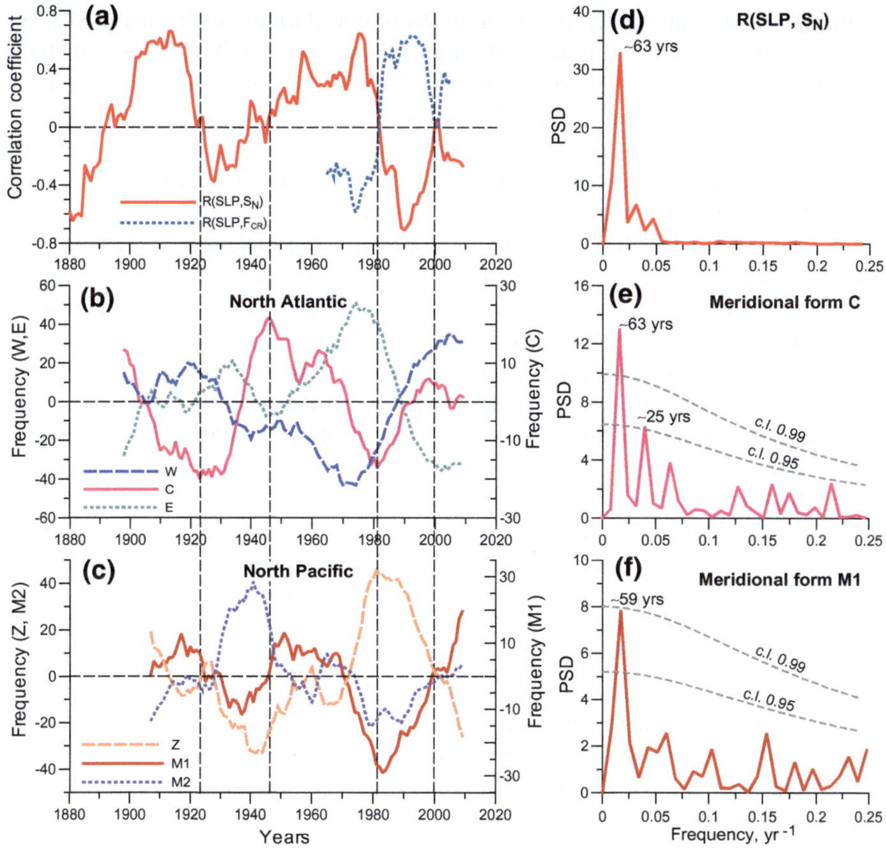

Fig. 1 Left: **a** Correlation coefficients between yearly values of sea-level pressure at high latitudes 60–85°N and sunspot numbers (solid red line), as well as GCR fluxes (dotted blue line) for sliding 15-year intervals; **b** frequencies of occurrence (number of days during a year) of the main forms of large-scale circulation for the North Atlantic region (15-year running averages); **c** the same for the North Pacific region. Vertical dashed lines show correlation reversals. Right: **d** Fourier spectrum (power spectral density) of sliding correlation coefficients R(SLP, S_N); **e** Fourier spectrum of frequency of occurrence of the meridional form C; **f** The same for the form M1. Dashed lines show the confidence levels calculated relative to the "red noise" with AR(1) coefficients $\alpha = 0.36$ (**e**) and 0.27 (**f**)

forms of circulation (zonal, meridional and easterly) in the Atlantic-Eurasian sector (W, C and E) and in the Pacific-American sector (Z, M1 and M2) of the Northern hemisphere. Under zonal circulation forms W and Z, small-amplitude waves moving quickly from west to east are observed in pressure field. Meridional forms C and M1 and easterly forms E and M2 characterize the development of meridional processes, stationary or slowly moving large-amplitude waves being observed in pressure field. The forms C and E, as well as M1 and M2 differ by the opposite location of troughs and crests.

In Fig. 1a one can see temporal variations of correlation coefficients R(SLP, S_N) between sea-level pressure in the high-latitudinal region (60°–85°N), the data being taken from [12] for 1873–2000 and from [13] after 2000, and sunspot numbers [14] for sliding 15-year intervals. Similarly, the correlation coefficients R(SLP, F_{CR}) were calculated between sea-level pressure and GCR fluxes in the stratosphere [15]. The time behavior of these correlation coefficients is compared with the evolution of annual frequencies of occurrence (number of days during a year) of the main circulation forms in the North Atlantic and North Pacific regions presented in Fig. 1b, c (the frequency data are shown after the removal of linear trends and smoothing by 15-year running averages). One can see that the correlation reversals took place in the end of the 19th century, the early 1920s, the 1950s, the early 1980s and near 2000. These reversals, as rule, are preceded or coincide with turning points in the development of all the circulation forms under study. The correlation sign depends, first of all, on the evolution of the meridional forms C and M1. When these forms are intensifying (their frequencies of occurrence are increasing), the correlation coefficients R(SLP, S_N) are negative which means an intensification of Arctic anticyclones (pressure increase) with S_N decreases (GCR increases). When these forms are weakening, one can see a positive correlation between sea-level pressure and sunspot numbers, i.e. a weakening of Arctic anticyclones (pressure decrease) with S_N decreases (GCR increases). The results of spectral analysis (Fig. 1, right panel) show that temporal variations of the correlation coefficients R(SLP, S_N) and frequencies of the meridional circulation forms are characterized by dominant harmonics with the periods close to 60 years. The statistical significance of these harmonics was estimated relative to the spectrum of the "red noise" using χ^2 statistics, with the "red noise" being modeled as the first-order autoregressive (AR(1)) process [16]. Thus, the results presented in Fig. 1 suggest a possible link between the character of solar activity (GCR) effects on troposphere pressure and the regime of the atmosphere circulation.

3 Stratospheric Polar Vortex and Its Possible Part in Formation of Solar Activity (GCR) Effects on Troposphere Pressure

Let us consider what may influence the formation of solar activity (GCR) effects on pressure variations (development of baric systems) under different epochs of large-scale circulation. An important part of atmospheric circulation is the polar vortex forming at polar latitudes above the level 500 hPa. The vortex manifests itself as an enhancement of zonal western winds in the stratosphere which decreases heat exchange between high and middle latitudes and, then, results in the formation of low temperature area with increased temperature gradients at its edges. A key feature of the polar vortex is its capability to influence the stratosphere-troposphere coupling via planetary waves. Under a strong vortex, when zonal wind velocity exceeds some critical value, planetary waves propagating upward are reflected back

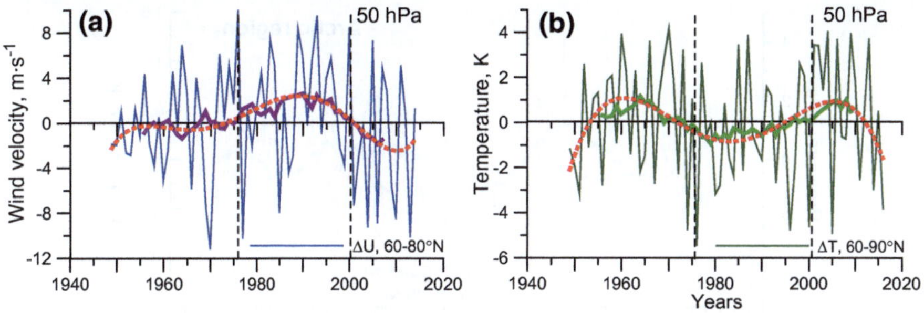

Fig. 2 Zonal wind velocity in the belt 60–80°N (**a**) and temperature in the region 60–90°N (**b**) at the stratospheric level 50 hPa for cold months (December–February) after the trend removal. Thick lines show 15-year running averages, dotted lines show the polynomial fits. Vertical dashed lines separate the periods of a strong and weak vortex

to the troposphere resulting in changes of pressure and temperature [17]. Under a weak vortex, planetary waves propagate in the stratosphere and higher. So, the stratosphere may influence the troposphere only under a strong vortex which seems to create favorable conditions for transferring solar signal from the stratosphere to the troposphere.

In Fig. 2 temporal variations of the polar vortex intensity are presented. To estimate the polar vortex strength, we used zonal western wind velocity and temperature variations at the stratospheric level 50 hPa (~20 km) according to the NCEP/NCAR reanalysis archive [18]. One can see that the period from the late 1970s to ~2000 was the period of a strong vortex: zonal winds were strongly enhanced and temperature was lowered in the vortex area. In the previous period from ~1950 to the late 1970s the polar vortex was noticeably weaker, with stratospheric temperature being increased.

To estimate the polar vortex intensity in the earlier period with no stratospheric data, we used oscillations of sea-level pressure and temperature at high latitudes as a proxy of the vortex state. According to Baldwin and Dunkerton [19], a strong vortex regime is characterized by negative SLP anomalies at high latitudes and positive ones at middle latitudes (the positive phase of the Arctic Oscillation), with cyclone tracks being shifted to the north. More intensive arrival of cyclones in the Arctic contributes to a warming in this region [7]. When the vortex is weak, SLP anomalies are positive at high latitudes and negative at middle ones (the negative phase of the Arctic Oscillation). In Fig. 3 oscillations of sea-level pressure and temperature in the Arctic are presented, the data from [12, 13, 20] being used. We can see that warm epochs accompanied by pressure decrease took place in the periods ~1920–1950 and ~1980–2000, with stratospheric wind enhancement being detected for the latest one (Fig. 2), providing evidence for a strong vortex conditions. The periods of cold epochs with pressure increases in the Arctic indicating a weakening of the vortex took place in ~1900–1920 and ~1950–1980. The above data suggest a roughly 60-year

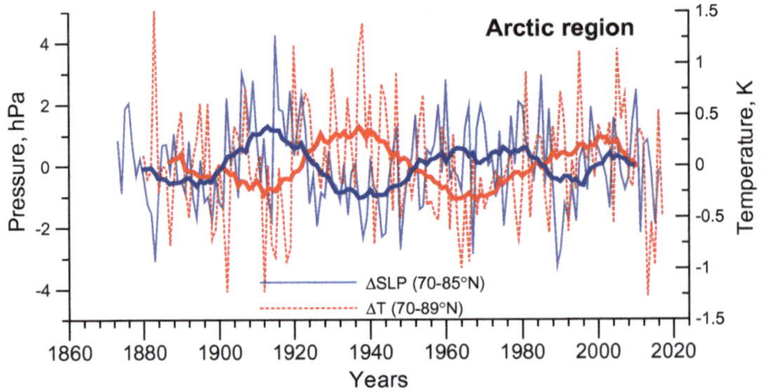

Fig. 3 Sea-level pressure and temperature (detrended yearly values) in the Arctic region (the Arctic Oscillation). Thick lines show 15-year running averages

periodicity in the polar vortex strength. The transitions between the different states of the vortex occurred near 1900, in the early 1920s, the 1950s, the early 1980s and, perhaps, near 2000 resulting in the changes of the large-scale circulation in these years (Fig. 1b, c).

Thus, the oscillations of the vortex strength at the multidecadal time scale can really be a reason for a ~60-year periodicity observed in solar activity effects on pressure changes at extratropical latitudes. As it was shown in [1, 2], under a strong vortex regime, increases of GCR intensity at minima of 11-year solar cycles are accompanied by an intensification both of cyclones (pressure decrease) at Polar fronts of middle latitudes and near-ground anticyclones (pressure increase) in the Arctic. Under a weak vortex, effects of GCR increases are quite opposite, a weakening both of mid-latitudinal cyclones (pressure increase at Polar fronts) and anticyclones in the Arctic (pressure decrease) being observed. The detected link between GCR influence on extratropical baric systems and the stratospheric polar vortex intensity allows suggesting an important part of the vortex in the formation of GCR effects at the multidecadal time scale. The observed reversals of correlation links between troposphere pressure and GCR intensity may be caused by changes of conditions for planetary wave propagation under a strong/weak vortex regime.

4 Variations of Solar Irradiance as a Possible Reason for Polar Vortex Evolution

Let us consider possible reasons for a roughly 60-year periodicity in the polar vortex strength and the corresponding changes in large-scale circulation epochs. A similar periodicity was earlier detected in global temperature anomalies (GTA) [3]. In Fig. 4a there are presented temporal variations of GTA according to the Climatic Research Unit data [21]. One can see a distinct ~60-year variation against the pos-

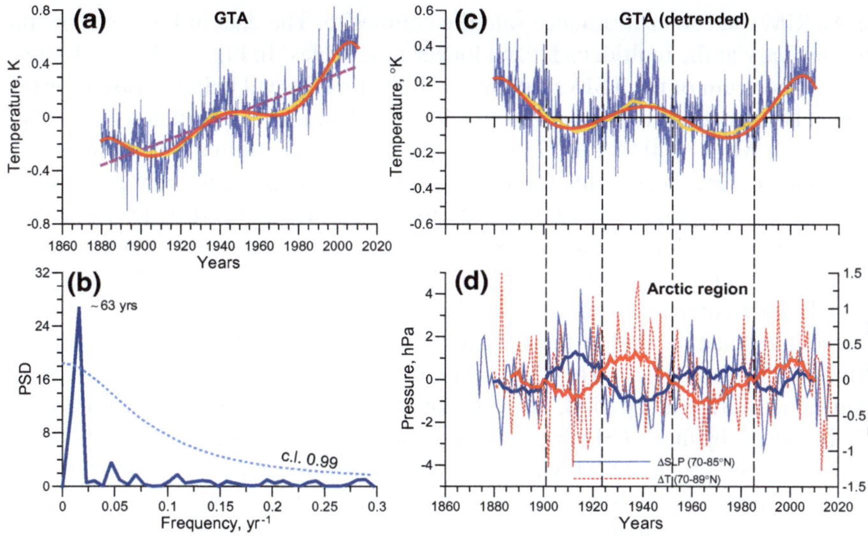

Fig. 4 **a** Global temperature anomalies (monthly values) [21]; **b** Fourier spectrum of GTA yearly values after the trend removal; **c** Detrended values of GTA; **d** Sea-level pressure and temperature variations in the Arctic (the Arctic Oscillation). Thick lines (**a**, **c**) show running averages of GTA over 241 months (~20 years) (yellow line) and the 6th order polynomial fit (red line). Dashed line (**b**) shows the confidence level calculated relative to the "red noise" with AR(1) coefficient $\alpha = 0.7$. Vertical dashed lines (**c, d**) separate the periods of positive and negative GTA deviations

itive linear trend which is confirmed by the results of spectral analysis (Fig. 4b). Comparing temporal variations of detrended GTA values and those of sea-level pressure and temperature in the Arctic (Fig. 4c, d), one can note that ~60-year oscillations of the circulation regime coincide well with similar GTA oscillations. When GTA deviations are positive, the positive phases of the Arctic Oscillation (a strong polar vortex) are observed, with pressure being lowered and temperature being increased in the Arctic. When GTA deviations are negative, the negative phases of the Arctic Oscillation take place (a weak polar vortex). The presented data suggest a possible link between the polar vortex state and global temperature variations contributing to the vortex intensification and the corresponding transformation of the circulation regime.

As the main energy source for the Earth's atmosphere is solar radiative energy, let us consider if long-term changes of total solar irradiance (TSI) could be a possible reason for ~60-year oscillations in GTA. In Fig. 5a one can see temporal variations of total solar irradiance according to the reconstruction created by Hoyt and Schatten [22] and recently updated by Scafetta and Wilson [23]. The Hoyt-Schatten reconstruction was based on the estimates of the Sun's convective energy transport, with five different solar indices (the length and the decay rate of the 11-year solar cycle, the mean level of solar activity, the fraction of penumbral sunspots and the equatorial rotation rate) being used. Scafetta and Wilson [23] updated this reconstruction using the ACRIM total solar irradiance satellite composite. The data in Fig. 5a show that TSI may vary at the multidecadal and longer time scales. In Fig. 5b the local Morlet wavelet spectrum of the TSI variations, with the linear trend being removed, is presented. The spectrum is normalized to 0.95 confidence level calculated for the "red noise" with AR(1) = 0.98 [16]. In Fig. 5c the Fourier spectrum of the reconstructed TSI is shown. One can see that the TSI spectra reveal highly significant periodicities ~80 and ~50–60 years. The ~80-year periodicity seems to be stronger in the 19th century, but in the 20th century the ~60-year periodicity starts dominating.

In Fig. 6 variations (deviations from the linear trends) of global temperature anomalies are compared with those of the reconstructed TSI taken since 1870. One can see that since the end of the 19th century the values under study vary approximately in the same phase. The TSI variations reach up to ± 1 W m^{-2} relative to the trend values during this period. Let us estimate changes of global temperature which may be caused by these TSI changes. The solar energy available for the Earth system per unit time is $(1 - A)\cdot TSI/4$, where A is the Earth's albedo ($A = 0.3$) and the factor of 4 arises, as the energy is averaged over the area of the Earth sphere (e.g., Gray et al. [24]). Taking into account the climate sensitivity to solar radiation changes which is characterized by the parameter λ ($\lambda = 0.5$ K (W $m^{-2})^{-1}$ according to [25]), we can estimate global temperature changes ΔT as

$$\Delta T = \lambda \times (1 - A) \times \Delta TSI/4$$

The estimates according to this formula show that global temperature changes associated with $\Delta TSI = \pm 1$ W m^{-2} amount to about ± 0.1 K which is consistent with GTA variations presented in Figs. 4c and 6.

Thus, variations of total solar radiation caused by long-term variations of solar activity may be one of possible reasons for ~60-year oscillations of global temperature, the polar vortex state and corresponding changes in the circulation regime of the lower atmosphere. This agrees with the suggestions made in [7, 26]. However, we should not deny other external reasons which may influence the state of the polar vortex. In particular, the vortex state may be influenced by electron precipitations associated with geomagnetic disturbances and resulting in ozone depletion in the polar stratosphere; this contributes to the stratosphere cooling and the vortex strengthening in cold months [27, 28]. So, ~60-year oscillations in the vortex intensity may be due to a combined influence of total solar irradiance and variable geomagnetic activity, as well as, possibly, other phenomena related to solar activity. Thus, the role of solar activity and related phenomena in the polar vortex evolution, as well as the role of the vortex in the formation of solar activity (GCR) effects on the development of extratropical baric systems need further investigations.

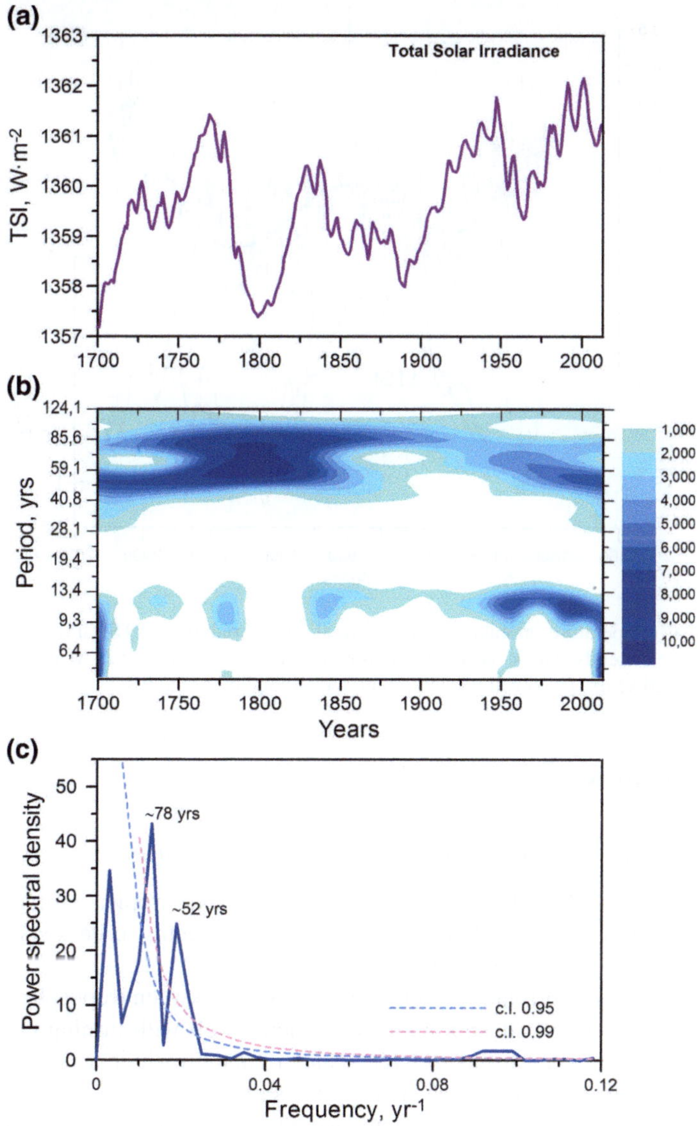

Fig. 5 **a** Long-term variations of total solar irradiance according to Scafetta and Willson [23]; **b** Local wavelet spectrum (Morlet basis) of TSI after the linear trend removal; **c** Fourier spectrum of TSI after the linear trend removal. Dashed lines show the confidence levels calculated relative to the "red noise" with AR(1) coefficient $\alpha = 0.98$

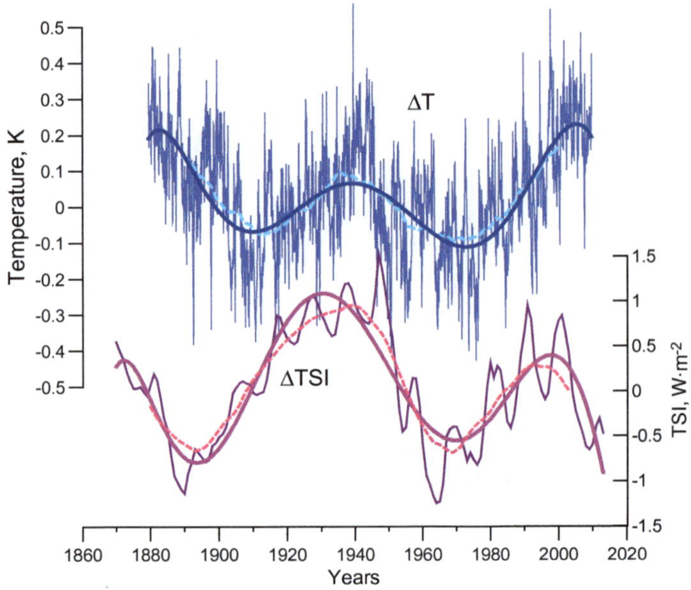

Fig. 6 Global temperature anomalies ΔT and total solar irradiance ΔTSI since 1870 after the removal of linear trends. Thick dashed lines show running averages over 21 years for ΔTSI and 241 months (~20 years) for ΔT, thick solid lines show the polynomial fits

5 Conclusions

The results of this study allow to draw the following conclusions:

1. Temporal variability of solar-atmospheric links at extratropical latitudes is characterized by a roughly 60-year periodicity which seems to be associated with changes of the regime of large-scale circulation.
2. Sign reversals of correlation coefficients detected between troposphere pressure at high latitudes and sunspot numbers coincide well with turning points in the evolution of the main forms of large-scale circulation which accompany transitions between strong and weak states of the stratospheric polar vortex.
3. The stratospheric polar vortex seems to play an important part in the mechanism of solar-atmospheric links due to its capability to influence troposphere-stratosphere coupling. The observed correlation reversals between troposphere pressure and solar activity (GCR) variations may be caused by changes of conditions for planetary wave propagation under a strong/weak vortex regime.
4. Global temperature variations may be considered as one of possible reasons for changes of the polar vortex state and the evolution of large-scale circulation. In turn, these temperature variations may be related to long-term changes of total solar irradiance.

References

1. Veretenenko, S., Ogurtsov, M.: Regional and temporal variability of solar activity and galactic cosmic ray effects on the lower atmosphere circulation. Adv. Space Res. **49**(4), 770–783 (2012)
2. Veretenenko, S., Ogurtsov, M.: Stratospheric polar vortex as a possible reason for temporal variations of solar activity and galactic cosmic ray effects on the lower atmosphere circulation. Adv. Space Res. **54**(12), 2467–2477 (2014)
3. Schlesinger, M.E., Ramankutty, N.: An oscillation in the global climate system of period 65–70 years. Nature **367**, 723–726 (1994)
4. Knudsen, M.F., Seidenkrantz, M.-S., Jacobsen, B.H., Kuijpers, A.: Tracking the Atlantic multidecadal oscillation through the last 8,000 years. Nat. Commun. **2**, 178 (2011). https://doi.org/10.1038/ncomms1186
5. Mantua, N.J., Hare, S.R.: The Pacific decadal oscillation. J. Oceanogr. **58**, 35–44 (2002)
6. Minobe, S.A.: 50–70 year oscillation over the North Pacific and North America. Geophys. Res. Lett. **24**(6), 683–686 (1997)
7. Gudkovich, Z.M., Karklin, V.P., Smolyanitsky, V.M., Frolov, I.E.: On the character and causes of the Earth's climate change. Problemy Arktiki i Antarktiki **1**(81), 15–23 (2009). (in Russian with English abstract)
8. Pudovkin, M.I., Raspopov, O.M.: The mechanism of action of solar activity on the state of the lower atmosphere and meteorological parameters: a review. Geomagn. Aeron. **32**(5), 593–608 (1992)
9. Svensmark, H.: Cosmic rays and clouds. Space Sci. Rev. **93**, 175–185 (2000)
10. Tinsley, B.A.: The global atmospheric electric circuit and its effects on cloud microphysics. Rep. Progr. Phys **71**(6), 066801 (2008)
11. Vangengeim, G.Ya.: Principles of the macro-circulation methods of long-term meteorological forecasts for the Arctic. Trudy AANII **34**, 11–66 (1952). (in Russian)
12. Climatic Research Unit. https://crudata.uea.ac.uk/cru/data/pressure
13. Earth System Research Laboratory, NOAA. https://www.esrl.noaa.gov/psd/repository
14. World Data Center WDC-SILSO. http://www.sidc.be/silso/datafiles
15. Stozhkov, YuI, Svirzhevsky, N.S., Bazilevskaya, G.A., Kvashnin, A.N., Makhmutov, V.S., Svirzhevskaya, A.K.: Long-term (50 years) measurements of cosmic ray fluxes in the atmosphere. Adv. Space Res. **44**, 1124–1137 (2009)
16. Torrence, C., Compo, G.P.: A practical guide to wavelet analyses. Bull. Amer. Meteorol. Soc. **79**, 61–78 (1998)
17. Avdyushin, S.I., Danilov, A.D.: The Sun, weather, and climate: a present-day view of the problem (Review). Geomagn. Aeron. **40**, 545–555 (2000)
18. Kalnay, E., Kanamitsu, M., Kistler, R., et al.: The NCEP/NCAR 40-year reanalysis project. Bull. Amer. Meteorol. Soc. **77**, 437–472 (1996)
19. Baldwin, M.P., Dunkerton, T.J.: Stratospheric harbingers of anomalous weather regimes. Science **294**, 581–584 (2001)
20. GISS Surface Temperature Analysis. http://data.giss.nasa.gov/gistemp/source
21. Climatic Research Unit. http://www.cru.uea.ac.uk/cru/info/warming
22. Hoyt, D.V., Shatten, K.H.: A discussion of plausible solar irradiance variations, 1700–1992. J. Geophys. Res. **98**(A11), 18895–18906 (1993)
23. Scafetta, N., Willson, R.C.: ACRIM total solar irradiance satellite composite validation versus TSI proxy models. Astrophys. Space Sci. **350**(2), 421–442 (2014)
24. Gray, L.J., Beer, J., Celler, M. et al.: Solar influences on climate. Rev. Geophys. **48**, RG4001 (2010). https://doi.org/10.1029/2009RG000282

25. Stocker, T. F., Qin, D., Plattner, G.-K. et al.: IPCC, 2013: Climate Change 2013: The Physical Science Basis. Contribution of Working Group I to the Fifth Assessment Report of the Intergovernmental Panel on Climate Change. Cambridge University Press. Cambridge, United Kingdom and New York, NY, USA (2013)
26. Soon, W.W.-H.: Variable solar irradiance as a plausible agent for multidecadal variations in the Arctic-wide surface air temperature record of the past 130 years. Geophys. Res. Lett. **32**, L16712 (2005). https://doi.org/10.1029/2005GL023429
27. Baumgaertner, A.J.G., Seppälä, A., Jöckel, P., Clilverd, M.A.: Geomagnetic activity related NO_x enhancements and polar surface air temperature variability in a chemistry climate model: Modulation of the NAM index. Atmos. Chem. Phys. **11**, 4521–4531 (2011)
28. Rozanov, E., Calisto, M., Egorova, T., Peter, T., Schmutz, W.: Influence of the precipitating energetic particles on atmospheric chemistry and climate. Surv. Geophys. **33**, 483–501 (2012)

MIX
Papier aus verantwortungsvollen Quellen
Paper from responsible sources
FSC® C105338

If you have any concerns about our products,
you can contact us on
ProductSafety@springernature.com

In case Publisher is established outside the EU,
the EU authorized representative is:
Springer Nature Customer Service Center GmbH
Europaplatz 3, 69115 Heidelberg, Germany

Printed by Libri Plureos GmbH
in Hamburg, Germany